Algorithms and Combinatorics 23

Springer
Berlin
Heidelberg
New York
Barcelona
Hong Kong
London
Milan
Paris
Tokyo

Michael Molloy
Bruce Reed

Graph Colouring and the Probabilistic Method

With 19 Figures

Springer

Michael Molloy
Department of Computer Science
University of Toronto
10 King's College Road
Toronto, Ontario, M5S 3G4, Canada
e-mail: molloy@cs.toronto.edu

Bruce Reed
Université de Paris VI, CNRS
4, Place Jussieu
75252 Paris Cedex 05, France
and
School of Computer Science, McGill University
3480 University Avenue
Montreal, Quebec, H3A 2A7, Canada
e-mail: breed@cs.mcgill.ca

Cataloging-in-Publication Data applied for.

Die Deutsche Bibliothek – CIP-Einheitsaufnahme

Molloy, Michael S.:
Graph colouring and the probabilistic method / Michael S. Molloy; Bruce Reed. –
Berlin; Heidelberg; New York; Barcelona; Hong Kong; London; Milan; Paris; Tokyo:
Springer, 2002
(Algorithms and combinatorics; 23)
ISBN 3-540-42139-4

Mathematics Subject Classification (2000): 5C15, 60C05

ISSN 0937-5511
ISBN 3-540-42139-4 Springer-Verlag Berlin Heidelberg New York

Springer-Verlag Berlin Heidelberg New York
a member of BertelsmannSpringer Science+Business Media GmbH

http://www.springer.de

Typesetting: Authors and LE-TEX Jelonek, Schmidt & Vöckler GbR
using a Springer LaTeX package

Cover design: *design & production* GmbH, Heidelberg

Printed on acid-free paper SPIN 10777803 46/3142/YL - 5 4 3 2 1 0

Preface

One technique for proving the existence of an object with certain properties is to show that a random object chosen from an appropriate probability distribution has the desired properties with positive probability. This approach is known as the *probabilistic method*; using it to find graph colourings with special properties is the topic of this book.

The probabilistic method was pioneered and championed by Paul Erdős who applied it mainly to problems in combinatorics and number theory from 1947 onwards. The authors have been told that at every combinatorics conference attended by Erdős in the 1960s and 1970s, there was at least one talk which concluded with Erdős informing the speaker that almost every graph was a counter-example to his conjecture. Although this story is apocryphal, it does illustrate three facts about the probabilistic method which are worth bearing in mind.

The first is that the probabilistic method allows us to consider graphs which are both *large* and *unstructured*. Even a modern computer looking for a counterexample via exhaustive search would not be able to run through all the graphs on twenty nodes in a reasonable amount of time. A researcher in the 60s certainly cannot be expected to have done so. In contrast, the counterexamples constructed using the probabilistic method routinely contain many, say 10^{10}, nodes.

Furthermore, any attempt to build large counterexamples via an explicit construction necessarily introduces some structuredness to the class of graphs built, which thus restricts the graphs considered. To illustrate this, we remark that even though the clique and stability numbers of a typical graph on n vertices are both $O(\log n)$, we do not know how to efficiently construct a graph with clique and stability numbers which are this small and for many years we could do no better than $o(\sqrt{n})$. Thus, the probabilistic method allows us to boldly go where no deterministic method has gone before, in our search for an object with the properties we desire. This accounts for its power.

The second moral of our story is that the method is not just powerful, it is also easy to use. Erdős would routinely perform the necessary calculations to disprove a conjecture in his head during a fifteen minute talk. Indeed, in the introduction to his Ph.D. thesis (which was published in 1970), Vasek Chvátal wrote *in this thesis we study hypergraphs using the probabilistic method where by the probabilistic method we mean the taking of sums in two different ways.* This *First Moment* approach was one of the main techniques used in the early days. For example, to bound the expected number of cliques of size k in a typical graph on n vertices, we consider each set S of k vertices and ask: *in what proportion of the graphs is S a clique?* Since this is roughly $2^{-\binom{k}{2}}$(as each of the $\binom{k}{2}$ edges will be present with probability $\frac{1}{2}$), there are on average $\binom{n}{k}2^{-\binom{k}{2}}$ cliques of size k in a graph on n vertices. This manner of calculating the sum is much easier than the alternative approach, which consists of determining the proportion of graphs with no cliques of size k, those with one clique of size k, etcetera. Furthermore, this one line calculation implies that the clique number (and in the same vein stability number) of a typical graph is $O(\log n)$, while, as mentioned above, we do not know how to prove such graphs exist without recourse to the probabilistic method.

The third and final moral of our story about Erdős and his counterexamples is that in classical applications of the probabilistic method, the result obtained was not just that a positive proportion of the random objects had the desired property, but rather that the overwhelming proportion of these objects did. E.g. Erdős did not say some graph is a counterexample to your conjecture, but rather *almost every* graph is a counter-example to your conjecture. It would be fair to say that the classical technique was typically limited to situations in which the probability that a randomly chosen object had the desired property was fairly large.

There were two motivating factors behind our decision to write this book. The first was to provide a gentle introduction to the probabilistic method, so that researchers who so desired could add this powerful and easy to use weapon to their arsenal. Thus, the book assumes little or no background in probability theory. Further, it contains an introductory chapter on probability, and each of the tools we use is introduced in a separate chapter containing examples and (at least in the early chapters) exercises illustrating how it is applied.

Of course, new probabilistic tools have been introduced in the last half-century which are more sophisticated than switching the order of two summation signs. Many of these focus on the typical difference between a random non-negative real-valued variable X and its average (or *expected*) value, denoted $\mathbf{E}(X)$. Since, $\mathbf{E}(X)$ is often easy to compute using the first moment method, any observations we can make about $|X - \mathbf{E}(X)|$ translate into statements about the value of X. Results which bound $|X - \mathbf{E}(X)|$ are called *concentration* results because they imply that independently sampled values of X tend to be concentrated around $\mathbf{E}(X)$ (the reader may find the image of darts concentrated around the bulls-eye of a target helpful here).

One classical approach to concentration is the Second Moment Method which bounds the expected value of $|X - \mathbf{E}(X)|$ using the First Moment Method applied to a related variable. It is this approach which allows us to tie down precisely the behaviour of the clique size of a typical graph on n vertices.

Certain random variables are better-behaved than others and for such random variables we can obtain much more information about $|X - \mathbf{E}(X)|$ than the relatively weak bounds on its expected value we obtain using the Second Moment Method. For example if Y is the number of heads obtained in n tosses of a *fair* coin then the probability that Y is k is simply $\binom{n}{k}2^{-n}$. Using this fact, and manipulating the terms involved, Chernoff obtained very strong bounds on the probability that $|Y - \mathbf{E}(Y)|$ exceeds t for a given real t. His bounds also apply to the number Y of heads obtained in n flips of a coin which comes up heads with some probability p.

Note that changing the outcome of any one coin flip can affect this number Y of heads by at most one. Recently, researchers have come up with more general tools which bound the concentration of random variables X which are determined by a sequence of n events such that changing the outcome of one event can affect the (expected) value of X by at most one. We present one such tool developed by Azuma, and another developed by Talagrand. Although the proofs that these bounds hold are non-trivial, as the examples given in the book attest, applying them is straightforward.

As mentioned above, using the classical probabilistic method researchers typically proved that a random object had certain desirable properties with probability near one. The Lovász Local Lemma is a very important tool which allows us to prove the existence of objects with certain properties when the probability that a random object has the desired properties is exponentially small. The proof of this lemma is short and it is very easy to use, as our examples once again attest. However, in contrast with e.g. the First Moment Method, it is difficult to construct efficient algorithms to find the objects the lemma guarantees exist. This is not too surprising as the number of such objects is so small. (In contrast, if almost every object has the property then to find such an object we can just pick one at random.) The last two chapters of the book discuss how to construct such algorithms.

Our second motivation for writing the book was to provide a unified treatment of a number of important results in graph colouring which have been proven using an iterative approach (the so-called semi-random method). These include results of Kim and Johansson on colouring triangle-free graphs with $O(\log n)$ colours, Kahn's results on $(1+o(1))\Delta$ edge list colouring linear hypergraphs, Kahn's proof that asymptotically the Goldberg-Seymour conjecture holds, and the fact that there is an absolute constant C such that every graph of maximum degree Δ has a $\Delta + C$ total colouring.

Our treatment of the results discussed above shows that each of them can be proven via an application of the semi-random method where we analyze an iteration using the Local Lemma and our concentration results. Although the proofs all have a similar flavour (indeed we reworked all the proofs to illustrate the unity of this body of work and hopefully to make them more accessible), some of them are more complicated than others, requiring auxiliary notions (such as entropy and hardcore distributions) and quite involved arguments. To ease the reader's burden, we present the easier results first, building up to more and more sophisticated variants of the technique. Fittingly, we end with Kahn's result on the list chromatic index of multigraphs which really is a tour-de-force.

Three researchers who have had a very important impact on the mathematical careers of the authors are Vasek Chvátal, Alan Frieze, and Colin McDiarmid. We both met all three of these researchers as graduate students. Vasek was the second author's Ph.D. supervisor and Alan supervised the first author. The first paper written by the second author on probabilistic combinatorics was joint with Alan and Colin. We are immensely grateful for all that we have learnt from these three researchers and the many enjoyable hours we have spent in their company.

We are very grateful to those researchers who read portions of the book, pointing out errors and suggesting reworkings. The following is a list of some of those researchers, we apologize to any we may have inadvertently left out.

Dimitris Achlioptas, Noga Alon, Spyros Angelopoulos, Etienne Birmele, Adrian Bondy, Babak Farzad, Nick Fountoulakis, Alan Frieze, Paul Gries, Jeff Kahn, Mark Kayll, Jeong Han Kim, Michael Krivelevich, Andre Kűndgen, Mohammad Mahdian, Colin McDiarmid, Steve Myers, Ioannis Papoutsakis, Ljubomir Perkovic, Mohammad Salavatipour, Benny Sudakov, Frank Van Bussel, Van Vu.

The authors are very grateful to Cynthia Pinheiros Santiago who prepared the figures and converted the manuscript into the Springer Latex format.

The second author would also like to thank the University of Toronto, McGill University, and the University of Sao Paulo, which all provided generous hospitality and support during the writing of the book. Both authors would like to thank the Federal University of Ceara in Fortaleza, Brazil where much of the polishing of this book took place.

Despite the many careful readings and rereadings, errors will inevitably remain, however we feel that the manuscript is now ready for publication and Boy! are we glad to be finished.

Montreal
September 2001

Michael Molloy
Bruce Reed

Contents

Part V. An Iterative Approach

Part VI. A Structural Decomposition

Part VII. Sharpening our Tools

Part VIII. Colour Assignment via Fractional Colouring

Part IX. Algorithmic Aspects

Part I

Preliminaries

> It was the afternoon of my eighty-first birthday, and I was in bed with my catamite when Ali announced that the archbishop had come to see me. ... you will be constrained to consider, if you know my work at all and take the trouble now to reread that first sentence, that I have lost none of my old cunning in the contrivance of what is known as an arresting opening.

These lines which open Burgess' book *Earthly Powers*, illustrate a huge advantage that novelists have over mathematicians. They can start with the juicy bits. Mathematicians usually have to plow through a list of definitions and well-known simple facts before they can discuss the developments they are really interested in presenting. This book is no exception. The first two chapters present the basic notions of graph colouring and probability theory.

In an attempt to liven up this a priori boring material, the results in the colouring chapter are presented in a way which emphasizes some of the central themes of the book. The only virtue we can claim for the probabilistic preliminaries chapter is brevity. These chapters may be skimmed quickly, or skipped altogether by the reader who is already familiar with these two areas.

1. Colouring Preliminaries

1.1 The Basic Definitions

We will be discussing colouring the vertices and edges of graphs. A *graph* G is a set $V = V(G)$ of *vertices* and a set $E = E(G)$ of *edges*, each linking a pair of vertices, its *endpoints* (formally, an edge is an unordered pair of vertices and thus our graphs have no loops or multiple edges), which are *adjacent*. We assume the reader has a basic knowledge of graph theory. A *k-colouring of the vertices* of a graph G is an assignment of k colours (often the integers $1, \ldots, k$) to the vertices of G so that no two adjacent vertices get the same colour. The *chromatic number* of G, denoted $\chi(G)$, is the minimum k for which there is a k-colouring of the vertices of G. The set S_j of vertices receiving colour j is a *colour class* and induces a graph with no edges, i.e. it is a *stable set* or *independent set*. So, a k-colouring of the vertices of G is simply a partition of $V(G)$ into k stable sets and the chromatic number of G is the minimum number of stable sets into which the vertices of G can be partitioned.

A *k-colouring of the edges* of a graph G is an assignment of k colours to the edges of G so that no two incident edges get the same colour. The *chromatic index* of G, denoted $\chi_e(G)$ is the minimum k for which there is a k-colouring of the edges of G. The set M_j of vertices receiving colour j is a *colour class*, and it is a set of edges no two of which share an endpoint, i.e. a *matching*. So, a k-colouring of the edges of G is simply a partition of $E(G)$ into k matchings, and the chromatic index of G is the minimum number of matchings into which the edges of G can be partitioned.

We sometimes want to colour both the edges and vertices of a graph. A *total k-colouring* of a graph G is an assignment of k colours to the vertices and edges of G so that no two adjacent vertices get the same colour, no two incident edges get the same colour, and no edge gets the same colour as one of its endpoints. The *total chromatic number* of G, denoted $\chi_T(G)$ is the minimum k for which there is a total k-colouring of G. The set T_j of vertices and edges receiving colour j is a *colour class* , and it consists of a stable set S_j and a matching M_j none of whose edges have endpoints in S_j. Such an object is called a *total stable set*. So, a k-colouring of $V(G) \cup E(G)$ is simply a partition of $V(G) \cup E(G)$ into k total stable sets and the total chromatic number of G is the minimum number of total stable sets required to partition $V(G) \cup E(G)$.

A *partial k-colouring* of a graph is an assignment of k colours (often the integers $1, \ldots, k$) to a (possibly empty) subset of the vertices of G so that no two adjacent vertices get the same colour. We *complete* a partial k-colouring by assigning colours to the uncoloured vertices to produce a k-colouring. (Of course, not every partial k-colouring can be completed, not even every partial k-colouring of a k-colourable graph.) These definitions extend in the obvious way to *partial edge colouring*, *partial total colouring*, etc.

For a graph G, the *line graph of* G, denoted $L(G)$ is the graph whose vertex set corresponds to the edge set of G and in which two vertices are adjacent precisely if the corresponding edges of G are incident. We note that the chromatic index of G is simply the chromatic number of $L(G)$.

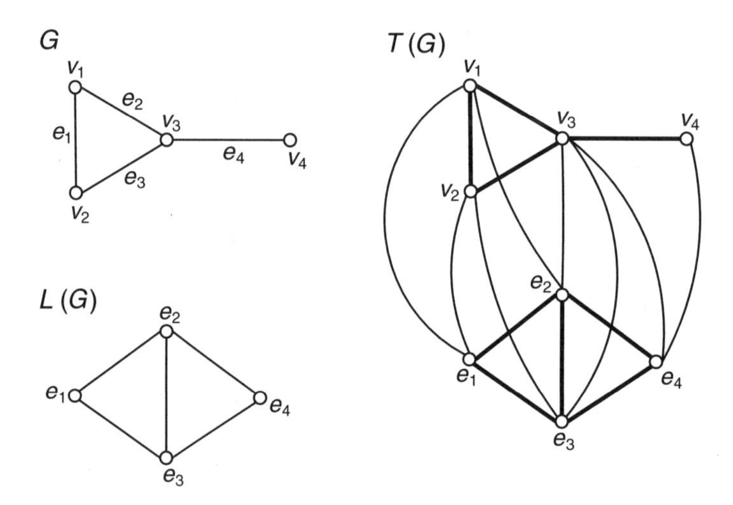

Fig. 1.1. A graph and its line and total graph

Similarly, for any graph G, we can construct a graph $T(G)$, *the total graph of* G, whose chromatic number is the total chromatic number of G. To obtain $T(G)$, we take a copy of G and a copy of $L(G)$ and add an edge between a vertex x of G and a vertex y of $L(G)$ precisely if x is an endpoint of the edge of G corresponding to y.

Thus, all the problems we discuss are really vertex colouring problems. When we talk about the chromatic index of graphs, we are just restricting our attention to vertex colouring line graphs. When we talk about the total chromatic index of graphs, we are just restricting our attention to vertex colouring total graphs.

1.2 Some Classical Results

Clearly a graph has chromatic number 0 precisely if it has no vertices and chromatic number 1 precisely if it has vertices but no edges. A well-known result states

Fact 1.1 *A graph has chromatic number at most 2,* i.e. *is* bipartite, *if and only if it contains no odd cycles.*

Proof It is easy to see that every odd cycle has chromatic number three. So, we need only show that any graph G without odd cycles is two colourable. Obviously, we can treat the components of G separately and hence can assume that G is connected. We choose some vertex v of G, assign it colour 1, and grow, vertex by vertex, a connected bipartite subgraph H of G containing v as follows. If H is G, then we are done. Otherwise, as G is connected, there is a vertex x in $G - H$ adjacent to a vertex in H. If x sees only vertices of colour 1, we colour it with colour 2 and add it to H. If x sees only vertices of colour 2, we colour it with colour 1 and add it to H. If x sees a vertex y of colour 1 in H and a vertex z of colour 2 in H then we let P be some yz path in the connected graph H. Since H is bipartite, P has an even number of vertices and so $P + x$ is an odd cycle. □

Remark This proof yields an ordering on $V(G)$ with certain properties. We highlight this fact, as we will find similar orderings useful later. If we label the vertices of H in the order in which we add them to H, then $v_1 = v$ and for all $j > 1$, v_j has a neighbour v_i with $j > i$. Similarly for any vertex x in a connected graph, taking the reverse of such an order yields an ordering $w_1, \ldots, w_n = x$ such that for $j < n$, w_j has a neighbour w_i with $i > j$.

The proof above yields an efficient (in fact, linear time) algorithm for determining if G is bipartite. On the other hand, it is NP-complete to determine if G has chromatic number three. In fact this was one of the first six problems which Karp, [93], proved NP-complete via a reduction, and it remains NP-complete for planar G (see [65]). Thus, determining the chromatic number of a graph G precisely seems difficult. There are however, a number of results which yield simple bounds on the chromatic number of G. Two of the most natural and important are:

Definition Recall that a *clique* is a set of pairwise adjacent vertices. We use $\omega(G)$ to denote the *clique number of* G, i.e. the number of vertices in the largest clique in G.

Observation 1.2 *The chromatic number of G is at least $\omega(G)$.*

Definitions The *degree* of a vertex v in a graph G, is the number of edges of G to which v is incident, and is denoted $d_G(v)$, or simply $d(v)$. We use $\Delta(G)$ or simply Δ to note the maximum vertex degree in G. We use $\delta(G)$ or simply δ to note the minimum vertex degree in G. The *neighbourhood* of a vertex v in a graph G, is the set of vertices of G to which v is adjacent, and is denoted $N_G(v)$, or simply $N(v)$. The members of $N(v)$ are referred to as *neighbours* of v.

Lemma 1.3 *For all G, $\chi(G) \leq \Delta(G) + 1$.*

Proof Arbitrarily order the vertices of G as $v_1, \ldots, v_n$. For each v_i in turn, colour v_i with the lowest positive integer not used on any of its neighbours appearing earlier in the ordering. Obviously, every vertex receives a colour between 1 and $\Delta + 1$. □

Now, both these bounds are tight for cliques, and the second is also tight for odd cycles. In 1941, Brooks [30] tightened Lemma 1.3 by showing that these two classes were essentially the only classes for which the second bound is tight. He proved (and we shall prove at the end of this chapter):

Theorem 1.4 (Brooks' Theorem) *$\chi(G) \leq \Delta$ unless some component of G is a clique with $\Delta + 1$ vertices or $\Delta = 2$ and some component of G is an odd cycle.*

Now, neither the bound of 1.2 nor the bound of 1.4 need be tight. In fact most graphs on n vertices satisfy $w(G) \leq 2\log(n)$, $\Delta(G) \geq \frac{n}{2}$, and $\chi(G) \approx \frac{n}{2\log n}$ (see, for example, [10]). Thus, usually, neither bound is a good approximation of the chromatic number. Moreover, recent very deep results [108] show that unless P=NP, χ cannot even be approximated to within a factor of $n^{1-\epsilon}$ for a particular small constant $\epsilon > 0$. (For similar further results, see [63, 57].)

The situation for line graphs is much different. Considering a maximum degree vertex of G, we see that $\chi_e(G) \geq \Delta(G)$. In fact $\chi_e(G) = \chi(L(G)) \geq \omega(L(G)) \geq \Delta(G)$, and the last inequality is tight here unless G is a graph of maximum degree two containing a triangle. Now, not all graphs have chromatic index Δ. Consider for example, a triangle. More generally, consider the odd clique K_{2l+1} on $2l+1$ nodes. It has $\Delta = 2l$, $|E(G)| = l(2l+1)$, and no matching with more than l edges. Thus, it cannot have a Δ colouring (unless $l = 0$). On the other hand, we do have:

Theorem 1.5 (Vizing's Theorem [154]) *For all G, $\chi_e(G) \leq \Delta(G) + 1$.*

Remark See [107], pp. 286–287 for a short sweet algorithmic proof of this theorem due to Ehrenfreucht, Faber, and Kierstead.

Thus, $\chi_e(G)$ is easy to approximate to within one, and determining $\chi_e(G)$ boils down to deciding if $\chi_e(G) = \Delta(G)$ or $\chi_e(G) = \Delta(G) + 1$. However, this problem also seems intractable (Holyer[82] has shown it is NP-complete to determine if $\chi_e(G) = 3$).

1.3 Fundamental Open Problems

The fundamental open problem concerning the total chromatic number is to decide if it can be approximated to within 1, as the chromatic index can. Considering a maximum degree vertex of G, we see that $\chi_T(G) \geq \Delta(G) + 1$. In fact $\chi_T(G) = \chi(T(G)) \geq \omega(T(G)) \geq \Delta(G) + 1$, and the second inequality is tight here unless G is a graph of maximum degree one. Now, $\chi_T(G)$ is not always $\Delta(G) + 1$. Consider, for example, the graph consisting of a single edge (what happens with larger even cliques?). Vizing[155] and Behzad[20] independently proposed what is undoubtedly the central open problem concerning total colourings:

Conjecture 1.6 (The Total Colouring Conjecture) $\chi_T(G) \leq \Delta(G)+2$.

We note that for a long period, the best bound on the total chromatic number of G was that given by Brooks' Theorem: $2\Delta(G)$. (The case $\Delta(G) < 2$ is trivial. For $\Delta(G) \geq 2$, the reader is invited to verify that $\Delta(T(G)) = 2\Delta(G)$ and that no component of $T(G)$ is a $(\Delta(T(G)) + 1)$-clique.) It took the probabilistic method to improve this far from optimal bound (see [27]).

The choice of the fundamental open problem concerning edge colourings is less clear. However, in the authors' opinion, the central question is to determine if we can still approximate the chromatic index to within one if we permit multiple edges.

A *multigraph* G is similar to a graph except that there may be more than one edge between the same pair of vertices. The definition of chromatic index and total chromatic number extend naturally to multigraphs (as does the notion of chromatic number, however to compute this parameter we need only consider one of the edges between each pair of adjacent vertices). The chromatic index of a multigraph can exceed its maximum degree by much more than one. Consider, for example, the multigraph obtained by taking k copies of each edge of a triangle. This multigraph has maximum degree $2k$ but its line graph is a clique of size $3k$ and so the multigraph has chromatic index $3k$. A multigraph's chromatic index can also significantly exceed the clique number of its line graph, as can be seen by replicating the edges of a cycle of length five. However, it still seems possible to approximate the chromatic index of multigraphs.

Definition Recall that a stable set is a set of pairwise non-adjacent vertices, we use $\alpha(G)$ to denote the *stability number of* G, i.e. the number of vertices in the largest stable set in G.

We define $\beta(G)$ to be $\lceil \frac{|V(G)|}{\alpha(G)} \rceil$. We define $\beta^*(G)$ to be the maximum of $\beta(H)$ over all subgraphs H of G. We have:

Observation 1.7 $\omega(G) \leq \beta^*(G) \leq \chi(G)$.

Proof The first inequality holds because cliques have stability number one. Now, $\chi(G) \geq \beta(G)$ because a colouring is a partition into stable sets. Since the chromatic number of G is at least as large as the chromatic number of any of its subgraphs, the second inequality follows. □

In the same spirit, for a multigraph G we define $\mu(G)$ to be the maximum over all subgraphs $H \subseteq G$ of $\lceil \frac{|E(H)|}{\lfloor |V(H)|/2 \rfloor} \rceil$. Clearly, $\beta^*(L(G))$ is at least $\mu(G)$. Since $\omega(L(G)) \geq \Delta(G)$, $\beta^*(L(G))$ is at least $\max(\Delta(G), \mu(G))$ (in fact, as we discuss in Chap. 21, seminal results of Edmonds[37] imply $\beta^*(L(G))$ is this maximum). So, $\max(\mu(G), \Delta(G))$ is a lower bound on the chromatic index of G. Our choice for the central conjecture about edge colourings states that this bound is always very close to correct.

Conjecture 1.8 (The Goldberg–Seymour Conjecture [67, 142]) *For every multigraph G, $\chi_e(G) \leq \max(\mu(G), \Delta + 1)$.*

There are even more choices for a central open problem in vertex colouring. (In fact Toft and Jensen have written an excellent book, *Graph Coloring* [sic] *Problems* [85] containing over 200 unsolved colouring problems, most of which deal with vertex colouring.) Two obvious candidates concern the relationship between χ and ω.

Definition A graph G is *perfect* if every induced subgraph H of G satisfies $\chi(H) = \omega(H)$.

Now a perfect graph cannot contain a subgraph H which is an induced odd cycle with more than five vertices for then $\chi(H) = 3$ and $\omega(H) = 2$. It is also straightforward to verify that the complement of an odd cycle on $2k+1 \geq 5$ nodes satisfies $\omega = k$ and $\chi = k+1$ (the colour classes in any colouring are one or two consecutive vertices of the cycle). Our first open problem is a possible characterization of perfect graphs in terms of these two families.

Conjecture 1.9 (The Strong Perfect Graph Conjecture(Berge [21])) *A graph is perfect if and only if it contains no induced subgraph isomorphic to either an odd cycle on at least five vertices or the complement of such a cycle.*

The second conjecture concerns minors. We say *K_k is a minor of G* if G contains k vertex disjoint connected subgraphs between every two of which there is an edge.

Conjecture 1.10 (Hadwiger's Conjecture [73]) *Every graph G satisfies: $\chi(G) \leq k_G = \max\{k | K_k$ is a minor of $G\}$.*

We can also define minors of graphs which are not cliques.

Definitions We *contract* an edge xy in a graph G to obtain a new graph G_{xy} with vertex set $V(G_{xy}) = V(G) - x - y + (x * y)$ and edge set $E(G_{xy}) =$

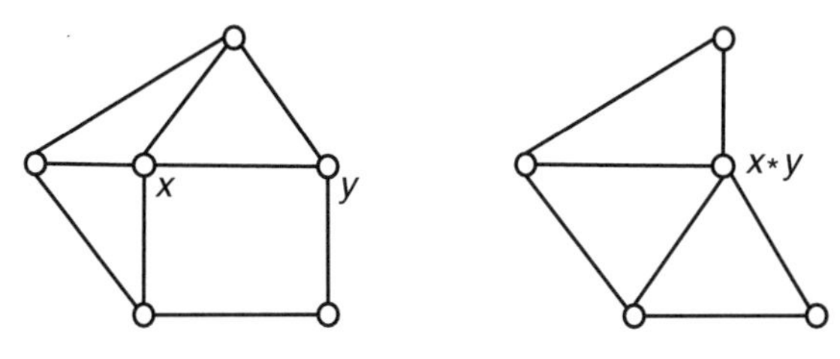

Fig. 1.2. Contracting an edge

$E(G-y-x) \cup \{(x*y)z | xz \text{ or } yz \in E(G)\}$ (see Fig. 1.2). H is a *minor* of G if H can be obtained from G via a sequence of edge deletions, edge contractions, and vertex deletions (we regard isomorphic graphs as being equal).

Note that our two definitions coincide for cliques.

Remark Note further that k_G is also $\max\{\omega(H)|\text{H is a minor of G}\}$.

Hadwiger's Conjecture for $k_G \leq 2$ is quite easy. Dirac [36] proved the conjecture for G with $k_G = 3$. The case $k_G = 4$ is significantly more challenging. The celebrated Four Colour Conjecture (that every planar graph G satisfies $\chi \leq 4$) is clearly implied by Hadwiger's Conjecture for $k_G = 4$, as it is well known that any graph with a K_5-minor is non-planar. In fact, Wagner [159] showed that this case of Hadwiger's Conjecture is equivalent to the Four Colour Conjecture. The Four Colour Conjecture remained open for over 100 years until being settled in the affirmative by Appel and Hakken in 1976 [12]. Hadwiger's Conjecture for $k_G = 5$ was recently resolved by Robertson, Seymour, and Thomas [137]. The conjecture remains open for all larger values of k_G.

1.4 A Point of View

We note that many of the problems above concern bounding the chromatic number of a graph in terms of its clique number.

The Total Colouring Conjecture states that $\chi(G) \leq \omega(G)+1$ if G is a total graph. The Strong Perfect Graph Conjecture involves characterising a special class of graphs all of whose members satisfy $\chi \leq \omega$. Hadwiger's Conjecture also involves using the clique number as an upper bound on the chromatic number, but now we consider the clique numbers of the minors of G rather than just G itself.

Many of the results we discuss in this book consist of proving that for some specially structured graphs, e.g. total graphs, we can bound χ by $(1+o(1))\omega$.

We will not have much to say about the Strong Perfect Graph Conjecture We shall however present some partial results on Hadwiger's Conjecture. We shall also present some results about bounding χ using ω in arbitrary graphs.

For example, we shall generalize Brooks' Theorem by proving that there is some $\epsilon > 0$ such that $\chi \leq \epsilon\omega + (1-\epsilon)(\Delta+1)$. We present this theorem as evidence for a stronger statement in the same vein:

Conjecture 1.11 (Reed [132]) *For any graph G, $\chi \leq \lceil \frac{1}{2}\omega + \frac{1}{2}(\Delta + 1) \rceil$.*

The proofs of all of the results mentioned above require only the simplest of probabilistic tools, and in most cases we analyze a very simple colouring procedure. Our intention is to stress the simplicity of the incredibly powerful techniques used, particularly early on in the book.

We also consider another graph invariant *the fractional chromatic number*. This number lies between ω and χ. We shall study bounding χ using the fractional chromatic number, possibly in combination with the maximum degree. As we shall see, the Goldberg–Seymour Conjecture implies that for any line graph G of a multigraph, $\chi(G)$ is at most one more than the fractional chromatic number of G. We shall present results of Kahn which show that for such G, χ does not exceed the fractional chromatic number significantly. These results require much more sophisticated probabilistic and combinatorial tools but the spirit of the arguments are the same as those introduced earlier in the book. This should aid the reader in understanding the more difficult material.

Recapitulating then, we view all these diverse conjectures as problems concerning bounding χ in terms of either ω or the fractional chromatic number and we discuss what light the probabilistic method sheds on this general problem. We will also discuss algorithms for obtaining the colourings that we prove exist. Of course, there are many graph colouring problems which do not fit this paradigm. We will also see how to treat some of these via the probabilistic method.

We complete this chapter with a few more technical results and definitions that we will need.

1.5 A Useful Technical Lemma

For ease of exposition, we often prefer to consider graphs in which all the vertices have the same degree. If this degree is d, then such a graph is called *d-regular.*

Fortunately, there is a simple construction which allows us to embed any graph of maximum degree Δ in a Δ-regular graph. This allows us to extend results on d-regular graphs to graphs with maximum degree d.

The construction proceeds as follows. We take two copies of G and join the two copies of any vertex not of maximum degree. If G is not already regular, this increases its minimum degree by one without changing the maximum degree. Iterating this procedure yields the desired result.

1.6 Constrained Colourings and the List Chromatic Number

Often colouring problems impose extra constraints which restrict the set of permissible colourings. A common side constraint is to restrict the colours permissible at each vertex.

Definitions Given a list L_v of colours for each vertex of G, we say that a vertex colouring is *acceptable* if every vertex is coloured with a colour on its list. The *list chromatic number* of a graph, denoted $\chi^\ell(G)$ is the minimum r which satisfies: if every list has at least r colours then there is an acceptable colouring. Given a list of colours for each edge of G, we say that an edge colouring is *acceptable* if every edge is coloured with a colour on its list. The *list chromatic index* of a graph, denoted $\chi^\ell_e(G)$ is the minimum r which satisfies: if every list has at least r members then there is an acceptable colouring.

Now, G is k vertex colourable if and only if there is an acceptable colouring when every vertex has the list $\{1,\ldots,k\}$. Thus, the list chromatic number is at least the chromatic number. Instinctively, one feels that making the list at different vertices dissimilar should make it easier to colour G. However this intuition is incorrect, as we see in Fig. 1.3. More generally, one can construct bipartite graphs with list chromatic number k for every k (see Exercise 1.8). Thus, the list chromatic number can be arbitrarily far away from the chromatic number. Once again, it is conjectured that line graphs are much better behaved. That is we have:

Conjecture 1.12 (The List Colouring Conjecture) *Every graph G satisfies $\chi^\ell_e(G) = \chi_e(G)$.*

A famous special case of this conjecture is due to Dinitz. It states that if we have n acceptable integers for each square of an n by n grid then we can

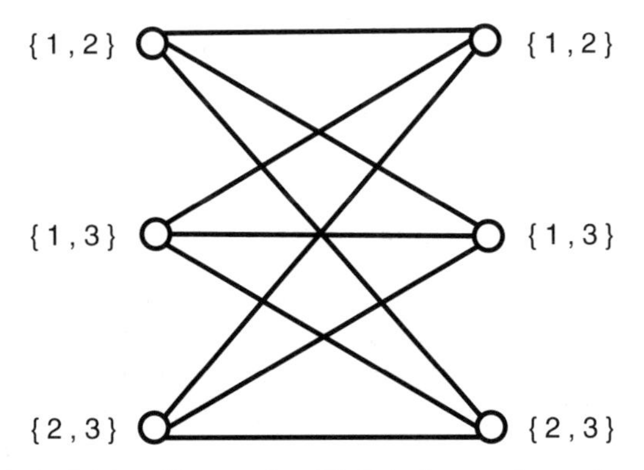

Fig. 1.3. A bipartite graph with list chromatic number three

fill in a Latin square on this grid so that each square of the grid contains an acceptable integer. Translated into the language of graphs, Dinitz's conjecture says that $\chi_e^\ell(K_{n,n})$ is n (where $K_{n,n}$ is the bipartite graph which has n vertices on each side and n^2 edges). Dinitz's conjecture was solved recently by Galvin [64]. In fact, Galvin proved the List Colouring Conjecture for all bipartite graphs (see [85] for more references).

1.7 Intelligent Greedy Colouring

The proof of Lemma 1.3, actually yields the following two stronger results:

Lemma 1.13 *Any partial $\Delta+1$ colouring of G can be extended to a $\Delta+1$ colouring of G.*

Proof Simply place the l coloured vertices at the beginning of the ordering under which we perform the colouring and begin colouring at vertex v_{l+1}. □

Definition The *colouring number* of G, denoted $col(G)$, is the maximum over all subgraphs H of G of $\delta(H)+1$

Lemma 1.14 *The chromatic number of G is at most the colouring number of G.*

Proof Choose an ordering of $V(G)$, beginning with v_n, by repeatedly setting v_i to be a minimum degree vertex of $H_i = G - \{v_{i+1}, \dots, v_n\}$. Then, when we come to colour v_i, we will use a colour in $\{1, \dots, \delta(H_i)+1\}$. □

Remark This lemma was proved independently by many people, see ([85], p 8) for a discussion.

Definition We use the term *greedy colouring algorithm* to refer to the process which colours $V(G)$ by colouring the vertices in some given order, always using the first available colour (under some arbitrary ordering of the colours). We also use the same term to refer to the analogous process for extending a partial colouring.

We can also use the greedy colouring algorithm to prove Brooks' Theorem. We end the chapter with this proof.

Proof of Brooks' Theorem. Since G is bipartite unless it contains an odd cycle, Brooks' Theorem holds for graphs with $\Delta = 2$. A minimal counterexample to Brooks' Theorem must be connected. In fact a minimal counterexample must be 2-connected. For if v were a cutvertex of G, and U a component of $G-v$ then we could colour $U+v$ and $G-U$ separately and obtain a colouring of G by relabeling so that v receives the same colour in these two colourings.

We need the following easy lemma whose short proof is left as an exercise (see Exercise 1.10).

Lemma 1.15 *Any 2-connected graph G of maximum degree at least three which is not a clique contains three vertices x, y, z such that $xy, xz \in E(G)$, $yz \notin E(G)$, and $G - y - z$ is connected.*

Corollary 1.16 *Let G be a 2-connected graph which has maximum degree $\Delta \geq 3$ and is not a clique. Then, we can order the vertices of G as $v_1, \ldots, v_n$ so that $v_1v_2 \notin E(G)$,$v_1v_n, v_2v_n \in E(G)$, and for all j between 3 and $n-1$, v_j has at most $\Delta - 1$ neighbours in $\{v_1, \ldots, v_{j-1}\}$.*

Proof Choose x, y, z as in the lemma. set $v_1 = y, v_2 = z$. As we remarked after the proof of Fact 1.1 we can order $G - y - z$ as $v_3, \ldots, v_n = x$ so that for each $i < n$, v_i has a neighbour v_j with $j > i$. □

Now, consider a graph G and ordering as in the corollary. If we apply our greedy colouring algorithm to G, we will assign colour 1 to both v_1 and v_2. Furthermore, for each i between 3 and $n-1$, v_i will receive a colour between 1 and Δ because it has a neighbour v_j with $j > i$ and so at most $\Delta - 1$ of its neighbours have been previously coloured. Finally, since v_n has two neighbours of colour 1, it will also be assigned a colour between 1 and Δ.

Thus, every 2-connected graph of maximum degree $\Delta \geq 3$ has a Δ-colouring and so Brooks' Theorem holds. □

Finding colourings using few colours by extending a colouring in which colours appear more than once in some vertices' neighbourhoods will be a key technique in this book.

Exercises

Exercise 1.1 Let G be a graph of maximum degree Δ and r an integer. Suppose that we can colour a subset of the vertices of G using $\Delta + 1 - r$ colours so that no two adjacent vertices are coloured with the same colour and for every $v \in V$ there are at least r colours which are used on two or more neighbours of v. Show that G has a $\Delta + 1 - r$ colouring.

Exercise 1.2 Euler's formula implies that every planar graph has a vertex of degree five.

(i) Show this implies that every planar graph has chromatic number at most six.

(ii) Suppose v is a vertex of degree five in a planar graph G and there is a colouring of $G - v$ with five colours. Let G_{ij} be the subgraph of G induced by those vertices coloured i or j in $G - v$. Show that if for some pair $\{i, j\}$ there is no path in $G_{i,j}$ between two vertices of $N(v)$ then $\chi(G) \leq 5$.

(iii) Use (ii) to show that $\chi(G) \leq 5$ for every planar graph G.

Exercise 1.3 We consider sets of rectilinear squares in the plane (i.e. the sides of the squares are parallel to the x and y axes). We say H is a *square graph* if its vertices correspond to such a set of squares and two vertices are adjacent precisely if the corresponding squares intersect.

(i) Show that if all the squares are the same size then for each vertex v in the corresponding square graph H, $N(v)$ can be partitioned into four or fewer cliques. Use this to show $\chi(H) \leq 4\omega(H) - 3$.
(ii) Show that if all the squares are the same size and H is finite then there is a vertex of H whose neighbourhood can be partitioned into two or fewer cliques. Use this to deduce that $\chi(H) \leq 2\omega(H) - 1$.
(iii) Show that *any* square graph H contains a vertex H whose neighbourhood can be partitioned into four or fewer cliques. Deduce $\chi(H) \leq 4\omega(H) - 3$.

Exercise 1.4 Use a variant of the greedy colouring procedure to show that every graph G contains a bipartite graph H with $|E(H)| \geq \frac{1}{2}|E(G)|$.

Exercise 1.5

(i) Show that if H is a bipartite subgraph of G with the maximum possible number of edges then $d_H(v) \geq \frac{d_G(v)}{2}$ for all v and in particular, $\delta(H) \geq \frac{\delta(G)}{2}$.
(ii) Show that we can construct a bipartite subgraph H of G with $d_H(v) \geq \frac{d_G(v)}{2}$ for all v in polynomial time.

Exercise 1.6 Show that if G has average degree d then it has a subgraph H with $\delta(H) \geq \frac{d}{2}$.

Exercise 1.7 Combine Exercises 1.5 and 1.6 to show that if G has average degree d then it has a bipartite subgraph with minimum degree at least $\frac{d}{4}$.

Exercise 1.8 Show that the list chromatic number of the complete bipartite graph $K_{\binom{2k-1}{k},\binom{2k-1}{k}}$ is at least $k+1$.

Exercise 1.9 Show inductively that if G contains 2^ℓ vertices then it contains a clique C and a stable set S such that $|C| + |S| = \ell + 1$. (Hint: put v_1 in $C \cup S$ and meditate on its degree.) Note that this implies that if G has 4^ℓ vertices then it contains either a clique of size ℓ or a stable set of size ℓ.

Exercise 1.10

(i) Prove that any connected graph G which is not a clique contains three vertices x, y, z such that $xy, xz \in E(G)$ and $yz \notin E(G)$.
(ii) Prove that if G is 2-connected and has minimum degree three, there is some such triple x, y, z, such that the removal of y, z leaves G connected. Note that this proves Lemma 1.15.

2. Probabilistic Preliminaries

2.1 Finite Probability Spaces

We consider experiments which have only a finite number of possible outcomes. We call the set of all possible outcomes, the *sample space* and denote it Ω. For example, our experiment may consist of rolling a six sided die and examining the top face, in which case $\Omega = \{1,2,3,4,5,6\}$. Alternatively, our experiment may consist of flipping a coin three times in a row, then $\Omega = \{HHH, HHT, HTH, THH, TTH, THT, HTT, TTT\}$ where H stands for heads and T for tails. The reader probably has an intuitive notion of what an event is, which corresponds to this word's use in everyday language. Formally, an *event* is a subset A of Ω. For example, we identify the event that the die roll is odd with the subset $(\{1,3,5\})$. Similarly, the event that the coin landed the same way up every time is the set $(\{HHH, TTT\})$.

A *finite probability space* $(\Omega, \mathbf{Pr})$ consists of a finite sample space Ω and a probability function $\mathbf{Pr} : \Omega \to [0..1]$ such that

2.1 $\sum_{x \in \Omega} \mathbf{Pr}(x) = 1$.

We often consider the *uniform distribution* in which $\mathbf{Pr}(x) = \frac{1}{|\Omega|}$ for each x in Ω. In other words a *uniformly chosen* element of Ω, or a *uniform element* of Ω is a random element where each possibility is equally likely. We extend $\mathbf{Pr}$ to the events in 2^{Ω} by setting

2.2 $\mathbf{Pr}(A) = \sum_{x \in A} \mathbf{Pr}(x)$.

So for example,

$$\mathbf{Pr}(\text{fair dice is odd}) = \frac{1}{2}.$$

If our coin is biased and lands heads two-thirds of the time then

$$\mathbf{Pr}(\text{all flips the same}) = \frac{8}{27} + \frac{1}{27} = \frac{1}{3}.$$

Letting $\overline{A}$ be the event that A does not occur, i.e. $\overline{A} = \Omega - A$, we have:

2.3 $\mathbf{Pr}(\overline{A}) = 1 - \mathbf{Pr}(A)$.

We also have:

2.4 $\mathbf{Pr}(A \cup B) = \mathbf{Pr}(A) + \mathbf{Pr}(B) - \mathbf{Pr}(A \cap B)$,

and for any partition of Ω into (disjoint) subsets $B_1, \ldots, B_l$:

2.5 $\mathbf{Pr}(A) = \sum_{i=1}^{l} \mathbf{Pr}(A \cap B_i)$.

Now, (2.4) implies:

2.6 $\mathbf{Pr}(A \cup B) \leq \mathbf{Pr}(A) + \mathbf{Pr}(B)$

and more generally:

$$\mathbf{Pr}(\cup_{i=1}^{l} A_i) \leq \sum_{i=1}^{l} \mathbf{Pr}(A_i).$$

This last fact we refer to as the *Subadditivity of Probabilities.*

We sometimes use $\mathbf{Pr}(A, B)$ for $\mathbf{Pr}(A \cap B)$.

For any two finite probability spaces $(\mathbf{Pr}_1, \Omega_1)$ and $(\mathbf{Pr}_2, \Omega_2)$, their *product space* has sample space $\Omega_1 \times \Omega_2$ and probability distribution $\mathbf{Pr}$ defined as follows: $\mathbf{Pr}(x \times y) = \mathbf{Pr}_1(x)\mathbf{Pr}_2(y)$. We can iteratively apply this definition to obtain the product of arbitrarily many component subspaces. Thus, we can consider a sequence of n independent coin flips as a product space. Another example of a product space is the random graph $G_{n,p}$ which has n vertices $\{1, \ldots, n\}$, and in which we choose each of the $\binom{n}{2}$ possible edges to be in the graph with probability p, where these choices are made independently. Thus, the probability that $G_{n,p}$ is any *specific* graph H on $1, \ldots, n$ is $p^{(|E(H)|)}(1-p)^{(\binom{n}{2} - |E(H)|)}$.

For any two events A and B, the *conditional probability* of A given B, denoted $\mathbf{Pr}(A|B)$ is, informally, the probability we would assign to A if we knew B occurred. Formally, $\mathbf{Pr}(A|B) = \frac{\mathbf{Pr}(A \cap B)}{\mathbf{Pr}(B)}$ and is read as the probability of A given B (if $\mathbf{Pr}(B) = 0$ we set $\mathbf{Pr}(A|B) = \mathbf{Pr}(A)$). Note how the formal definition agrees with the informal one presented first.

For example if A is the event that the die roll is odd and B is the event that the die roll is at most three, then $\mathbf{Pr}(A|B) = \frac{2}{3}$.

Intuitively, A is independent of B if knowing that B occurs yields no hint as to whether or not A occurs. Formally, A is *independent* of B if $\mathbf{Pr}(A|B) = \mathbf{Pr}(A)$ or equivalently $\mathbf{Pr}(A \cap B) = \mathbf{Pr}(A)\mathbf{Pr}(B)$. Note that this implies that $\mathbf{Pr}(B|A) = \mathbf{Pr}(B)$, i.e. A is independent of B if and only if B is independent of A. Thus, we may speak of a pair of *independent events.* Note that A is independent of B if and only if it is independent of $\overline{B}$, by (2.5).

A set of events is *pairwise independent* if every pair of events in the set is independent. The set is *mutually independent* if for any subset$\{A_0, A_1, \ldots, A_l\}$ of events we have:

$$\mathbf{Pr}(A_0 | \cap_{i=1}^{l} A_i) = \mathbf{Pr}(A_0).$$

Note that a set of events may be pairwise independent without being mutually independent, consider for example tossing a fair coin three times and the three events: the number of heads is even, the first two flips are the same, the second two flips are heads.

The reader should verify that if $A_1, \ldots, A_t$ are mutually independent then $\mathbf{Pr}(A_1 \cap \ldots \cap A_t) = \mathbf{Pr}(A_1) \times \ldots \times \mathbf{Pr}(A_t)$.

An event A is *mutually independent* of a set of events $\mathcal{E}$ if for every $B_1, \ldots, B_r \in \mathcal{E}$,

$$\mathbf{Pr}(A|B_1 \cap \ldots \cap B_r) = \mathbf{Pr}(A).$$

It is straightforward to show that the mutual independence condition implies a more general condition: for every $B_1, \ldots, B_t, C_1, \ldots, C_s \in \mathcal{E}$,

$$\mathbf{Pr}(A|B_1 \cap \ldots \cap B_r \cap \overline{C_1} \cap \ldots \cap \overline{C_s}) = \mathbf{Pr}(A).$$

2.2 Random Variables and Their Expectations

A *random variable* defined on a finite probability space $(\mathbf{Pr}, \Omega)$ is a function X from Ω to the reals. For example: the sum of the top faces of two rolls of a fair die, the number of heads in N flips of a biased coin which comes up heads two-thirds of the time. The random variable X naturally defines a probability space $(\mathbf{Pr}_X, \Omega_X)$ on its range Ω_X. The corresponding probability function $\mathbf{Pr}_X$ is called the mass function for X and satisfies:

$$\text{for each} \quad x \in \Omega_X, \mathbf{Pr}_X(x) = \sum \{\mathbf{Pr}(w) | w \in \Omega, X(w) = x\}.$$

For example, if we roll a fair die twice and let X be the sum of the two rolls then $\Omega_X = \{2, 3, \ldots, 12\}$ and

$$\mathbf{Pr}_X(2) = \mathbf{Pr}_X(12) = \frac{1}{36}, \ \mathbf{Pr}_X(3) = \mathbf{Pr}_X(11) = \frac{2}{36},$$

$$\mathbf{Pr}_X(4) = \mathbf{Pr}_X(10) = \frac{3}{36}, \ \mathbf{Pr}_X(5) = \mathbf{Pr}_X(9) = \frac{4}{36},$$

$$\mathbf{Pr}_X(6) = \mathbf{Pr}_X(8) = \frac{5}{36}, \ \text{and } \mathbf{Pr}_X(7) = \frac{6}{36}.$$

The *expected value* of a random variable X is:

$$\mathbf{E}(X) = \sum_{w \in \Omega} \mathbf{Pr}(w) X(w).$$

Intuitively, $\mathbf{E}(X)$ is the value we would expect to obtain if we repeated a random experiment several times and took the average of the outcomes of X.

Thus, if X is a random variable which is 1 with probability p and 0 with probability $1-p$ then the expected value of X is $p \times 1 + (1-p) \times 0 = p$. In the same vein, the expected number of pips on the top face of a fair die after it is rolled is $\sum_{i=1}^{6} i \times \frac{1}{6} = 3.5$.

We remark that

$$\begin{aligned} \mathbf{E}(X) &= \sum_{w \in \Omega} \mathbf{Pr}(w) X(w) = \sum_{x \in \Omega_X} \sum_{w \in \Omega, X(w)=x} \mathbf{Pr}(w) x \\ &= \sum_{x \in \Omega_X} x \sum_{w \in \Omega, X(w)=x} \mathbf{Pr}(w) = \sum_{x \in \Omega_X} x \mathbf{Pr}_X(x) . \end{aligned}$$

Thus, if X is the sum of two die rolls then our calculation of $\mathbf{Pr}_X$ above yields:

$$\begin{aligned} \mathbf{E}(X) = \; & 2 \times \frac{1}{36} + 3 \times \frac{2}{36} + 4 \times \frac{3}{36} + 5 \times \frac{4}{36} + 6 \times \frac{5}{36} + 7 \times \frac{6}{36} \\ & + 8 \times \frac{5}{36} + 9 \times \frac{4}{36} + 10 \times \frac{3}{36} + 11 \times \frac{2}{36} + 12 \times \frac{1}{36} = 7 . \end{aligned}$$

There is another equation (whose proof we leave as an exercise) that can be obtained by swapping the order of a double sum, which will prove much more useful in computing expected values. It is:

Linearity of Expectation $\mathbf{E}\left(\sum_{i=1}^{\ell} X_i\right) = \sum_{i=1}^{\ell} \mathbf{E}(X_i)$.

Thus, if X is the total rolled using n fair dice then letting X_i be the number on the *ith* dice we have that $\mathbf{E}(X_i) = 3.5$ and since X is the sum of the X_i, Linearity of Expectation implies $\mathbf{E}(X) = 3.5n$. Similarly, since there are $\binom{n}{2}$ possible edges in $G_{n,p}$, each of which is present with probability p, the expected number of edges of $G_{n,p}$ is $p\binom{n}{2}$. As a final example, consider a random permutation π of $1, \ldots, n$ where each permutation is equally likely (thus Ω is the set of all $n!$ permutations of $\{1, \ldots, n\}$ and $\mathbf{Pr}$ is the uniform distribution on Ω). Let $X = X(\pi)$ be the number of i for which $\pi(i) = i$, that is the number of fixed points of π. We see that setting X_i to be 1 if $\pi(i) = i$ and 0 otherwise, we have $X = \sum_{i=1}^{n} X_i$. It is easy to show that $\mathbf{E}(X_i) = \frac{1}{n}$, hence by the Linearity of Expectation, $\mathbf{E}(X) = 1$. Note that we have computed $\mathbf{E}(X)$ without computing, for example, $\mathbf{Pr}(X = 0)$ which is much more difficult.

These last two examples are instances of a general paradigm which will reoccur frequently. Many variables are the sum of 0-1 variables and hence their expected value is easily computed by computing the sum of the expected values of the 0-1 variables. We will see several examples of this approach in the next chapter. We use $BIN(n,p)$ to denote the variable which is the sum of n 0-1 variables each of which is 1 with probability p (the BIN here stands for Binomial which stresses the fact that each of the n simple variables can take one of two values).

The *conditional expectation of X given B* , denoted $\mathbf{E}(X|B)$ is equal to $\sum_{x\in\omega_X} x\mathbf{Pr}(X=x|B)$. For example, if X is the value of a fair die roll:

$$\mathbf{E}(X|\text{ X is odd}) = 1\times\frac{1}{3}+2\times 0+3\times\frac{1}{3}+4\times 0+5\times\frac{1}{3}+6\times 0 = 3\,.$$

Similarly,

$$\mathbf{E}(X|\text{ X is even}) = 4\,.$$

Linearity of Expectation generalizes to conditional expectation, we leave it as an exercise for the reader to obtain:

2.7 *If* $X=\sum_{i=1}^{l} X_i$, *then* $\mathbf{E}(X|B)=\sum_{i=1}^{l}\mathbf{E}(X_i|B)$.

Thus, we see that if X is the number of heads in n flips of a coin that comes up heads with probability p then considering each flip separately and applying (2.7), we have:

$$\mathbf{E}(X|\text{ first flip is a head}) = 1+p(n-1)\,.$$

Similarly, if $X=X(\pi)$ denotes the number of fixed points of π as above, then we can obtain (another exercise): $\mathbf{E}(X|\pi(1)=1)=1+(n-1)\frac{1}{n-1}=2$, $\mathbf{E}(X|\pi(1)=3)=0+0+(n-2)\frac{1}{n-1}$, and by symmetry: $\mathbf{E}(X|\pi(1)\neq 1)=\frac{n-2}{n-1}$.

Finally, we note that $\mathbf{E}(X)=\mathbf{E}(X|B)\mathbf{Pr}(B)+\mathbf{E}(X|\overline{B})\mathbf{Pr}(\overline{B})$ by the definition of $P(A|B)$ and (2.5). That is, $\mathbf{E}(X)$ is a convex combination of $\mathbf{E}(X|B)$ and $\mathbf{E}(X|\overline{B})$. Thus, one of the latter will be at most $\mathbf{E}(X)$ and the other will be at least $\mathbf{E}(X)$.

2.3 One Last Definition

The *median* of a random variable X, denoted $\mathbf{Med}(x)$, is defined to be the minimum real number m such that $\mathbf{Pr}(X\leq m)\geq\frac{1}{2}$. By symmetry if X is the number of heads in n tosses of a fair coin then $\mathbf{Med}(X)=\lfloor\frac{n}{2}\rfloor$. Thus, the median of X, in this case, is either exactly $\mathbf{E}(X)$ or $\mathbf{E}(X)-\frac{1}{2}$.

Intuitively, as long as a variable has a "nice" distribution, the median should be very close to the expected value. However this is not always the case (see Exercise 2.4). Due mainly to the Linearity of Expectation, expected values are usually much easier to compute than medians, and so we tend to focus far more on expected values. In fact, medians only show their heads twice in this book, and even then they only appear in situations where they are in fact close enough to the expected value that the two are virtually interchangeable.

2.4 The Method of Deferred Decisions

We frequently consider probability spaces which are the product of independent subspaces. One example mentioned earlier is $G_{n,\frac{1}{2}}$. In analyzing such an object, we often find it convenient to expose the choices made in some of the subspaces and use the outcomes exposed to determine how to continue our analysis. This is the simplest case of *The Method of Deferred Decisions*. To illustrate this technique, we will show that the probability that $G_{2l,\frac{1}{2}}$ has a perfect matching is at least $\frac{1}{3}$. Exercise 2.24 asks you to improve this result by showing that the probability $G_{2l,\frac{1}{2}}$ has a perfect matching (that is a matching containing l edges) goes to 1 as l goes to infinity.

We label the vertices $v_1, .., v_{2l}$, We will try to find a perfect matching by repeatedly matching the lowest indexed unmatched vertex to some other vertex using the following procedure:

Step 1. Set $M = \emptyset$ and $S = V$.
Step 2. For $i = 1, \ldots, l$ do:
(2.1) Let $j(i)$ be the smallest integer such that $v_{j(i)}$ is in S. Consider all the pairs $v_{j(i)}, y$ with $y \in S$ and determine which are edges.
(2.2) If $\exists y \in S$ s.t. $v_{j(i)}y \in E(G)$ then choose y_i in S s.t. $v_{j(i)}y_i \in E(G)$ and set $M = M + v_{j(i)}y_i$; else choose any $y_i \in S$.
(2.3) Set $S = S - x_{j(i)} - y_i$.

Note that for each pair x, z of vertices, we examine whether there is an edge between this pair in Step 2.1 in at most iteration. Specifically the first iteration i for which one of x or z is $v_{j(i)}$. So we can think of performing the coin flip which determines if $v_{j(i)}, y$ is an edge and exposing the result in iteration i. Since these coin flips are independent, for every (x, y) considered in Step 2.1, the probability xy is an edge is $\frac{1}{2}$.

Now, we let A_i be the event that we fail to find an edge between $v_{j(i)}$ and $S - v_{j(i)}$ in iteration i. We want to show that with probability at least $\frac{1}{3}$ none of the A_i occur and hence M is a perfect matching.

By the above remarks, $\mathbf{Pr}(A_i) = 2^{-|S-v_{j(i)}|} = 2^{2i-1-2l}$ irregardless of what has happened in the earlier iterations. Thus, the probability that we fail in any iteration between 1 and l is, by Linearity of Expectation, at most $\sum_{i=1}^{l} 2^{2i-1-2l} = \frac{1}{2}\sum_{j=0}^{l-1} 4^{-j} < \frac{2}{3}$.

Note here that we cannot determine whether A_i holds until we have performed the first $i - 1$ iterations. For the outcome of the earlier random choices will determine which vertices remain in S, and hence which edges we consider in the ith iteration. The key fact that allows us to perform our analysis is that regardless of what happens before iteration i, the probability of A_i is the same.

All applications of the Method of Deferred Decisions involve iterative analysis. For the simplest variant of the method, in each iteration we expose the outcome of the choices for some of the random subspaces of a product space. Which outcomes we expose depends on the results of the previous iterations. Note that we may not have complete independence between iterations, as we do above. (For instance, one can imagine a procedure which is similar to the one above but in which the number of vertices in S in an iteration may vary depending on the results of earlier iterations.) Rather, we obtain upper and/or lower bounds on the probability that certain events hold which are valid regardless of the outcome of previous choices.

In more sophisticated variants of the method, we may expose whether or not a certain event holds in an iteration and then in future iterations condition on this event holding or not as the case may be. For example, we may need to expose whether or not a die roll is odd in one iteration and use, in a subsequent iteration, the fact that regardless of the result, we still have that the expected value of the die roll is at least 3.

Exercises

Most of the exercises presented here are from two excellent introductory textbooks on probability theory, one written by Grimmett and Stirzaker[69], the other by Grimmett and Welsh[71]. These exercises are marked by a [GS] or a [GW] to indicate the source. We thank the authors for allowing us to include them. Further exercises can be found in [70].

Exercise 2.1 [GS] A traditional fair die is thrown twice. What is the probability that

(a) a six turns up exactly once,
(b) both numbers are odd,
(c) the sum of the scores is 4,
(d) the sum of the scores is divisible by three.

Exercise 2.2 [GS] Six cups and saucers come in pairs. There are two cups and saucers of each of the three colours in the tricolour: red, white and blue. If the cups are placed randomly on the saucers find the probability that no cup is on a saucer of the same colour.

Exercise 2.3 [GW] Show that the probability that a hand in bridge contains 6 spades, 3 hearts, 2 diamonds, and 2 clubs is

$$\frac{\binom{13}{6}\binom{13}{3}\binom{13}{2}^2}{\binom{52}{13}}.$$

(In bridge the 52 cards are dealt out so that each player has 13 cards).

Exercise 2.4 [GS] Airlines find that each passenger who reserves a seat fails to turn up with probability $\frac{1}{10}$ independently of the other passengers. So Teeny Weeny Airlines always sells 10 tickets for their 9 seat airplane and Blockbuster Airways sells 20 tickets for their 18 seat airplane. Who is more frequently overbooked?

Exercise 2.5 Adapted from [GW]. You are traveling on a train with your sister. Neither of you has a valid ticket and the inspector has caught you both. He is authorized to administer a special punishment for this offence. He holds a box containing nine apparently identical chocolates, but three of these are contaminated with a deadly poison. He makes each of you, in turn, choose and immediately eat a single chocolate.

(a) What is the probability you both survive?
(b) What is the probability you both die?
(c) If you choose first what is the probability that you survive but your sister dies?

Exercise 2.6 [GS] **transitive coins** Three coins each show heads with probability $\frac{3}{5}$ and tails otherwise. The first counts 10 points for a head and 2 for a tail. The second counts 4 points for a head and 4 for a tail. The third counts 3 points for a head and 20 for a tail.

You and your opponent each choose a coin; you cannot choose the same coin. Each of you tosses and the person with the larger score wins ten million francs. Would you prefer to be the first to pick a coin or the second?

Exercise 2.7 [GS] Eight rooks are placed randomly on a chessboard, no more than one to a square. What is the probability that: (a) they are in a straight line (do not forget the diagonals)? (b) no two are in the same row or column?

Exercise 2.8 What is the probability that two dice sum to at least seven given that the first is even.

Exercise 2.9 Under what conditions will $\mathbf{Pr}(A_1 \cup \ldots \cup A_n) = \sum_{i=1}^{n} \mathbf{Pr}(A_i)$?

Exercise 2.10 Adapted from [GS]. You have once again been caught by the ticket inspector of Exercise 2.5:

(a) If you choose first and survive what is the conditional probability your sister survives?
(b) If you choose first and die what is the conditional probability your sister survives?
(c) Is it in your best interests to persuade your sister to choose first? Could you? Discuss.
(d) If you choose first, what is the probability you survive given your sister survives?

Exercise 2.11

(a) Prove the Linearity of Expectation.
(b) Prove that $\mathbf{E}(BIN(n,p)) = np$.

Exercise 2.12 Prove that for any c, k, n we have

$$\mathbf{Pr}\left(BIN\left(n, \frac{c}{n}\right) \geq k\right) \leq \frac{c^k}{k!}.$$

Exercise 2.13 Adapted from [GS]. Anne, Brian, Chloe, and David were all friends at school. Subsequently each of the $\binom{4}{2} = 6$ pairs met up; at each of the six meetings, the pair involved quarrel with some probability p, or become firm friends with probability $1-p$. Quarrels take place independently of each other. In future, if any of the four hears a rumour they tell it to all their firm friends and only their firm friends. If Anne hears a rumour what is the probability that:

(a) David hears it?
(b) David hears it if Anne and Brian have quarreled?
(c) David hears it if Brian and Chloe have quarreled?
(d) David hears it if he has quarreled with Anne?

Exercise 2.14 Adapted from [GS]. A bowl contains twenty cherries, exactly fifteen of which have had their stones removed. A greedy pig eats five whole cherries, picked at random, without remarking on the presence or absence of stones. Subsequently a cherry is picked randomly from the remaining fifteen

(a) What is the probability that this cherry contains a stone?
(b) Given that this cherry contained a stone what is the probability that the pig consumed at least one stone?
(c) Given that the pig consumed at least two stones, what is the probability that this cherry contains a stone?

Exercise 2.15 Consider a random $n \times n$ matrix whose entries are 0 and 1. Suppose each entry is chosen to be a one with probability $\frac{1}{2}$ independently of the other choices. Let A_i be the event that all the entries in row i are the same. Let B_i be the event that all the entries in column i are the same. Show that the set of events $\{A_1, \ldots, A_n, B_1, \ldots, B_n\}$ is pairwise independent but not mutually independent.

Exercise 2.16 [GS] On your desk there is a very special fair die, which has a prime number p of faces, and you throw this die once. Show that no two non-empty events A and B can be independent unless one of A or B is the whole space.

Exercise 2.17 [GW] If $A_1, \ldots, A_m$ are mutually independent and $P(A_i) = p$ for $i = 1, \ldots, m$ then find the probability that (a) none of the As occur, (b) an even number of As occur.

Exercise 2.18 Calculate the expected number of singleton vertices (incident to no edges) in $G_{n,p}$.

Exercise 2.19 [GS] Of the $2n$ people in a given collection of n couples, exactly m die. Assuming that the m have been picked at random, find the mean number of surviving couples. This problem was formulated by Bernoulli in 1768.

Exercise 2.20 Give an example of a random variable X for which $\mathbf{E}(X)$ is far from $\mathbf{Med}(X)$.

Exercise 2.21 Suppose you perform a sequence of n coin flips, the first with a fair coin. Then if the *ith* flip is a head (resp. tails), for the $i+1st$ flip you use a coin which comes up heads with probability $\frac{2}{3}$ (respectively $\frac{1}{3}$). What is the expected total number of heads?

Exercise 2.22 Let X_n be the number of fixed points in a random permutation of $\{1,\ldots,n\}$. Show that for $n \geq 2$, we have:
$\mathbf{Pr}(X_n = 1) \leq \frac{n}{n-1}\mathbf{Pr}(X_n = 0)$.

Exercise 2.23 Use the result of Exercise 2.22 and the fact that the expected value of $\mathbf{E}(X_n)$ is 1 to show that for n at least 3, $\mathbf{Pr}(X_n = 0) \geq \frac{2}{7}$. Can you calculate this probability more precisely?

Exercise 2.24 (Hard) Show that the probability that $G_{2l,\frac{1}{2}}$ has a perfect matching is $1 - o(1)$.

Part II

Basic Probabilistic Tools

In these chapters, we present three fundamental tools of the probabilistic method. We begin with the First Moment Method, which was for many years synonymous with the probabilistic method. While the underlying mathematics is very elementary, the technique is surprisingly powerful. We then turn to the celebrated Lovász Local Lemma, which has been and continues to be one of the most useful tools in probabilistic combinatorics. Finally, we discuss the simplest of several concentration tools that will be used throughout this book, the Chernoff Bound.

As we wish to emphasize that the probabilistic method can be mastered with a very basic understanding of probability, we omit the proofs of these tools for now, stressing that the reader need only understand how they are used.

3. The First Moment Method

In this chapter, we introduce the First Moment Method[1], which is the most fundamental tool of the probabilistic method. The essence of the first moment method can be summarized in this simple and surprisingly powerful statement:

The First Moment Principle: If $\mathbf{E}(X) \leq t$ then $\mathbf{Pr}(X \leq t) > 0$.

We leave the proof of this easy fact to the reader.

Applying the first moment method requires a judicious choice of the random variable X, along with a (usually straightforward) expected value computation. Most often X is positive integer-valued and $\mathbf{E}(X)$ is shown to be less than 1, thus proving that $\mathbf{Pr}(X = 0)$ is positive.

Recalling that, for such X, $\mathbf{E}(X) = \sum_i i \times \mathbf{Pr}(X = i)$, it may seem at first glance that one cannot compute $\mathbf{E}(X)$ without first computing $\mathbf{Pr}(X = i)$ for every value of i, which is in itself at least as difficult a task as computing $\mathbf{Pr}(X \leq t)$ directly. Herein lies the power of the linearity of expectation, which allows us to compute $\mathbf{E}(X)$ without computing $\mathbf{Pr}(X = i)$ for any value of i, in effect by computing a different sum which has the same total! This is the central principle behind virtually every application of the First Moment Method, and we will apply it many times throughout this book, starting with the example in the next section.

Another way of stating the First Moment Principle is to say that the probability that X is larger than $\mathbf{E}(X)$ is less than 1. Our next tool bounds the probability that X is much larger than $\mathbf{E}(X)$.

Markov's Inequality: For any positive random variable X,

$$\mathbf{Pr}(X \geq t) \leq \mathbf{E}(X)/t.$$

Again, we leave the proof as an exercise.

Markov's Inequality is frequently used when X is positive integer-valued and $\mathbf{E}(X)$ is less than 1, in which case we have a bound on the probability which the First Moment Principle guarantees to be positive:

[1] For positive integral k, the kth moment of a real-valued variable X is defined to be $\mathbf{E}(X^k)$, and so the first moment is simply the expected value.

$$\mathbf{Pr}(X > 0) \leq \mathbf{E}(X).$$

The First Moment Method is the name usually used to describe applications of these two tools. In the remainder of this chapter, we will illustrate this method with four examples.

3.1 2-Colouring Hypergraphs

A *hypergraph* is a generalized graph, where an edge may have more than 2 vertices, i.e. a *hyperedge* is any subset of the vertices. If every hyperedge has the same size, k, then we say that the hypergraph is k*-uniform*. Thus, a 2-uniform hypergraph is simply a graph.

By a *proper 2-colouring* of a hypergraph, we mean an assignment of one of 2 colours to each of the vertices, such that no hyperedge is *monochromatic*, i.e. has the same colour assigned to each of its vertices. Note that this is a natural generalization of a proper 2-colouring of a graph. As we saw in Chap. 1, those graphs which can be properly 2-coloured have a precise simple structure which is easily recognized. On the other hand, determining whether a hypergraph is 2-colourable is NP-complete, even for 3-uniform hypergraphs [152], and so characterizing 2-colourable hypergraphs is a very difficult task. In this section we provide a simple proof of a sufficient condition for a hypergraph to be 2-colourable.

Theorem 3.1 *If $\mathcal{H}$ is a hypergraph with fewer than 2^{k-1} hyperedges, each of size at least k, then $\mathcal{H}$ is 2-colourable.*

Proof Colour the vertices at random, assigning to each vertex the colour red with probability $\frac{1}{2}$ and blue otherwise, and making each such choice independently of the choices for all other vertices. In other words, choose a uniformly random 2-colouring of the vertices. For each hyperedge e, define the random variable X_e to be 1 if e is monochromatic and 0 otherwise. (X_e is called an *indicator variable.*) Let $X = \sum_{e \in \mathcal{H}} X_e$, and note that X is the number of monochromatic edges. Any one hyperedge, e, is monochromatic with probability at most $2^{-(k-1)}$, and so $\mathbf{E}(X_e) \leq 2^{-(k-1)}$. Therefore, by the Linearity of Expectation, $\mathbf{E}(X) = \sum_{e \in \mathcal{H}} \mathbf{E}(X_e) \leq |E(\mathcal{H})| \times 2^{-(k-1)} < 1$. Therefore, the probability that $X = 0$, i.e. that there are no monochromatic hyperedges, is positive. □

Remark An alternative proof, avoiding the First Moment Principle, would have been as follows: For each hyperedge e, we denote by A_e the event that e is monochromatic. By the Subadditivity of Probabilities, the probability that at least one such event holds is at most $\sum_{e \in \mathcal{H}} \mathbf{Pr}(A_e) \leq |E(\mathcal{H})| \times 2^{k-1} < 1$, and so the probability that no edge is monochromatic is positive. Note that the calculations are identical.

As the reader will discover in Exercise 3.3, the Subadditivity of Probabilities is a special case of the First Moment Principle. In fact, in many applications of the First Moment Method including the one we have just presented, one can actually get away with using the Subadditivity of Probabilities, which some might prefer since it is a slightly less advanced probabilistic tool. However, in our discussion we will always present our proofs in terms of the First Moment Method, and will not bother to point out the cases where the Subadditivity of Probabilities would have sufficed.

Theorem 3.1 has been improved twice. First, by Beck [18] who replaced 2^{k-1} by $\Omega(k^{1/3-o(1)}2^k)^2$, and then further by Radhakrishnanz and Srivinasan [131] who increased the bound to $\Omega(k^{1/2-o(1)}2^k)$. Both of these proofs involve a much more complicated application of the First Moment Method. Erdős [40] showed that this bound must be at most $O(k^2 2^k)$, again by using the First Moment Method.

3.2 Triangle-Free Graphs with High Chromatic Number

A fundamental question discussed in this book is: What can we say about $\chi(G)$ if we bound $\omega(G)$? In this section we show that bounding $\omega(G)$ alone does not allow us to say much about $\chi(G)$ because there are graphs with no triangles, i.e. with $\omega \leq 2$, and χ arbitrarily high:

Theorem 3.2 *For any $k \geq 1$ there exist triangle-free graphs with chromatic number greater than k.*

We prove this theorem using a probabilistic construction due to Erdős [39], which was one of the first noteworthy applications of the probabilistic method. Actually, Theorem 3.2 had been proved many years before Erdős' construction by Zykov [160], but this construction yields the much stronger generalization that there are graphs with arbitrarily high girth and arbitrarily high chromatic number (see Exercise 3.5). The fact that no one was able to produce a non-probabilistic construction of such graphs for more than 10 years [106, 126] is a testament to the power of the First Moment Method.

This will be our first use of $G_{n,p}$, the random graph defined in Chap. 2. One of the many reasons why this model is fascinating is that it can often be used to prove the existence of large graphs with interesting properties, as we will see here.

Proof of Theorem 3.2. Choose a random graph G from the model $G_{n,p}$ with $p = n^{-\frac{2}{3}}$.

In order to prove that $\chi(G) > k$, it suffices to prove that G has no stable sets of size $\lceil \frac{n}{k} \rceil$. In fact, we will show that with high probability, G does not

[2] $a = \Omega(b)$ means the same thing as $b = O(a)$, i.e. the asymptotic order of a is at least that of b

even have any stable sets of size $\lceil \frac{n}{2k} \rceil$, since we will require this stronger fact to apply an elegant trick in a few paragraphs.

We prove the bound via a simple expected number calculation. Let I be the number of stable sets of size $\lceil \frac{n}{2k} \rceil$. For each subset S of $\lceil \frac{n}{2k} \rceil$ vertices, we define the random variable I_S to be 1 if S is a stable set and 0 otherwise. $\mathbf{E}(I_S)$ is simply the probability that S is a stable set, which is $(1-p)^{\binom{\lceil n/2k \rceil}{2}}$. Therefore by Linearity of Expectation:

$$\begin{aligned}\mathbf{E}(I) &= \sum_S \mathbf{E}(I_S) \\ &= \binom{n}{\lceil n/2k \rceil}(1-p)^{\binom{\lceil n/2k \rceil}{2}} \\ &< \binom{n}{\lceil n/2k \rceil}(1-p)^{\binom{n/2k}{2}}\end{aligned}$$

Since, for positive x, $1-x < e^{-x}$, this yields:

$$\begin{aligned}\mathbf{E}(I) &< 2^n \times \mathrm{e}^{-pn(n-2k)/8k^2} \\ &< 2^n \times \mathrm{e}^{-n^{4/3}/16k^2} \\ &< \frac{1}{2}\end{aligned}$$

for $n \geq 2^{12}k^6$. Therefore, by Markov's Inequality, $\mathbf{Pr}(I > 0) < \frac{1}{2}$ for large n.

Our next step should be to show that the expected number of triangles is also much less than one. Unfortunately, this is not true. However, as we will see, by applying the clever trick alluded to earlier it will suffice to show that with high enough probability the number of triangles is at most $\frac{n}{2}$.

To do this, we compute the expected value of T, the number of triangles. Each of the $\binom{n}{3}$ sets of 3 vertices forms a triangle with probability p^3. Therefore, by applying Linearity of Expectation as in the previous example,

$$\begin{aligned}\mathbf{E}(T) &= \binom{n}{3}p^3 \\ &< \frac{n^3}{3!}\left(n^{-2/3}\right)^3 \\ &= \frac{n}{6}.\end{aligned}$$

Therefore, by Markov's Inequality, $\mathbf{Pr}(T \geq \frac{n}{2}) < \frac{1}{3}$ for large n.

Since $\mathbf{Pr}(I \geq 1) + \mathbf{Pr}(T \geq \frac{n}{2}) < 1$, the probability that $I = 0$ and $T < \frac{n}{2}$ is positive. Therefore, there exists a graph G for which $I = 0$ and $T < \frac{n}{2}$.

And now for the trick that we promised. Choose a set of at most $\frac{n}{2}$ vertices, with at least one from each triangle of G, and delete them to leave the subgraph G'. Clearly G' is triangle-free, and $|G'| \geq \frac{n}{2}$. Furthermore, G' has no stable set of size $\lceil \frac{n}{2k} \rceil \leq \lceil \frac{|G'|}{k} \rceil$, and so $\chi(G') > k$ as desired! □

3.3 Bounding the List Chromatic Number as a Function of the Colouring Number

We now discuss the list chromatic number, defined in Chap. 1. We recall that the list chromatic number lies between the chromatic number and the colouring number. A fundamental though perhaps not well-known result, due to Alon [4], implies that a graph's list chromatic number is much more closely tied to its colouring number than to its chromatic number.

Theorem 3.3 *There exists a function $g(d)$ tending to infinity with d, such that if the minimum degree of a bipartite graph G is at least d then the list chromatic number of G is at least $g(d)$.*

It follows immediately from Exercise 1.5 that if G has colouring number r, then it has a bipartite subgraph of minimum degree at least $\frac{r-1}{2}$. Hence, Theorem 3.3 immediately implies:

Corollary 3.4 *There exists a function $g'(r)$ tending to infinity with r, such that if the colouring number of a graph G is at least r then the list chromatic number of G is at least $g'(r)$.*

The bipartite examples of Exercise 1.8 show that, in contrast, the list chromatic number may be arbitrarily large, even if the chromatic number is only 2. I.e., there is no analogue of Corollary 3.4 bounding the list chromatic number by a function of the chromatic number.

In this section, we prove Theorem 3.3 using Markov's Inequality and Linearity of Expectation.

Consider any bipartite graph H with bipartition (A, B) where $|A| \geq |B|$. In what follows, we will only consider lists of size s, each drawn from the colours $\{1, \ldots, s^4\}$. We will show that if H has minimum degree at least $d = s^4\binom{s^4}{s}$ then there is an assignment of lists to $A \cup B$ such that H does not have an acceptable colouring. This clearly proves the theorem.

We call a set of lists, one for each vertex of A, an *A-set* and we call a set of lists, one for each vertex of B, a *B-set.* Our strategy will be to first fix a B-set $\mathcal{B}$ with a certain property, and then try to find an A-set $\mathcal{A}$ such that there is no acceptable colouring for $\mathcal{A} \cup \mathcal{B}$.

Consider any particular B-set of lists, $\mathcal{B}$. We say that a vertex $v \in A$ is *surrounded* by $\mathcal{B}$ if each of the $\binom{s^4}{s}$ possible lists appears on at least one neighbour of v. We say a B-set $\mathcal{B}$ is *bad* if at least half the vertices of A are surrounded by $\mathcal{B}$. Our proof falls into the following two parts:

Lemma 3.5 *If H has minimum degree at least $d = s^4\binom{s^4}{s}$ then there is a bad B-set of lists.*

Lemma 3.6 *For any bad B-set $\mathcal{B}$, there is an A-set $\mathcal{A}$ such that H does not have an acceptable colouring with respect to the lists $\mathcal{A} \cup \mathcal{B}$.*

Proof of Lemma 3.5: We choose a random B-set by choosing for each $w \in B$ one of the $\binom{s^4}{s}$ possible lists, with each list equally likely to be chosen. Consider a particular vertex $v \in A$. We will bound the probability that v is surrounded.

Let X denote the number of subsets of $\{1, \ldots, s^4\}$ of size s which do not appear as a list on a neighbour of v. There are $\binom{s^4}{s}$ possibilities for such a subset, and the probability that one particular subset does not appear on any neighbour of v is at most $(1 - 1/\binom{s^4}{s})^d$. Therefore, by Linearity of Expectation, $\mathbf{E}(X) \leq \binom{s^4}{s}(1 - 1/\binom{s^4}{s})^d$, and so by Markov's Inequality, the probability that v is not surrounded, i.e. the probability that $X > 0$ is at most $\binom{s^4}{s}(1 - 1/\binom{s^4}{s})^d$. Since $1 - x < e^{-x}$ for positive x, this yields $\mathbf{Pr}(X > 0) < 2^{s^4} \times e^{-s^4} < \frac{1}{2}$.

Let Y denote the number of vertices in A which are surrounded by $\mathcal{B}$. We have already shown that the probability that a particular vertex is surrounded is at least $\frac{1}{2}$. Therefore, by Linearity of Expectation, $\mathbf{E}(Y) \geq \frac{1}{2}|A|$. By the First Moment Principle, the probability that $Y \geq \frac{1}{2}|A|$, i.e. that $\mathcal{B}$ is bad, is positive. Therefore, there must be at least one bad set. □

Proof of Lemma 3.6: Consider any bad set $\mathcal{B}$ and let $\mathcal{A}$ be a random A-set of lists, formed by choosing for each $v \in A$ one of the $\binom{s^4}{s}$ possible lists, with each list equally likely to be chosen. We define the random variable Z to be the number of acceptable colourings of B which extend to an acceptable colouring of $A \cup B$, where of course we mean acceptable with respect to the lists $\mathcal{A} \cup \mathcal{B}$. We start by bounding $\mathbf{E}(Z)$.

There are exactly $s^{|B|}$ acceptable colourings of B. Consider any particular one of them, C. We will bound the probability that C can be extended to an acceptable colouring of $A \cup B$ when $\mathcal{A}$ is chosen. We say that a colour is *available* for a vertex $v \in A$ if it does not appear on a neighbour of v under C. C will be extendible iff for every vertex $v \in A$, the list chosen for v includes at least one available colour. The key observation is that if $v \in A$ is surrounded by $\mathcal{B}$, then v has at most $s - 1$ available colours from $\{1, \ldots, s^4\}$. To see this, note that if v had s available colours, then that subset of colours must form the list of at least one neighbour of v, and so one of those colours must appear on that neighbour under C – contradicting the assertion that they are all available!

Thus, for any surrounded vertex v, the probability that a random colour is available for v is at most $\frac{s-1}{s^4}$, and so the probability that the list chosen for v contains an available colour is at most $s \times \frac{s-1}{s^4} < \frac{1}{s^2}$. Since there are at least $\frac{1}{2}|A|$ such vertices, and their lists are chosen independently, the probability that every surrounded vertex has an acceptable colour in its list is less than $(\frac{1}{s^2})^{\frac{1}{2}|A|} = s^{-|A|} \leq s^{-|B|}$ (where the last inequality is due to the fact that we chose A so that $|A| \geq |B|$).

Therefore, by the Linearity of Expectation, $\mathbf{E}(Z) < s^{|B|} \times s^{-|B|} = 1$, and so with positive probability $Z = 0$. Therefore there is at least one A-set $\mathcal{A}$ for which $Z = 0$. Clearly if we assign the list $\mathcal{A} \cup \mathcal{B}$ to H then there is no acceptable colouring. □

In summary, our proof essentially consisted of the following argument: We first considered a uniformly random B-set of lists in order to prove the existence of a bad B-set, $\mathcal{B}$. We then showed that the expected number of acceptable colourings of H for a list formed by taking the union of $\mathcal{B}$ and a uniformly random A-set of lists is less than 1. (We actually showed that the expected value of a slightly smaller variable is less than 1, but with a little more care, we could also have shown that the expected number of acceptable colourings of H is also less than 1.)

At first glance, it may appear that this was just an awkward way of considering the union of a uniformly random B-set and a uniformly random A-set, and showing that the expected number of acceptable colourings for such a set of lists is less than 1. It is important to understand that this is not at all what happened. In fact, it is not even true that the expected number of such colourings is less than 1 (see Exercise 3.8), and this is the reason that we had to generate our random set of lists in a 2-step process.

The reason that we were not simply analyzing the expected number of acceptable colourings in a uniformly random set of lists, is that the B-set that we chose was *not* a uniformly random B-set. Rather, we merely *considered* a uniformly random B-set to prove the existence of a B-set which was not random and which had some very specific properties, namely it was bad.

This idea of choosing a combinatorial object one step at a time, at each step considering a random choice to prove that we can make a choice which has very specific properties, and thus which might not at all resemble a random choice, will recur frequently in this book. We call it the pseudo-random method.

3.3.1 An Open Problem

By examining our calculations more closely, the reader will see that we proved Theorem 3.3 with $g(d) = \Theta(\log d / \log\log d)$. Alon [6] has improved this to $g(d) = \Theta(\log d)$, and so, for example, there is a constant A such that every bipartite graph G with minimum degree δ has $\chi_\ell(G) \geq A \log \delta$. This is best possible, up to the value of A, since $\chi_\ell(K_{n,n}) = O(\log n)$ (see [45]).

So we understand the minimum possible list chromatic number of, say, a Δ-regular bipartite graph fairly well. The same is far from true about the maximum possible list chromatic number.

Open Question: *Determine the smallest function $h(\Delta)$ such that every bipartite graph G with maximum degree Δ has $\chi_\ell(G) \leq h(\Delta)$.*

The best known upper bound on h is $h(\Delta) \le O(\Delta/\log \Delta)$. This comes from Johansson's bound on the list chromatic number of triangle-free graphs (see Chap. 13). For all we know, it might be the case that $h(\Delta) = O(\log \Delta)$. We don't even have good bounds on the list chromatic number of some fairly common bipartite graphs. For example, Alon[6] has asked what the asymptotic value is of the list chromatic number of the n-cube.

3.4 The Cochromatic Number

We close this chapter with an elegant application of the first moment method due to Alon, Krivelevich and Sudakov [7].

Recall that the chromatic number of a graph G can be thought of as the smallest number t such that $V(G)$ can be partitioned into t stable sets. The chromatic number of $\overline{G}$ is the smallest number t such that $V(G)$ can be partitioned into t cliques. The *cochromatic number* of G, $z(G)$, is the smallest number t such that $V(G)$ can be partitioned into t sets each of which is either a stable set or a clique. This notion was introduced by Lesniak and Straight [104] and inspired in part by its relationship to Ramsey Theory. One appealing property of the cochromatic number which distinguishes it from the chromatic number is that for any graph G, $z(G) = z(\overline{G})$.

Clearly $z(G) \le \chi(G)$, and so it is natural to ask how much smaller $z(G)$ can be than $\chi(G)$. The answer is trivial: the extreme case is the complete graph, as $z(K_n) = 1$ and $\chi(K_n) = n$. Noting that the cochromatic number is nonmonotonic (i.e. it is not necessarily true that for any subgraph $H \subseteq G$, $z(H) \le z(G)$), we can ask a more interesting question: Define $z^*(G) = \max_{H \subseteq G} z(H)$, and note that $z^*(G) \le \chi(G)$. How much smaller can $z^*(G)$ be than $\chi(G)$?

A straightforward argument (see Exercise 3.9) shows that every graph on at most t vertices has cochromatic number at most $(2 + o(1))\frac{t}{\log_2 t}$, and so $z^*(K_t) \le (2 + o(1))\frac{t}{\log_2 t}$ while $\chi(K_t) = t$ (see [43]). Here we will see that, up to a constant multiple, this is the smallest possible value of $z^*(G)$ for any graph with chromatic number t by showing:

Theorem 3.7 [7] *If* $\chi(G) = t$ *then* $z^*(G) \ge \frac{t}{4\log_2 t}(1 + o(1))$.

This improves a result of Erdős and Gimbel[42] who proved a lower bound of $O(\sqrt{\frac{t}{\log_2 t}})$, and answers Question 17.3 of [85], posed by Gimbel in 1990. We first prove that our theorem is true for $|G| \le t^2$:

Lemma 3.8 *If* $\chi(G) = t$ *and* G *has at most* t^2 *vertices, then* $z^*(G) \ge \frac{t}{4\log_2 t}(1 + o(1))$.

Proof Because our desired bound on $z^*(G)$ is an asymptotic one, we can assume that t is large enough to satisfy a few inequalities implicit in our proof. We must show that G has a subgraph H such that $z(H) \geq \frac{t}{4\log_2 t}$. To do this, we choose H at random as follows: $V(H) = V(G)$ and for each edge $e \in E(G)$ we keep e in H with probability $\frac{1}{2}$. This is very reminiscent of the random graph model $G_{n,\frac{1}{2}}$, and in fact, if G were the complete graph on n vertices, then our random subgraph H would be distributed exactly as $G_{n,\frac{1}{2}}$. We will show that with positive probability, $z(H)$ is sufficiently high.

Let X denote the number of cliques in H of size greater than $4\log_2 t$. Let Y denote the number of stable sets in H which induce subgraphs with chromatic number greater than $4\log_2 t$ in G. We will see that with positive probability both X and Y are equal to 0, and so there is at least one subgraph H_1 for which X and Y are 0. This implies that each colour class of any cocolouring of H_1 induces a subgraph of G with chromatic number at most $4\log_2 t$. Thus, $\chi(G) \leq z(H_1) \times 4\log_2 t$ and so $z(H_1) \geq \frac{t}{4\log_2 t}$.

To bound the probability that $Y = 0$, we will focus on Y', the number of stable sets in H which induce subgraphs with minimum degree at least $4\log_2 t - 1$ in G. Note that if $Y' = 0$ then $Y = 0$, by Lemma 1.14. We will now see that $\mathbf{E}(X) < \frac{1}{2}$ and $\mathbf{E}(Y') < \frac{1}{2}$, from which it follows by Markov's Inequality and the Subadditivity of Probabilities that

$$\mathbf{Pr}(X = 0, Y' = 0) > 1 - (\mathbf{Pr}(X > 0) + \mathbf{Pr}(Y > 0)) > 1 - \left(\frac{1}{2} + \frac{1}{2}\right) = 0 .$$

Enumerate the $a \leq \binom{t^2}{\lceil 4\log_2 t\rceil}$ cliques of size $\lceil 4\log_2 t\rceil$ in G: $C_1, \ldots, C_a$. For each i, we define the random variable X_i to be 1 if C_i also induces a clique in H and 0 otherwise. By the Linearity of Expectation, $\mathbf{E}(X) = \mathbf{E}(X_1) + \ldots + \mathbf{E}(X_a)$. Furthermore, for each i, $\mathbf{E}(X_i) = (\frac{1}{2})^{\binom{\lceil 4\log_2 t\rceil}{2}}$. Therefore,

$$\begin{aligned}
\mathbf{E}(X) &\leq \binom{t^2}{\lceil 4\log_2 t\rceil}\left(\frac{1}{2}\right)^{\binom{\lceil 4\log_2 t\rceil}{2}} \\
&\leq \frac{(t^2)^{\lceil 4\log_2 t\rceil}}{(\lceil 4\log_2 t\rceil)!}\left(\frac{1}{2}\right)^{2(\log_2 t)(4\log_2 t-1)} \\
&= \frac{(t^2)^{\lceil 4\log_2 t\rceil}}{(\lceil 4\log_2 t\rceil)!}\left(\frac{1}{t^2}\right)^{(4\log_2 t-1)} \\
&< \frac{t^4}{(4\log_2 t)!} ,
\end{aligned}$$

which is less than $\frac{1}{2}$ for t sufficiently large.

If the subgraph induced by some subset U of V has minimum degree at least $4\log_2 t - 1$ in G, then $|U| \geq 4\log_2 t$ and the number of edges in the subgraph is at least $\frac{1}{2}|U|(4\log_2 t - 1)$. Therefore, each subset of size $r \geq 4\log_2 t$ has a probability of at most $(\frac{1}{2})^{2r\log_2 t - \frac{r}{2}}$ of being counted towards Y'.

Therefore,

$$\begin{aligned}\mathbf{E}(Y') &\leq \sum_{r=\lceil 4\log_2 t\rceil}^{t^2} \binom{t^2}{r}\left(\frac{1}{2}\right)^{2r\log_2 t-\frac{r}{2}} \\ &\leq \sum_{r=\lceil 4\log_2 t\lceil}^{t^2} \frac{t^{2r}}{r!}\times 2^{\frac{r}{2}}\times\frac{1}{t^{2r}} \\ &\leq \sum_{r\geq 4\log_2 t}\frac{2^{\frac{r}{2}}}{r!},\end{aligned}$$

which is less than $\frac{1}{2}$ for t sufficiently large. Note how focusing on Y' rather than Y made our expected value computation easier.

Therefore, with positive probability, we have both $X = 0$ and $Y' = 0$, and so $Y = 0$ as well. □

And now our main theorem follows:

Proof of Theorem 3.7. Again, we can assume that t is large. By Lemma 3.8, we can also assume that $|V(G)| > t^2$. If $z(G)$ is at least $\frac{t}{\log_2 t}$ then so is $z^*(G)$. Otherwise, we will see that G contains a subgraph G' such that $|G'| \leq \chi(G')^2$ and $\chi(G') \geq t - t/\log_2 t$. By Lemma 3.8,

$$z^*(G') \geq \frac{\chi(G')}{4\log_2(\chi(G'))}(1+o(1)) \geq \frac{t}{4\log_2 t}(1+o(1)).$$

Our theorem follows immediately since $z^*(G) \geq z^*(G')$.

If $z(G) < \frac{t}{\log_2 t}$ then $V(G)$ can be partitioned into k stable sets $U_1, \ldots, U_k$ and ℓ cliques $W_1, \ldots, W_\ell$ for some $k + \ell < \frac{t}{\log_2 t}$. We will take G' to be the subgraph induced by $W_1 \cup \ldots \cup W_\ell$. Clearly $\chi(G') \geq \chi(G) - k \geq t - t/\log_2 t$ as any proper colouring of G', along with a new colour for each of $U_1, \ldots, U_k$ provides a proper colouring of G. Furthermore, by our bound on $z(G)$, G contains no cliques of size greater than $\chi(G) = t$, and so $|G'| \leq \ell t < t^2/\log_2 t$ which is less than $\chi(G')^2$ for t sufficiently large. □

Exercises

Exercise 3.1 Prove the First Moment Principle.

Exercise 3.2 Prove Markov's Inequality.

Exercise 3.3 Show how to use the First Moment Method to prove the Subadditivity of Probabilities .

Exercise 3.4

(i) Prove that if H is a hypergraph with every edge of size at least r, then H can be 2-coloured such that at most $\frac{|E(H)|}{2^{r-1}}$ of its hyperedges are monochromatic.

(ii) Prove that if H is a hypergraph with every edge of size at least r, then H can be 2-coloured such that *fewer than* $\frac{|E(H)|}{2^{r-1}}$ of its hyperedges are monochromatic.

(iii) Show that for each $N, r \geq 2$, there exists an r-uniform hypergraph H with $E(H) \geq N$ such that every 2-colouring of H produces at least $\frac{|E(H)|}{2^{r-1}}(1 + o(1))$ monochromatic hyperedges.

Exercise 3.5 Modify the proof of Theorem 3.2 to show that for every g, k there exists a graph G without any cycles of length at most g and with $\chi(G) > k$.

Exercise 3.6 Show that for any n sufficiently large, there exists a graph G on n vertices with chromatic number at least $\frac{n}{2}$ and with clique number at most $n^{3/4}$. (Hint: What can you say about the chromatic number of the complement of a triangle-free graph?)

Exercise 3.7 Use the First Moment Method to prove that every graph G containing a matching with M edges has a bipartite subgraph with at least $\frac{1}{2}(|E(G)| + M)$ edges. (Hint: think of a way to choose a random bipartition such that the M edges of the matching are guaranteed to each have endpoints on opposite sides of the bipartition.)

Exercise 3.8 Consider any bipartite graph H (with no restrictions whatsoever on the degrees of the vertices in H), and any integer $t \geq 3$. For each vertex $v \in H$, choose for v a uniformly random list of size 3 from amongst the colours $\{1, \ldots, 2t\}$. Show that the expected number of acceptable colourings for H with this random set of lists grows very quickly with the number of vertices. What are the implications of this fact, in light of the discussion preceding Subsection 3.3.1? (Hint: show that even the expected number of acceptable colourings in which one part of the bipartition uses only colours from $\{1, \ldots, t\}$ and the other side uses colours from $\{t + 1, \ldots, 2t\}$ is high.)

Exercise 3.9 Use Exercise 1.9 to prove that any graph on t vertices has cochromatic number at most $(2 + o(1))\frac{t}{\log_2 t}$.

4. The Lovász Local Lemma

In this chapter, we introduce one of the most powerful tools of the probabilistic method: The Lovász Local Lemma . We present the Local Lemma by reconsidering the problem of 2-colouring a hypergraph.

Recall that in Sect. 3.1 we showed that any hypergraph with fewer than 2^{k-1} hyperedges, each of size at least k, has a proper 2-colouring because the expected number of monochromatic edges in a uniformly random 2-colouring of the vertices is less than 1.

Now suppose that a k-uniform hypergraph has many more than 2^{k-1} hyperedges, say 2^{2^k} hyperedges. Obviously, the First Moment Method will fail in this case. In fact, at first glance it appears that any attempt to apply the probabilistic method by simply selecting a uniformly random 2-colouring is doomed since the chances of it being a proper 2-colouring are typically very remote indeed. Fortunately however, for the probabilistic method to succeed we don't require a high probability of success, just a positive probability of success.

To be more precise, we will choose a uniformly random 2-colouring of the vertices, and for each hyperedge e, we denote by A_e the event that e is monochromatic. Suppose, for example, that our k-uniform hypergraph consisted of m completely disjoint hyperedges. In this case, the events A_e are mutually independent, and so the probability that none of them hold is exactly $(1-2^{-(k-1)})^m$ which is positive no matter how large m is. Therefore, the hypergraph is 2-colourable.[1]

Of course for a general hypergraph, H, the events $\{A_e | e \in E(\mathcal{H})\}$ are not independent as many pairs of hyperedges intersect. The Lovász Local Lemma is a remarkably powerful tool which says that in such situations, as long as there is a sufficiently limited amount of dependency, we can still claim a positive probability of success.

Here, we state the Lovász Local Lemma in its simplest form. We omit the proof for now, as we will prove it in a more general form in Chap. 19.

The Lovász Local Lemma [44]: *Consider a set $\mathcal{E}$ of (typically bad) events such that for each $A \in \mathcal{E}$*

(a) $\mathbf{Pr}(A) \le p < 1$, *and*

[1] The astute reader may have found an alternate proof of this fact.

(b) A is mutually independent of a set of all but at most d of the other events.

If $4pd \leq 1$ *then with positive probability, none of the events in* $\mathcal{E}$ *occur.*

Remark 4.1 *The inequality* $4pd \leq 1$ *can be replaced by* $\mathrm{e}p(d+1) < 1$*, which typically yields a slightly sharper result. (Here* $\mathrm{e} = 2.71\ldots$*) Only rarely do we desire such precision so we usually use the first form. Shearer [143] proved that we cannot replace "e" by any smaller constant.*

Our first application of the Lovász Local Lemma is the following:

Theorem 4.2 *If* $\mathcal{H}$ *is a hypergraph such that each hyperedge has size at least k and intersects at most* 2^{k-3} *other hyperedges, then* $\mathcal{H}$ *is 2-colourable.*

Remark This application is the one used in virtually every introduction to the Lovász Local Lemma. The authors do not apologize for using it again here, because it is by far the best example. We refer the reader who for once would like to see a different first example to [125] where this application is disguised as a satisfiability problem.

Proof We will select a uniformly random 2-colouring of the vertices. For each hyperedge e, we define A_e to be the event that e is monochromatic. We also define N_e to be the set of edges which e intersects (i.e. its neighbourhood in the line graph of $\mathcal{H}$). Recall that $|N_e| < 2^{k-3}$ by assumption. We shall apply the Local Lemma to the set of events $\mathcal{E} = \{A_e | e \in E(\mathcal{H})\}$.

Claim: Each event A_e is mutually independent of the set of events $\{A_f : f \notin N_e\} \cup A_e$.

The proof follows easily from this claim and the Lovász Local Lemma, as $\mathbf{Pr}(A_e) \leq 2^{-(k-1)}$ and $4 \times 2^{-(k-1)} \times 2^{k-3} \leq 1$. The claim seems intuitively clear, but we should take care to prove it, as looks can often be deceiving in this field.

Suppose that the vertices are ordered $v_1, \ldots, v_n$ where $e = \{v_1, \ldots, v_t\}$. Consider any edges $f_1, \ldots, f_r, g_1, \ldots, g_s \notin N_e$. Let Υ be the set of 2-colourings for which the event $B = A_{f_1} \cap \ldots \cap A_{f_r} \cap \overline{A_{g_1}} \cap \ldots \cap \overline{A_{g_s}}$ holds.

For any 2-colouring ρ of $G - V(e)$, define T_ρ to be the set of the 2^t different 2-colourings of G which extend ρ. It is straightforward to verify that for each ρ, Υ contains either all of T_ρ or none of T_ρ. In other words, there is an ℓ such that Υ is the disjoint union $T_{\rho_1} \cup \ldots \cup T_{\rho_\ell}$ for some $\rho_1, \ldots, \rho_\ell$. Thus, $\mathbf{Pr}(B) = \frac{2^t \ell}{2^n}$.

Within each T_{ρ_i}, there are exactly two 2-colourings in which e is monochromatic, and so $\mathbf{Pr}(A_e \cap B) = \frac{2\ell}{2^n}$. Thus, $\mathbf{Pr}(A_e | B) = (\frac{2\ell}{2^n}) / \mathbf{Pr}(B) = 2^{-(t-1)} = \mathbf{Pr}(A_e)$ as claimed. □

The claim in the preceding proof is a special case of a very useful principle concerning mutual independence. In fact, we appeal to the following fact nearly every time we wish to establish mutual independence in this book.

The Mutual Independence Principle *Suppose that* $\mathcal{X} = X_1, \ldots, X_m$ *is a sequence of independent random experiments. Suppose further that* $A_1, \ldots, A_n$ *is a set of events, where each* A_i *is determined by* $F_i \subseteq \mathcal{X}$. *If* $F_i \cap (F_{i_1}, \ldots, F_{i_k}) = \emptyset$ *then* A_i *is mutually independent of* $\{A_{i_1}, \ldots, A_{i_k}\}$.

The proof follows along the lines of that of the preceding claim, and we leave the details as an exercise.

We end this chapter with another application of the Local Lemma.

4.1 Constrained Colourings and the List Chromatic Number

As discussed in Sect. 3.3, Alon has shown that a graph has bounded list chromatic number if and only if it has bounded colouring number. Thus, if we impose no extra conditions on our lists, to approximately determine how big our lists must be to ensure that an acceptable colouring exists, we need only consider the colouring number. In this section, we show that we can ensure the existence of acceptable colourings for much shorter lists, if we impose a (natural) constraint on the ways in which the lists can intersect. The results discussed in this section first appeared in [133]. For further discussion, including some conjectures, the reader should consult that paper.

As we mentioned in Chap. 1, the greedy colouring procedure yields a bound of $\Delta + 1$ on $\chi_l(G)$. The following theorem suggests that a much stronger result, stated as a conjecture below, may be true. The theorem is quite powerful in its own right and will be used repeatedly throughout the book.

Theorem 4.3 *If there are at least* ℓ *acceptable colours for each vertex, and each colour is acceptable for at most* $\frac{\ell}{8}$ *of the neighbours of any one vertex, then there there is an acceptable colouring.*

Conjecture 4.4 *The* $\frac{\ell}{8}$ *in the above theorem can be replaced by* $\ell - 1$.

Remark The $\frac{\ell}{8}$ in the above theorem can be replaced by $\frac{\ell}{2e}$ by using the more precise version of the Local Lemma. Furthermore, using different techniques, Haxell [77] has proven that the result holds if the value is $\frac{\ell}{2}$, and by iteratively applying the Local Lemma, Reed and Sudakov [134] have shown that $\ell - o(\ell)$ is sufficient.

We now prove Theorem 4.3, which requires an application of the Local Lemma.

Proof of Theorem 4.3. Fix a graph G and an acceptable list of colours L_v for each vertex v, which satisfy the conditions of the theorem. For ease of exposition, we truncate each L_v so that it has exactly ℓ colours.

Now, we consider the random colour assignment in which each vertex is independently assigned a uniform element of L_v. For each edge $e = xy$ and colour $i \in L_x \cap L_y$, we let $A_{i,e}$ be the event that both x and y are coloured with i. We let $\mathcal{E}$ be the set of all such events. We use the Local Lemma to show that with positive probability none of the events in $\mathcal{E}$ occur, i.e. the colouring obtained is acceptable.

Consider first the probability of $A_{i,e}$, clearly this is $\left(\frac{1}{\ell}\right)^2$. Consider next the dependency between events. If e has endpoints x and y, then $A_{i,e}$ depends only on the colours assigned to x and y. Thus, letting $E_x = \{A_{j,f} | j \in L_x, x$ is an endpoint of $f\}$ and letting $E_y = \{A_{j,f} | j \in L_y, y$ is an endpoint of $f\}$, we see that $A_{i,e}$ is mutually independent of $\mathcal{E} - E_x - E_y$. Now, since L_x has exactly ℓ elements, and x has at most $\frac{\ell}{8}$ neighbours of colour i for each $i \in L_x$, we see that $|E_x| \leq \frac{\ell^2}{8}$. Similarly, $|E_y| \leq \frac{\ell^2}{8}$. Thus, setting $d = \frac{\ell^2}{4}$, we see that each $A_{e,i}$ is mutually independent of a set of all but at most d of the other events in $\mathcal{E}$. Since $\left(\frac{1}{\ell}\right)^2 \times \frac{\ell^2}{4} \leq \frac{1}{4}$, the Local Lemma implies that an acceptable colouring exists. This yields the desired result. □

As is often the case with the Local Lemma, once we choose our bad events, the proof is straightforward. However, choosing the good bad events can sometimes be a bit tricky. For example, in Exercise 4.2, we see that two natural attempts at defining the bad events for this application do not lead to proofs.

Exercises

Exercise 4.1 Prove the Mutual Independence Principle.

Exercise 4.2 Show what would go wrong if you attempted to prove Theorem 4.3 by giving each vertex a uniformly random colour from its list and applying the Local Lemma to either of the following sets of bad events.

1. For each vertex v, A_v is the event that v receives the same colour as one of its neighbours.
2. For each edge e, A_e is the event that the endpoints of e both receive the same colour.

Exercise 4.3 Consider a graph G with maximum degree Δ where every vertex v of G has a list L_v of acceptable colours. Each colour $c \in L_v$ has a weight $w_v(c)$ such that $\sum_{c \in L_v} w_v(c) = 1$. Prove that if for every edge uv we have $\sum_{c \in L_u \cap L_v} w_u(c) w_v(c) \leq \frac{1}{8\Delta}$ then G has an acceptable colouring.

5. The Chernoff Bound

The First Moment Principle states that a random variable X is at most $\mathbf{E}(X)$ with positive probability. Often we require that X is near $\mathbf{E}(X)$ with very high probability. When this is the case, we say that X is *concentrated.* In this book, we will see a number of tools for proving that a random variable is concentrated, including Talagrand's Inequality and Azuma's Inequality. In this chapter, we begin with the simplest such tool, the Chernoff Bound.

Recall that $\text{BIN}(n,p)$ is the sum of n independent variables, each equal to 1 with probability p and 0 otherwise. The Chernoff Bound bounds the probability that $\text{BIN}(n,p)$ is far from np, its expected value. For a proof, we refer the reader to [10] or [112].

The Chernoff Bound *For any* $0 \le t \le np$*:*

$$\mathbf{Pr}(|\text{BIN}(n,p) - np| > t) < 2\mathrm{e}^{-t^2/3np}.$$

Remark In the case that $t > np$, it is usually sufficient for our purposes to use the bound

$$\mathbf{Pr}(|\text{BIN}(n,p) - np| > t) < \mathbf{Pr}(|\text{BIN}(n,p) - np| > np) < 2\mathrm{e}^{-np/3}.$$

Only occasionally in applications of the probabilistic method, does one have to resort to the stronger but somewhat more unwieldy bound:

$$\mathbf{Pr}(|\text{BIN}(n,p) - np| > t) < 2\mathrm{e}^{-((1+\frac{t}{np})\ln(1+\frac{t}{np})-\frac{t}{np})np},$$

which holds for all t.

To see how strong the Chernoff Bound is, it is instructive to compare it with the bound obtained from Markov's Inequality:

$$\mathbf{Pr}(\text{BIN}(n,p) - np > t) < \frac{1}{1+\frac{t}{np}}.$$

We will illustrate the Chernoff Bound, with yet another application to 2-colouring hypergraphs. This time, instead of merely ensuring that no hyperedge is monochromatic, we must ensure that no hyperedge has many more vertices of one colour than of the other.

Given a hypergraph H, and a 2-colouring $\mathcal{C}$ of its vertices, for each hyperedge e, we define the *discrepancy* of e to be the absolute value of the difference between the number of vertices of e in each colour class. The *discrepancy of H, with respect to $\mathcal{C}$*, is defined to be the maximum of the discrepancies of the edges of H. The *discrepancy* of H, *disc*(H), is the minimum over all 2-colourings of the vertex set of H, of the discrepancy of H with respect to the 2-colouring.

For example, if H is k-uniform, then $\text{disc}(H) < k$ iff H is 2-colourable. Furthermore, $\text{disc}(H)$ in a sense measures how "good" a 2-colouring we can obtain for H.

Theorem 5.1 *If H is a k-uniform hypergraph with $k > 1$ edges, then* $\text{disc}(H) \leq \sqrt{8k \ln k}$.

Proof We can assume $k \geq 9$ as if $k \leq 8$ then there is nothing to prove as $\text{disc}(H) \leq k < \sqrt{8k \ln k}$.

We two colour the vertex set of H by assigning to each vertex a random colour where each colour is equally likely to be chosen, and where the choices corresponding to different vertices are independent. For any edge e, the number of vertices in e which get colour 1, is distributed precisely like $\text{BIN}(k, \frac{1}{2})$, and so by applying the Chernoff Bound with $t = \sqrt{2k \ln k}$ we find that the probability that the discrepancy of e is greater than $2t$ is at most:

$$\mathbf{Pr}(|\text{BIN}(k, \frac{1}{2}) - \frac{1}{2}k| > t) < 2\mathrm{e}^{-t^2/3(\frac{1}{2}k)} = 2k^{-\frac{4}{3}} < \frac{1}{k}.$$

Therefore, by Linearity of Expectation, the expected number of edges with discrepancy greater than $2t$ is less than 1, and so with positive probability the number of such edges is 0. Thus, the desired two colouring of H exists. $\square$

We refer the reader to [10] for a more thorough discussion of discrepancy, including a deep result of Spencer [144], improving Theorem 5.1:

Theorem 5.2 *If H is a hypergraph with k vertices and k edges, then* $\text{disc}(H) \leq 6\sqrt{k}$.

5.1 Hajós's Conjecture

In this section, we will see how the Chernoff Bound was applied to resolve a strengthening of Hadwiger's Conjecture. First we present a formulation of the conjecture which is slightly different from that given in Sect. 1.5:

Hadwiger's Conjecture *If $\chi(G) \geq k$ then G contains K_k as a minor.*

A *K_k-subdivision* in a graph G, is a subgraph $H \subseteq G$ consisting of k vertices $v_1, \ldots, v_k$, its *centres*, along with a collection of $\binom{k}{2}$ paths, one joining each pair of these vertices, such that the internal vertices of each path do not lie on any other path. (It is possible that some of these paths are merely edges.) Clearly if G contains a K_k-subdivision then it also contains K_k as a minor.

Hajós proposed the following strengthening of Hadwiger's Conjecture: if $\chi(G) \geq k$ then G contains a K_k-subdivision. Here we will show that the uniform random graph usually provides a counterexample, as proved by Erdős and Fajtlowicz [41]. We remark that Caitlin [32] was the first to construct a counterexample to Hajós' conjecture. The result of Erdős and Fajtlowicz is interesting because it shows that almost every large graph is a counterexample.

Theorem 5.3 *For n sufficiently large, there exist graphs with chromatic number at least $\frac{n}{2\log_2 n}$ and with no $K_{8\sqrt{n}}$-subdivision.*

Proof We will choose a graph uniformly at random from all graphs on n vertices. Note that this is equivalent to choosing $G_{n,p}$ where $p = \frac{1}{2}$. We first saw this random graph model in Chap. 2, and then again in our proof of Theorem 3.2.

Our first step is to show that with high probability $\chi(G_{n,\frac{1}{2}}) \geq \frac{n}{2\log_2 n}$. In the proof of Theorem 3.2, we bounded the chromatic number of $G_{n,p}$ by bounding the independence number of $G_{n,p}$. We take the same approach here. In particular, we will show that with high probability $\alpha(G_{n,\frac{1}{2}}) \leq \lceil 2\log_2 n \rceil$, and apply the basic inequality $\chi(G) \geq |G|/\alpha(G)$. Let X denote the number of stable sets of size $a = \lceil 2\log_2 n \rceil$.

$$\begin{aligned}
\mathbf{E}(X) &= \binom{n}{a}\left(\frac{1}{2}\right)^{\binom{a}{2}} \\
&\leq \frac{n^a}{a!}\left(\left(\frac{1}{2}\right)^{\frac{a}{2}}\right)^{a-1} \\
&\leq \frac{n}{a!} \\
&< \frac{1}{n},
\end{aligned}$$

for n sufficiently large. Thus, by Markov's Inequality, $\mathbf{Pr}(X > 0) < \frac{1}{n}$, and so $\mathbf{Pr}\left(\chi(G_{n,\frac{1}{2}}) < \frac{n}{2\log_2 n}\right) < \frac{1}{n}$.

Next, we must show that with high probability $G_{n,\frac{1}{2}}$ has no K_ℓ-subdivision, where $\ell = \lceil 8\sqrt{n} \rceil$. We say that $U \subset V(G)$ is a $\frac{3}{4}$-clique if the subgraph of G induced by U has at least $\frac{3}{4}\binom{|U|}{2}$ edges. We will show that with high probability $G_{n,\frac{1}{2}}$ has no $\frac{3}{4}$-clique of size ℓ. Note that this will be sufficient for our needs as it is easy to show that for any r, the centres of every K_r-subdivision on at most $\frac{r^2}{8}$ vertices must induce a $\frac{3}{4}$-clique of size r.

Consider any subset of ℓ vertices and let Y denote the number of edges between these vertices. Note that Y is distributed exactly like $BIN\left(\binom{\ell}{2}, \frac{1}{2}\right)$. Therefore by the Chernoff Bound,

$$\begin{aligned}\mathbf{Pr}\left(Y \geq \frac{3}{4}\binom{\ell}{2}\right) &\leq 2\mathrm{e}^{-\frac{1}{24}\binom{\ell}{2}} \\ &< 2\mathrm{e}^{-\frac{5}{4}n},\end{aligned}$$

for n sufficiently large. Thus, the expected number of $\frac{3}{4}$-cliques of size ℓ is less than

$$\binom{n}{\ell} 2\mathrm{e}^{-\frac{5}{4}n} < 2^n \mathrm{e}^{-\frac{5}{4}n} < \mathrm{e}^{-\frac{n}{4}},$$

and so by Markov's Inequality, the probability that $G_{n,\frac{1}{2}}$ has a $\frac{3}{4}$-clique of size ℓ is at most $\mathrm{e}^{-\frac{n}{4}}$.

Therefore, for n sufficiently large, the probability that $G_{n,\frac{1}{2}}$ has chromatic number at least $\frac{n}{2\log_2 n}$ and has no K_ℓ-subdivision is, by the Subadditivity of Probabilities, at least $1 - \frac{1}{n} - \mathrm{e}^{-\frac{n}{4}} > 0$, and so such a graph must exist. □

Remark Key to this proof was showing that $\mathbf{Pr}(\chi(G_{n,\frac{1}{2}}) < \frac{n}{2\log_2 n}) < \frac{1}{n}$. It is worth noting that Bollobás [25] and Matula and Kučera [111] independently proved that with high probability $\chi(G_{n,\frac{1}{2}}) = \frac{n}{2\log_2 n}(1 + \mathrm{o}(1))$.

As mentioned earlier, this proof does not just provide a counterexample to Hajós' Conjecture, but it shows that the vast majority of graphs on any large number of vertices are counterexamples. Note however, that for $n = 2^{30}$, $8\sqrt{n}$ is still bigger than $\frac{n}{2\log_2 n}$, and so this proof says nothing about graphs on a smaller number of vertices. So perhaps Hajós can be forgiven for missing this plethora of counterexamples.

Exercises

Exercise 5.1 Consider any hypergraph H where every edge has size at least k and intersects at most d other edges. Use the Chernoff Bound and the Local Lemma to prove that if $d \leq \frac{1}{8}\mathrm{e}^{\ell^2/6k}$ then $\mathrm{disc}(H) \leq \ell$.

Exercise 5.2 Use the Chernoff Bound and Linearity of Expectation to prove that the probability that $G_{n,\frac{1}{2}}$ contains a bipartite subgraph with more than $\frac{n^2}{8} + n^{\frac{3}{2}}$ edges is $o(1)$. Thus, the result in Exercise 1.4 is nearly best possible.

Part III

Vertex Partitions

In each of the next four chapters, an important step of the argument presented is to find a partition of the vertices with certain desirable properties. To find these partitions, we apply the tools of the previous chapters to show that the probability that a randomly chosen partition has the desired properties is positive, thereby proving that such a partition exists.

The next two chapters contain fairly straightforward applications of this technique. In the third chapter, we will have to obtain our partition iteratively – we repeatedly split the vertex set into two parts to obtain finer and finer partitions. In the final chapter of the section, we again consider an iterative procedure. This time, there are two extra twists. The first is that there is some interaction between the random choices in the different iterations. The second is that we will have to combine our probabilistic approach with some structural combinatorics (to wit, Tutte's characterization of those graphs with perfect matchings). These are both complications which will arise in similar settings several times throughout the book.

6. Hadwiger's Conjecture

Kostochka [98] and Thomason [149] have shown that if $\chi(G) \geq \Omega(k\sqrt{\ln k})$ then G has a K_k-minor. To date, this is the best asymptotic progress towards Hadwiger's Conjecture. Actually, they proved the stronger result that if $col(G) \geq \Omega(k\sqrt{\ln k})$ then G has a K_k-minor, and in [150], Thomason determined the best possible multiplicative constant for this bound. In this chapter, we prove the slightly weaker result, originally shown by Mader [109], that if $col(G) \geq \Omega(k \ln k)$ then G has a K_k-minor. Our main theorem is:

Theorem 6.1 *For k sufficiently large, if G has average degree at least $100k \ln k$ then G has a K_k-minor.*

Corollary 6.2 *For k sufficiently large, if $col(G) \geq 100k \ln k + 1$ then G has a K_k-minor.*

Proof If $col(G) \geq 100k \ln k + 1$ then G has a subgraph with minimum degree at least $100k \ln k$. That subgraph certainly has average degree at least $100k \ln k$, and so by Theorem 6.1, it contains a K_k-minor. □

It is worth remarking, that Theorem 6.1 would hold for all values of k, if we replaced 100 by a much larger constant.

In Exercise 6.2, you will prove that Theorem 6.1 remains true when we replace $100k \ln k$ by $Ck\sqrt{\ln k}$, for some constant C. This is best possible up to a constant multiple since for k arbitrarily large, there are graphs with average degree at least $\frac{1}{2}k\sqrt{\ln k}$ which do not contain a K_k-minor (see [98, 55, 26]).

Before proceeding with the proof, we will need a few definitions. Recall that H is a minor of G if H can be obtained from G by a series of vertex-deletions, edge-deletions and edge-contractions. It follows that K_k is a minor of G iff there is a collection of disjoint subsets $V_1, \ldots, V_k \subset V(G)$ such that (a) the subgraph induced by each V_i is connected, and (b) for each i, j, there is an edge from V_i to V_j. We call this collection of subsets a *K_k-minor.*

A *K_k-split-minor* of G is a collection of disjoint subsets $V_1, \ldots, V_k \subset V(G)$ such that (a) the subgraph induced by each V_i *has at most two components*, and (b) for each i, j, there is an edge from *every component of* V_i to V_j.

A graph G is said to be *minor-balanced* if every minor of G has smaller average degree than G.

Our proof will consist of 3 steps:

Lemma 6.3 *If G is minor-balanced with average degree d then G has a subgraph H with at most d vertices and with minimum degree at least $\frac{1}{2}d - 1$.*

Lemma 6.4 *If $|H| \leq d$ and $\delta(H) \geq \frac{1}{2}d - 1$ for some $d \geq 100k \ln k$, then H has a K_{4k}-split-minor.*

Lemma 6.5 *Every K_{4k}-split-minor has a K_k-minor.*

Theorem 6.1 follows immediately from these three lemmas since it is easy to verify that every graph G has a minor-balanced minor with average degree at least that of G – simply repeatedly replace the graph with a minor of at least the same average degree, until no such minor exists. Of these three lemmas, only Lemma 6.4 has a probabilistic proof.

6.1 Step 1: Finding a Dense Subgraph

In this section, we prove Lemma 6.3. Our main tool is the following:

Claim 6.6 *If G is minor-balanced with average degree d then for each vertex v in G and each $u \in N(v)$, we have $|N(u) \cap N(v)| > \frac{1}{2}d - 1$.*

Proof Since G has average degree d, $|E(G)| = \frac{1}{2}d|V(G)|$. Furthermore, since G is minor-balanced, the minor formed by contracting the edge uv has average degree less than d. Therefore, $|E(G)| - |N(u) \cap N(v)| - 1 < \frac{1}{2}d(|V(G)| - 1)$ and the claim follows. □

And now the proof of Lemma 6.3 is quite short:

Proof of Lemma 6.3. Since G has average degree d, it must have a vertex v of degree at most d. Set H to be the subgraph induced by $N(v)$. Clearly $|H| \leq d$, and for any $u \in N(v)$, Claim 6.6 implies that u has at least $\frac{1}{2}d - 1$ neighbours in $N(v)$, and so H has minimum degree at least $\frac{1}{2}d - 1$. □

6.2 Step 2: Finding a Split Minor

Our main step is to show that the H of Lemma 6.3 has a K_{4k}-split-minor, i.e. to prove Lemma 6.4. We will find this split-minor by randomly partitioning H into $4k$ parts, $H_1, \ldots, H_{4k}$, which with positive probability will form the split-minor. We take our random partition in the most natural way possible: for each $v \in V(H)$, we place v into a uniformly chosen part. We will require the following two properties of our partition:

Property 1 *Each vertex has a neighbour in every part.*

Property 2 *For each i, any pair $u, v \in H_i$ such that $|N(u) \cap N(v)| \geq \frac{1}{6}|H|$ has a common neighbour in H_i.*

More precisely, we need:

Claim 6.7 *With positive probability, Properties 1 and 2 both hold.*

Lemma 6.4 follows immediately:

Proof of Lemma 6.4. By Claim 6.7, there is a partition $H_1, \ldots, H_{4k}$ for which Properties 1 and 2 both hold. We claim that such a partition forms a K_{4k}-split-minor. Property 1 clearly ensures that for each i, j there is an edge from every component of H_i to H_j, so all we need to show is that each H_i has at most 2 components.

Suppose that H_i has 3 vertices, x, y, z, in different components of H_i. $|N(x) \cup N(y) \cup N(z)| \leq |H| - 3$, since neither of x, y, z lies in the neighbourhood of another. Also, $|N(x)| + |N(y)| + |N(z)| \geq \frac{3}{2}|H| - 3$, since, by hypothesis, the minimum degree in H is at least $\frac{1}{2}|H| - 1$. Therefore, $|N(x) \cap N(y)| + |N(x) \cap N(z)| + |N(y) \cap N(z)| \geq \frac{1}{2}|H|$, and so at least one of these intersections has size at least $\frac{1}{6}|H|$. But no two of x, y, z have a common neighbour in H_i as they lie in different components, and so this contradicts Property 2. Therefore, no three such vertices exist and so H_i has at most 2 components. □

And now we just need to prove Claim 6.7. This follows from a straightforward application of the First Moment Method.

Proof of Claim 6.7. Let X be the number of vertices v which violate Property 1 and let Y be the number of pairs u, v which violate Property 2.

For any i, the probability that no neighbour of v is placed into H_i is $(1 - 1/4k)^{\deg(v)}$ and so by the Subadditivity of Probabilities, the probability that for at least one i, v has no edge to H_i, is at most $4k \times (1 - 1/4k)^{\frac{1}{2}d-1}$. Therefore,

$$\begin{aligned} \mathbf{E}(X) &\leq |H| \times 4k \times (1 - 1/4k)^{\frac{1}{2}d-1} \\ &< d \times 4k \times \mathrm{e}^{-(d-2)/8k} \end{aligned}$$

Since $(1 - x) < e^{-x}$ for positive x, $d \geq 100k \ln k$, and $d\mathrm{e}^{-(d-2)/8k}$ decreases as d increases above $8k$, we have:

$$\begin{aligned} \mathbf{E}(X) &\leq 100k \ln k \times 4k \times \mathrm{e}^{-12 \ln k} \\ &< \frac{1}{2} \end{aligned}$$

for k sufficiently large.

We turn now to computing $\mathbf{E}(Y)$. The probability that u, v both lie in some H_i is $1/4k$, and if they do then the probability that they do not have a common neighbour in H_i is $(1-1/4k)^{|N_u \cap N_v|}$. Therefore, since $(1-x) < e^{-x}$ for positive x, and $d^2 e^{-d/24k}$ decreases as d increases above $50k \ln k$, we have:

$$\begin{aligned}
\mathbf{E}(Y) &\leq \binom{|H|}{2} \frac{1}{4k} \left(1 - \frac{1}{4k}\right)^{\frac{1}{6}|H|} \\
&< \frac{|H|^2}{2} \times \frac{1}{4k} \times e^{-|H|/24k} \\
&< 1250k(\ln k)^2 e^{-2\ln k} \qquad \text{since } |H| \geq \frac{1}{2}d \geq 50k \ln k \\
&< \frac{1}{2}
\end{aligned}$$

for k sufficiently large.

Therefore, by Markov's Inequality, $\mathbf{Pr}(X > 0) < \frac{1}{2}$ and $\mathbf{Pr}(Y > 0) < \frac{1}{2}$ and so by the Subadditivity of Probabilities, the desired result holds. □

6.3 Step 3: Finding the Minor

We now complete our proof of Theorem 6.1 by showing that every K_{4k}-split-minor, $H_1, \ldots, H_{4k}$, contains a K_k-minor. First, for ease of discussion, we will simplify the structure of our split-minor in two respects:

1. By contracting each component of each part into a vertex, we will assume that each H_i consists of either 1 or 2 vertices.
2. We can assume that in fact each part H_i has exactly 2 vertices, To see this, note that if ℓ parts, say $H_1, .., H_\ell$, each have one vertex and if $H_{\ell+1}, \ldots, H_{4k}$ contains a $K_{k-\ell}$-minor $\{V_{\ell+1}, \ldots, V_k\}$, then $\{H_1, \ldots, H_\ell, V_{\ell+1}, \ldots, V_k\}$ is a K_k-minor. Hence, because $4k - \ell > 4(k - \ell)$, our assumption is justified.

So to prove Lemma 6.5, it suffices to prove the following.

Claim 6.8 *Suppose that the vertices of a graph S are paired $\{a_1, b_1\}, \ldots, \{a_{4k}, b_{4k}\}$, and suppose further that every vertex has a neighbour in every pair but its own. Then S has a K_k-minor.*

Intuitively, if the pairs can be labeled so that most of the a_i's are adjacent to each other then S has a large clique. Otherwise, the edges must "cross" in such a way that by adding only a few extra vertices to a pair we can "connect" it, thus creating many disjoint connected subgraphs each containing a pair, and thereby forming a large clique-minor. To prove our claim, we show that one of these two situations must indeed occur.

Proof of Claim 6.8. We say that a triple of pairs $H_{i_1}, H_{i_2}, H_{i_3}$ is a *connected triple* if their union induces a connected subgraph.

The subgraph induced by any triple of pairs has 6 vertices and minimum degree at least 2. It follows easily that every triple of pairs either induces two disjoint triangles, or is a connected triple. Therefore, any collection of ℓ pairs which does not contain a connected triple, must induce two disjoint ℓ-cliques. By considering a maximum-sized collection of disjoint connected triples, one can easily show that S has either a k-clique or a collection of k disjoint connected triples.

If S has a k-clique, then S certainly has a K_k-minor. On the other hand, if S has a collection of k disjoint connected triples, then they form a collection of disjoint connected subgraphs $C_1, \ldots, C_k$ of S such that each C_i contains both vertices of a pair (in fact, 3 pairs) and so has a neighbour in every other C_j. These subgraphs form a K_k-minor. □

Exercises

Exercise 6.1 Mader [109] proved the following strengthening of Lemma 6.3: *If G is minor-minimal with average degree d sufficiently large then G has a subgraph H with at most d vertices and with minimum degree at least $\frac{3}{5}d$.*

Show that by applying this lemma we can modify the proof of Theorem 6.1 by avoiding the use of split-minors and thus eliminating Lemma 6.5.

Exercise 6.2 Use Mader's lemma from Exercise 6.1 to find a strengthening of Theorem 6.1 to Kostochka and Thomason's result that if G has average degree at least $O(k\sqrt{\ln k})$ then G has a K_k-minor. The outline of your proof might be as follows:

Suppose k is sufficiently large, and take H as guaranteed by Mader's lemma. Randomly partition $V(H)$ into $2k$ subsets $H_1, \ldots, H_{2k}$, each of size exactly $\lfloor\frac{|H|}{2k}\rfloor$, along with a few extra vertices if $2k$ does not divide H evenly. (where, of course, every such partition is equally likely).

For each i, define M_i to be the set of vertices in $H - H_i$ which do not have a neighbour in H_i. Call a set H_i *good* if $|M_i| \leq \left(\frac{1}{100}\right)^{\sqrt{\ln k}} \times |H|$. Call a set H_i *nasty* if there is at least one H_j, $j \neq i$, such that there are no edges between H_i and H_j.

Find a constant C such that if $|H| \leq Ck\sqrt{\ln k}$ and $\delta(H) \geq \frac{3}{5}Ck\sqrt{\ln k}$ then:

1. The expected number of disconnected H_i is at most $\frac{1}{3}k$. (Hint: consider the probability that H_i contains two vertices which do not have a common neighbour in H_i.)
2. Use Markov's Inequality to show that the expected number of sets which are not good is at most $\frac{1}{3}k$.
3. The expected number of sets which are good and nasty is at most $\frac{1}{3}k$. (Hint: bound the probability that there is no edge from H_i to H_j conditional on the event that H_i is good.)

Observe that this implies that G has a K_k-minor.

7. A First Glimpse of Total Colouring

In Part II, we introduced three probabilistic tools and saw an application of each of them. In the last chapter, we saw a more complicated application of one of them, the First Moment Method. In this chapter, we will illustrate the power of combining the other two, the Local Lemma and the Chernoff Bound, by discussing their application to total colouring.

Recall that a total colouring of a graph G consists of a colouring of the vertices and the edges so that:

(i) no two adjacent vertices receive the same colour,
(ii) no two incident edges receive the same colour,
(iii) no edge receives the same colour as one of its endpoints.

The total chromatic number of G, denoted $\chi_T(G)$, is the minimum k for which G has a total colouring using k colours. As mentioned in Chap. 1, Behzad and Vizing independently conjectured that every graph G has a total colouring using $\Delta(G)+2$ colours.

Now, finding a $\Delta+2$ vertex colouring presents no difficulty, as the greedy colouring procedure discussed in Sect. 1.7 will generate one for us. Colouring the edges with $\Delta+2$ colours is also straightforward, for we can apply Vizing's Theorem which ensures that an edge colouring using $\Delta+1$ colours exists. Complications arise when we try to put two such colourings together, as an edge may receive the same colour as one of its endpoints. The crux of the matter is to pair a $\Delta+2$ vertex colouring with a $\Delta+2$ edge colouring so that no such conflicts arise.

Actually, provided we can find a pairing which generates only a few conflicts then we can find a total colouring using not many more than $\Delta+2$ colours. For example, if only r conflicts arise then we can recolour the r edges involved in conflicts with r new colours to generate a $\Delta+r+2$ total colouring. Of course, we may be able to use fewer than r new colours. For example, if the edges involved in conflicts form a matching then we need only one new colour. More generally, if we let R be the graph formed by all those edges whose colour is rejected because they are involved in a conflict, then we can recolour the edges of R with $\chi_e(R) \leq \Delta(R)+1$ new colours to obtain a $\Delta+\Delta(R)+3$ total colouring of G. This is the approach we take in this chapter.

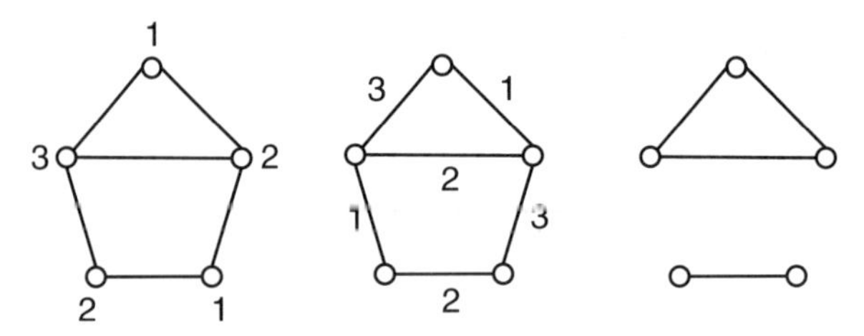

Fig. 7.1. An edge colouring, a vertex colouring and the resultant reject edges

To warm up, we present a result from [117] (the proof also appears in [74]) which uses the Chernoff Bound and the First Moment Method to show:

Theorem 7.1 *Every graph G satisfies: $\chi_T(G) \le \Delta + \lceil \log(|V(G)|) \rceil + 3$.*

Proof We assume $|V(G)|$ is at least three, as otherwise the theorem is trivial. let $l = \lceil \log(|V(G)|) \rceil + 2$. Consider an arbitrary $\Delta + 1$ vertex colouring $C = \{S_1, \ldots, S_{\Delta+1}\}$ and an arbitrary $\Delta + 1$ edge colouring $D = \{M_1, \ldots, M_{\Delta+1}\}$ of G. Let $C_1, \ldots, C_{(\Delta+1)!}$ be the $(\Delta + 1)!$ vertex colourings which are obtained by permuting the colour class names of C. Note that if some of the colour classes are empty then some vertex colourings may appear more than once on this list. We show that for some i, combining C_i with D yields a reject graph R_i with $\Delta(R_i) \le l - 1$ (to be precise, $R_i = \cup_{j=1}^{\Delta+1}\{xy | xy \in M_j,\ x \text{ or } y \text{ receives colour } j \text{ under } C_i\}$). Thus, we can edge colour R_i using l matchings, thereby completing the desired $\Delta + l + 1$ total colouring of G.

To do so, we consider picking a C_i uniformly at random and let $R = R_i$ be the random reject graph thereby obtained. We show that the expected number of vertices of degree at least l in R is less than one and thereby prove that there exists an R_i with maximum degree less than l.

By the Linearity of Expectation, to show that the expected number of vertices of degree at least l is less than 1, it is enough to show that for each vertex v, $\mathbf{Pr}(d_R(v) \ge l) < \frac{1}{n}$.

Now, at most one edge incident to v is in R because it conflicts with v. So we consider the event that there are $l - 1$ edges incident to v which conflict with their other endpoint. We need only show that the probability of this event is less than $\frac{1}{n}$.

We actually show that for any vertex v, the expected number of sets of $l-1$ edges incident to v, all of which are in R because they conflict with their other endpoint is less than $\frac{1}{n}$. Applying Markov's Inequality, we obtain the desired result. To this end, we first compute the probability that a particular set $\{vu_1, .., vu_{l-1}\}$ of $l-1$ edges incident to v are all in R because they conflict with their other endpoint. We let α_i be the colour of vu_i. We let β_i be the colour that u_i is assigned under C. We are computing the probability that our random permutation takes β_i to α_i for $1 \le i \le l - 1$. This probability is zero if the β_i are not distinct. Otherwise, the probability that the permutation does indeed take each of the $l-1$ colours β_i to the corresponding α_i is: $\frac{(\Delta+1-(l-1))!}{(\Delta+1)!}$.

Now, there are at most $\binom{\Delta}{l-1}$ sets of $l-1$ edges incident to v in G. So the expected number of sets of $l-1$ edges incident with v which conflict with their other endpoint is at most:

$$\binom{\Delta}{l-1}\frac{(\Delta+1-(l-1))!}{(\Delta+1)!} < \frac{1}{(l-1)!}.$$

It is easy to see that $(\lceil \log n \rceil + 1)!$ is greater than n provided n is at least three, so the result holds. □

We now want to apply the same technique to obtain a bound on $\chi_T(G)$ which is independent of $|V(G)|$. To do so, we wish to apply the Local Lemma. However, the Local Lemma will only work if we are analyzing a random procedure for which the conflicts in distant parts of the graph occur independently. One way of ensuring that this is true is to assign each vertex a uniformly random colour without considering the colours assigned to the other vertices. Our bad events would each be determined only by the colours on a cluster of vertices which are all very close together, and so events corresponding to clusters in distant parts of the graph would occur independently.

The problem with this approach is that it is very unlikely to generate a proper vertex colouring. To overcome this problem, we will consider a two phase procedure, consisting of a random initial phase which retains the flavour of the random procedure proposed in the preceding paragraph, followed by a deterministic phase which ensures that we have a proper total colouring. We first randomly partition V into k sets $V_1, \dots, V_k$ such that for each i, the graph H_i induced by V_i has maximum degree at most $l-1$ with l near $\frac{\Delta}{k}$. We then greedily colour the vertices of each H_i using the colours in $C_i = \{(i-1)l, \dots, il-1\}$. This yields a kl colouring of $V(G)$.

We fix any $\Delta+1$ edge colouring $\{M_1, \dots, M_{\Delta+1}\}$ before performing this process. We say that an edge xy *conflicts* with the endpoint x if xy is coloured with a colour in C_i and x is assigned to V_i. We note that if e does not conflict with x then in the second phase, the colour assigned to x will be different from that used on e. The advantage to widening our definition of conflict in this way is that now the conflicts depend only on the random phase of the procedure, and this allows us to apply the Local Lemma. Forthwith the details.

Theorem 7.2 *For any graph G with maximum degree Δ sufficiently large, $\chi_T(G) \le \Delta + 2\Delta^{\frac{3}{4}}$*

Proof As usual, we can assume that G is Δ-regular by the construction in Sect. 1.5. Set $k = k_\Delta = \lceil \Delta^{\frac{1}{3}} \rceil$ and $l = l_\Delta = \lfloor \frac{\Delta+\Delta^{\frac{3}{4}}}{k} \rfloor$. We fix an arbitrary edge colouring of G using the colours $1, \dots, \Delta+1$. We then specify a vertex colouring of G using the colours $0, \dots, kl-1 \le \Delta + \Delta^{\frac{3}{4}} - 1$, as follows.

We first partition $V(G)$ into $V_1, \ldots, V_k$ such that

(i) for each vertex v and part i, $|N_v \cap V_i| \leq l-1$,
(ii) For each vertex v, there are at most $\Delta^{\frac{3}{4}} - 3$ edges $e = (u,v)$ such that $u \in V_i$ and e has a colour in C_i.

Our next step will be to refine this partition into a proper colouring, colouring the vertices of V_i using the colours in C_i.

By (i), we can do so using the simple greedy procedure of Lemma 1.3, since the subgraph induced by V_i has maximum degree $l-1$. By (ii), the reject graph formed has maximum degree at most $\Delta^{\frac{3}{4}} - 2$ (there is a 2 and not a 3 here because we may reject an edge incident to v because it has the same colour as v). Recolouring these edges with at most $\Delta^{\frac{3}{4}} - 1$ new colours yields the desired total colouring of G.

It only remains to show that we can actually partition the vertices so that (i) and (ii) hold. To do so, we simply assign each vertex to a uniformly random part (where of course, these choices are made independently). For each v, i we let $A_{v,i}$ be the event that (i) fails to hold for $\{v, i\}$ and B_v be the event that (ii) fails to hold for v. We will use the Local Lemma to prove that with positive probability none of these bad events occur. B_v and $A_{v,i}$ are determined by the colours of the vertices adjacent to v. Thus, by the Mutual Independence Principal, they are mutually independent of all events concerning vertices which are at distance more than 2 from v, and so every event is mutually independent of all but at most $(k+1)\Delta^2 < \Delta^3$ other events. We will show that the probability that any particular bad event holds is much less than $\frac{1}{4\Delta^3}$. Thus, by the Local Lemma, there exists a colouring satisfying (i) and (ii).

Consider first the event B_v. Let Rej_v be the set of edges $e = (u,v)$ with the property that e has a colour in C_i and $u \in V_i$. Since there are k parts, the probability that this occurs for a given e is exactly $\frac{1}{k}$. Furthermore, as the choices of the parts are independent, the size of Rej_v is just the sum of Δ independent 0-1 variables each of which is 1 with probability $p = \frac{1}{k}$. Applying the Chernoff Bound for $\mathrm{BIN}(\Delta, p)$ we obtain:

$$\mathbf{Pr}\left(\left||\mathrm{Rej}_v| - \frac{\Delta}{k}\right| > \frac{\Delta}{k}\right) \leq 2e^{-\frac{\Delta}{3k}},$$

Since $k = \lceil \Delta^{\frac{1}{3}} \rceil$ and $\frac{\Delta^{\frac{3}{4}}}{2} > \frac{\Delta}{k}$, it follows that for Δ sufficiently large,

$$\mathbf{Pr}(B_v) \leq 2e^{-\Delta^{1/2}}.$$

The size of $N_v \cap V_i$ is just the sum of Δ independent 0-1 variables each of which is 1 with probability $\frac{1}{k}$, and so applying the Chernoff Bound as above we obtain that for large Δ,

$$\mathbf{Pr}(A_{v,i}) \leq \mathbf{Pr}\left(\left||N_v \cap V_i| - \frac{\Delta}{k}\right| > \frac{\Delta^{\frac{3}{4}}}{2}\right) \leq 2e^{-\Delta^{1/2}}$$

□

Remark Actually, we can obtain a $\Delta + \mathrm{O}(\Delta^{\frac{2}{3}} \log \Delta)$ total colouring using exactly the same technique, but the computations are slightly more complicated.

8. The Strong Chromatic Number

Consider a graph G on kr vertices. We say that G is *strongly r-colourable* if for any partition of $V(G)$ into parts $V_1, \ldots, V_k$, each of size r, G has a r-colouring such that every colour class contains exactly one vertex from each part (and so every part contains exactly one vertex of each colour). Equivalently, G is strongly r-colourable if for any graph G' which is the union of k disjoint r-cliques on the same vertex set, $\chi(G \cup G') = r$. A well-known conjecture of Erdős, recently proven by Fleischner and Steibitz [58], states that the union of a Hamilton cycle on $3n$ vertices and n vertex disjoint triangles on the same vertex set has chromatic number 3. In other words, C_{3n} is strongly 3-colourable. Strongly r-colourable graphs are of interest partially because of their relationship to this problem, and also because they have other applications (see for example, Exercise 8.1).

To generalize our definition to the case when r does not divide $|V(G)|$, we set $k = \lceil \frac{|V(G)|}{r} \rceil$ and we say that G is *strongly r-colourable* if for any partition of $V(G)$ into parts $V_1, \ldots, V_k$, each of size at most r, G has a r-colouring such that each colour class contains at most one vertex from each part. Equivalently, G is strongly r-colourable if the graph obtained by adding $rk - |V(G)|$ isolated vertices to G is strongly r-colourable.

The *strong chromatic number* of G, $s_\chi(G)$, is defined to be the minimum r such that G is strongly r-colourable. Fellows [54] showed that for every $r \geq s_\chi(G)$, G is strongly r-colourable.

It is not surprising that the strong chromatic number of a graph is often larger than its chromatic number. In fact, there are bipartite graphs with arbitrarily large strong chromatic numbers since $s_\chi(K_{n,n}) = 2n$. To see this, note that $K_{n,n}$ is not strongly $(2n-1)$-colourable since if we set V_1, V_2 to be the two sides of the bipartition, then it is impossible to $(2n-1)$-colour the vertices so that each colour is used at most once in each V_i; the fact that there are only $2n$ vertices implies that $K_{n,n}$ is strongly $2n$-colourable.

Thus, $K_{n,n}$ has strong chromatic number twice as high as its maximum degree, and so it is natural to ask if there is any relation between the strong chromatic number and the maximum degree of a graph. In this chapter, we will see that the strong chromatic number grows at worst linearly with the maximum degree by proving the following theorem, due to Alon [3]:

Theorem 8.1 *If G has maximum degree Δ then $s_\chi(G) \leq 2^{20{,}000}\Delta$.*

It would be interesting to determine the minimum value of c such that every graph with maximum degree Δ, has strong chromatic number at most $c\Delta$ (this is problem 4.14 of [85]). $K_{n,n}$ demonstrates that c must be at least 2. By applying the proof of this chapter more carefully, c can be reduced below 10^{10}.

The first step in the proof of Theorem 8.1 is to show that the strong chromatic number grows at most exponentially with the maximum degree:

Lemma 8.2 *If G has maximum degree Δ, then $s_\chi(G) \leq 2^{\Delta+1}$.*

Proof We can assume $|V(G)| = k2^{\Delta+1}$ for some integer k, as otherwise, we can simply add some isolated vertices to G. We first note that since by Vizing's Theorem $\chi_e(G) \leq \Delta + 1$, $E(G)$ is the union of $\Delta + 1$ edge-disjoint matchings $M_1, \ldots, M_{\Delta+1}$. We wish to show that if H is the union of G and any set of k vertex disjoint $2^{\Delta+1}$-cliques $C_1, \ldots, C_k$ on $V(G)$, then $\chi(H) = 2^{\Delta+1}$.

We will show that we can partition each C_i into two 2^{Δ}-cliques, C_i^1, C_i^2 such that every edge of $M_{\Delta+1}$ has one endpoint in some C_i^1 and the other endpoint in some C_j^2 (where perhaps $i = j$). Therefore, for $\ell = 1, 2$, the subgraph H_ℓ induced by $C_1^\ell, \ldots, C_k^\ell$ is the union of Δ edge-disjoint matchings and k disjoint 2^{Δ}-cliques. Thus, we can repeat the operation again and again, so that at iteration j, H is partitioned into 2^j subgraphs, each of which is the union of $\Delta+1-j$ edge-disjoint matchings and k disjoint $2^{\Delta+1-j}$-cliques. After $\Delta+1$ iterations, we have partitioned H into $2^{\Delta+1}$ independent sets of size k, i.e. we have found our desired colouring.

It only remains to see how to obtain our initial partition. We form a perfect matching M in H by taking an arbitrary perfect matching in each clique C_i. Now, $\chi(M_{\Delta+1} \cup M) = 2$ as it is the union of 2 (not necessarily disjoint) matchings. Take any 2-colouring of $M_{\Delta+1} \cup M$ and note that this colouring induces a partition of each C_i into two parts C_i^1 and C_i^2 with the desired properties. □

Let's look at the proof of Lemma 8.2 in another way. We say that the *G-degree* of a vertex v in any subgraph of H, is the number of edges of G incident to v in that subgraph, i.e. the degree of v not counting the edges added by the cliques. The main idea behind our proof is that we repeatedly split the cliques in half, each time (essentially) reducing the maximum G-degree of the subgraphs by 1 (this is not exactly what happens – we really reduce the chromatic index of the subgraphs of G each time, not their maximum degrees – but it is close enough). Thus, as long as the size of the original cliques are exponential in Δ, we will succeed.

Reflecting on this, it seems remarkably inefficient that upon splitting the vertex set of H in half, we can only reduce the maximum G-degree by 1. A more reasonable goal would be to cut the maximum degree in half. Note that if we could achieve this more ambitious goal at every iteration then we would obtain a linear bound on the strong chromatic number.

We essentially prove Theorem 8.1 in this manner, by considering random partitions. After j iterations, we will have partitioned H into 2^j subgraphs, each of whose vertex sets is the union of cliques of size $k/2^j$. Unfortunately, the maximum G-degrees of the parts are slightly higher than half those in the previous iteration. Furthermore, we cannot continue to reduce the maximum G-degrees all the way to 0. Instead, at some point we will appeal to Lemma 8.2 to complete the colouring. This is why the constant term in Theorem 8.1 is so high.

We actually have to stop splitting before the maximum G-degrees ever drop below 5000. At this point, we will appeal to Lemma 8.2 to show that as long as the size of the cliques at this point is exponential in the maximum G-degree of the subgraphs (and thus is a very large constant), we can colour the subgraphs.

More precisely, we express $\Delta = a2^j$ where a is some real number satisfying $5000 \leq a \leq 10,000$. We will prove that $s_{\chi}(G) \leq 2^{2a+1} \times 2^j$, thus proving Theorem 8.1, as $2^{2a+1}/a < 2^{20,000}$. Again, we can add isolated vertices so that $|V(G)| = k \times 2^{2a+1} \times 2^j$ for some integer k, and so our goal is to prove that if G has maximum degree Δ and if H is the union of G and k disjoint $(2^{2a+1} \times 2^j)$-cliques on the same vertex set then $\chi(H) = 2^{2a+1} \times 2^j$. To do this, we proceed for j iterations, splitting H into 2^j subgraphs, each of which has G-degree at most $2a+1$. Applying Lemma 8.2, each subgraph has chromatic number at most 2^{2a+1} and so H has chromatic number at most $2^{2a+1} \times 2^j$ as required.

Our main lemma is the following:

Lemma 8.3 *Suppose H is the union of a graph G with maximum degree $\Delta \geq 10,000$ and k disjoint $2R$-cliques $C_1, \ldots, C_k$ on the same vertex set. Then we can partition each C_j into two R-cliques C_j^1, C_j^2 such that for $i = 1, 2$, each vertex has at most $\frac{1}{2}\Delta + 2\sqrt{\Delta \ln \Delta}$ G-neighbours in H_i, the subgraph induced by $C_1^i, \ldots, C_k^i$.*

Proof We will choose a random partition of each C_j into 2 equal parts, in a manner to be described later. For each $v \in G$ and $i = 1, 2$, we will denote by A_v^i the event that v has more than $\frac{1}{2}\Delta + 2\sqrt{\Delta \ln \Delta}$ G-neighbours in H_i. We will apply the Local Lemma to show that with positive probability, none of these events occur. Our first step will be to prove that $\mathbf{Pr}(A_v^i) \leq \frac{1}{8\Delta^2}$ for each v, i.

The most natural way to choose our random partition would be to simply take a uniformly random partition. Suppose that we do so. If v has at most one G-neighbour in each C_j, then the number of its G-neighbours in H_i is distributed like $BIN(\deg_G(v), \frac{1}{2})$ and so our desired bound on $\mathbf{Pr}(A_v^i)$ follows from a straightforward application of the Chernoff Bound. If, on the other hand, v has many G-neighbours in some C_j, then this is no longer true as the events that those neighbours land in H_i are not independent, and so the Chernoff Bound does not apply.

This problem can be overcome by applying some of the concentration bounds from later chapters, so it is more of a bother than a serious problem. There is however, a second problem which is less easily overcome. This arises when we try to bound the dependency of the events. A^i_v is clearly dependent on every event $A^{i'}_u$ such that u and v have neighbours lying in the same clique. Thus, the number of events upon which A^i_v is dependent is a function of R, which can be much higher than Δ, and this makes it impossible for us to apply the Local Lemma.

Fortunately, we can overcome both of these problems by not picking our random partition uniformly, but rather by using a matching in a manner reminiscent of the proof of Lemma 8.2, and of Exercise 3.7. Specifically, we first arbitrarily match the vertices of each C_j into pairs. Next, independently for each pair, we put the two vertices into different parts, where each of the two possible choices is equally likely.

For any v we let t be the total number of pairs which contain exactly one neighbour of v, and note that A^i_v holds iff the number of such pairs for which that neighbour is placed into H_i is more than $\frac{1}{2}t + 2\sqrt{\Delta \ln \Delta}$. I.e., the choices for all other pairs, including those containing two neighbours of v, are irrelevant! Thus, $\mathbf{Pr}(A^i_v) = \mathbf{Pr}(BIN(t, \frac{1}{2}) > \frac{1}{2}t + 2\sqrt{\Delta \ln \Delta})$ which by the Chernoff Bound is less than $2e^{-(2\sqrt{\Delta \ln \Delta})^2/3(\frac{t}{2})} < 1/8\Delta^2$, since $t \le \Delta$. (Note that we can assume $2\sqrt{\Delta \ln \Delta} \le \frac{1}{2}t$ here, as otherwise the probability is 0. Therefore, the Chernoff Bound applies.)

Now we turn to bounding the dependency amongst our events. By the Mutual Independence Principle, each event A^i_v is mutually independent of all events $A^{i'}_u$ other than those for which $N_u \cap N_v \neq \emptyset$ or for which a vertex in N_u and a vertex in N_v form one of the pairs used to partition a clique. In other words, dependency is not created merely by having neighbours which both lie in the same clique – they must also both lie on one of these pairs! Thus, A^i_v is mutually independent of all but at most $2\Delta^2$ other events. Therefore, our Lemma follows from the Local Lemma since $2\Delta^2 \times (1/8\Delta^2) = \frac{1}{4}$. □

And now we complete the proof of Theorem 8.1:

Proof of Theorem 8.1. Set $d_0 = \Delta$ and $d_{i+1} = \frac{1}{2}d_i + 2\sqrt{d_i \ln d_i}$. By repeated applications of Lemma 8.3, we can partition H into 2^j subgraphs each of which is the union of k disjoint 2^{2a+1}-cliques and a graph of maximum degree d_j.

Note that $d_j \ge \Delta/2^j = a$. We will show that $d_j \le 2a$. In fact, we show by induction that for $i \le j$, we have: $d_i \le \frac{\Delta}{2^i} + 16\sqrt{d_i \ln d_i}$. This is clearly true for $i = 0$. For higher $i \le j$,

$$d_i \le \frac{1}{2}\left(\frac{\Delta}{2^{i-1}} + 16\sqrt{d_{i-1}\ln d_{i-1}}\right) + 2\sqrt{d_{i-1}\ln d_{i-1}}$$

$$= \frac{\Delta}{2^i} + 10\sqrt{d_{i-1}\ln d_{i-1}}$$

$$\leq \frac{\Delta}{2^i} + 10\sqrt{2d_i \ln 2d_i}$$
$$\leq \frac{\Delta}{2^i} + 16\sqrt{d_i \ln d_i}, \qquad \text{since } d_i \geq a \geq 5000.$$

Thus the desired result does indeed follow by induction. In particular, $d_j \leq \frac{\Delta}{2^j} + 16\sqrt{d_j \ln d_j}$.

Now, $\frac{\Delta}{2^j} = a$. Furthermore, since $a \geq 5000$, we have that every $b \geq 2a$ satisfies $a + 16\sqrt{b \ln(b)} < b$. This implies $d_j \leq 2a$. Therefore, by Lemma 8.2 each of the 2^j subgraphs can be 2^{2a+1}-coloured. By using a different set of colours on each subgraph, we obtain our $2^{2a+1} \times 2^j$-colouring of H. □

Exercises

Exercise 8.1 Show that Theorem 4.3, with the constant 8 increased to $2^{20,000}$ is a corollary of Theorem 8.1. In fact, prove the stronger result that we can find a set ℓ different acceptable colourings such that for each vertex v and color $c \in L(v)$, v receives c in exactly one of the colourings.
(Hint: given a graph G, with lists of size ℓ, consider forming a new graph G' which has exactly ℓ copies of each vertex of G, and where each such collection of copies forms an ℓ-clique.)

9. Total Colouring Revisited

9.1 The Idea

In Chap. 7, we constructed total colourings by first choosing an edge colouring and then choosing a vertex colouring which didn't significantly conflict with it. We then obtained a total colouring by modifying the edge colouring so as to eliminate the conflicts. In this chapter, we take the opposite approach, first choosing a vertex colouring and then choosing an edge colouring which does not conflict *at all* with the vertex colouring, thereby obtaining a total colouring.

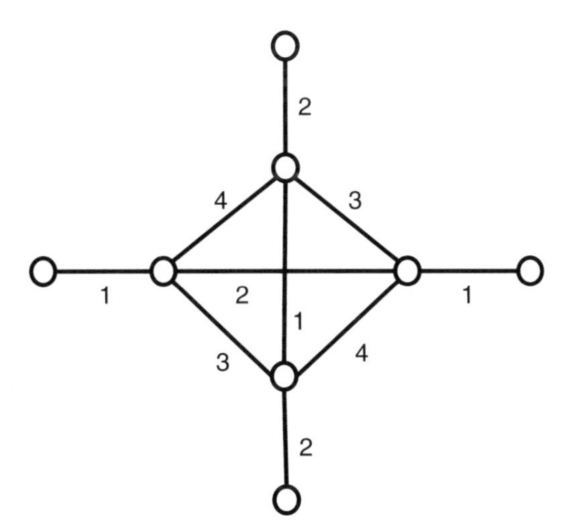

Fig. 9.1. A non-extendible Δ edge colouring

It is believed that for every $(\Delta+3)$-vertex colouring, there is some edge colouring using the same colours with which it does not conflict. We note that the analogous statement does not hold for edge colouring, even if we replace the $\Delta+3$ with $2\Delta-1$. To see this, consider the graph obtained from a clique of order Δ by adding a pendant edge at each vertex (the case $\Delta=4$ is illustrated in Fig. 9.1). The clique has maximum degree $\Delta-1$ and hence has a Δ-edge colouring. We can extend this to a Δ-edge colouring of the whole graph by colouring the pendant edge from a vertex v of the clique with the colour which does not appear on any of the other edges incident with v. But now Δ new colours are needed to colour the vertices of the clique if we are to avoid conflicts.

Given that we cannot always extend $(\Delta + 3)$-edge colourings without introducing new colours, why should we expect to be able to extend $(\Delta + 3)$-vertex colourings? The answer is that colouring the vertices places very few restrictions on the edge colouring. Specifically, consider fixing a vertex colouring $\mathcal{C}$ which uses the colours $\{1, \ldots, \Delta + 3\}$. Then, for each edge e there is a list of $\Delta + 3 - 2 = \Delta + 1$ acceptable colours whose assignment to e will not generate a conflict. Thus, if the List Colouring Conjecture is correct there is an edge colouring in which each edge receives an acceptable colour and hence which does not conflict with $\mathcal{C}$. We note that $\Delta + 3$ is best possible here, as Hind [79] has given examples of $(\Delta + 2)$-vertex colourings which cannot be extended to $(\Delta + 2)$-total colourings.

Although we believe that every $(\Delta + 3)$-vertex colouring can be extended to a $(\Delta + 3)$-total colouring, we will consider only vertex colourings with a special property which makes them easier to extend. Before defining the vertex colourings we are interested in, we sketch our approach.

Definition A total colouring with colour classes $\{T_1, \ldots, T_\ell\}$ is an *extension* of a vertex colouring with colour classes $\{S_1, \ldots, S_m\}$ if $m \leq \ell$ and for j between 1 and m, there is a (possibly empty) matching M_j such that $T_j = S_j \cup M_j$ whilst for $j > m$, T_j is a matching. In this case, we say $\{S_1 \ldots, S_m\}$ can be *extended* to an ℓ-total colouring.

Definition A matching M *misses* a set of vertices A if no vertex in A is the endpoint of an edge in M. M *hits* A if every vertex in A is the endpoint of an edge of M.

As discussed in Sect. 1.5, we can assume that G is Δ-regular. We consider some $(\Delta + 1)$-vertex colouring of G and for $1 \leq i \leq \Delta + 1$, we let S_i be the set of vertices of colour i. Suppose that we could choose disjoint matchings $M_1, \ldots, M_{\Delta+1}$ such that each M_i is a perfect matching in $G - S_i$. Then the M_i would form a partition of $E(G)$ because each vertex is in exactly one S_i, and hence is incident to an edge in exactly Δ of these matchings. Thus, giving the edges of M_i colour i yields a $(\Delta + 1)-$total colouring of G.

Of course, the approach sketched in the above paragraph will not always work, as some Δ-regular graphs have total chromatic number $\Delta + 2$. To see what can go wrong, let us consider even cliques. The only $\Delta + 1$ vertex colouring of an even clique assigns every vertex a different colour. Now, for each S_i, $G - S_i$ is odd and hence has no perfect matching.

Fortunately, this is not a significant problem. For example, for each i we can find a matching missing only one vertex of $G - S_i$ and such that each vertex is missed by exactly two of these matchings, one of which corresponds to the colour assigned to it. Thus, the edges remaining uncoloured after deleting $M_1, \ldots, M_{\Delta+1}$ form a 1-regular graph, i.e. a matching which can be coloured with one new colour. Hence every even clique has total chromatic number $\Delta + 2$.

This notion generalizes to other regular graphs G. For each i, we choose some set X_i of vertices in $G - S_i$ which we allow M_i to miss, such that every vertex of G is in a small number of these X_i. Note that if we start with a $(\Delta + 1)$-vertex colouring and insist that each vertex lies in at most ℓ of the X_i, then the graph $G - \cup\{M_i | 1 \leq i \leq \Delta + 1\}$ has maximum degree at most ℓ and hence its edges can be $(\ell + 1)-$coloured, yielding a $(\Delta + \ell + 2)$-total colouring of G. By considering large values of ℓ, we deal with the parity problem mentioned above, as well as with other problems of a similar nature.

A more significant difficulty arises if we are careless in choosing the original vertex colouring. Suppose for example that all the neighbours of some vertex v are assigned the colour 1. In this case, M_1 must miss v. More generally, if a vertex v has a large number of neighbours in each of many different S_i then it may prove very difficult to find the desired total colouring. To avoid this problem, we choose a vertex colouring for which the number of times any colour appears in a neighbourhood is bounded.

Definition A *k-frugal colouring* is a proper vertex colouring in which no colour appears more than k times in any one neighbourhood.

Hind, Molloy and Reed [80] proved the following:

Theorem 9.1 *Every graph G with maximum degree Δ sufficiently large, has a $\log^8 \Delta$-frugal $(\Delta + 1)$-colouring.*

In this chapter, we will prove:

Theorem 9.2 *There exists a Δ_0 such that for $\Delta \geq \Delta_0$ every $\log^8 \Delta-$frugal $\Delta + 1$ colouring of a graph G with maximum degree Δ can be extended to a $(\Delta + 2\log^{10} \Delta + 2)$-total colouring of G.*

Combining these two results yields the main result of [81]:

Theorem 9.3 *If G has maximum degree Δ then $\chi_T(G) \leq \Delta + O(\log^{10} \Delta)$.*

Proof For $\Delta < \Delta_0$, the result is true since G has a $2\Delta_0 + 1$ total colouring, and for $\Delta \geq \Delta_0$, the result is true by Theorems 9.1 and 9.2. □

It remains to give more details of the proof of Theorem 9.2 which follows the lines sketched above. In doing so, the following well-known extension of Tutte's Theorem characterizing which subsets of $V(G)$ can be hit by a matching will be useful (see Exercise 3.1.8 on page 88 of [107]).

Lemma 9.4 *For any $R \subseteq V(G)$, there is a matching hitting R if and only if there is no set T of vertices of G such that $G - T$ contains $|T| + 1$ odd components which are completely contained in R.*

To prove Theorem 9.2 using the Local Lemma, we need to find a local condition which ensures that the global condition of Lemma 9.4 holds. This is provided by the following corollary of Lemma 9.4 whose derivation we leave as an exercise (see Exercise 9.1).

Corollary 9.5 *Let R be a subset of the vertices of a graph G such that for some positive integer ℓ:*

(i) for each $u \in R$, $d(u) \geq \Delta(G) - \ell$, and
(ii) for each $v \in V(G)$, if $d(v) \geq \Delta(G) - \ell$, then $|N(v) - R| > \ell$.

Then G has a matching hitting R.

9.2 Some Details

To recap, we are given a $\log^8 \Delta$-frugal vertex colouring of G with colour classes $\{S_1, \ldots, S_{\Delta+1}\}$, which we wish to extend to a total colouring. One approach is to find a sequence of vertex sets $X_1, \ldots, X_{\Delta+1}$ such that (i) each vertex is only in a small number of these sets, and (ii) we can find a sequence of disjoint matchings $M_1, \ldots, M_{\Delta+1}$ such that each M_i misses S_i and hits $V(G) - S_i - X_i$. In fact, as we will see, we only need to require M_i to hit all vertices in $V(G) - S_i - X_i$ which have nearly maximum degree. Since M_i must miss every vertex in S_i, it makes sense to pick X_i disjoint from S_i.

Thus, in choosing our X_i, we require two properties. The first is that no vertex lies in too many of these sets:

(P1) Each vertex lies in at most $\log^{10} \Delta$ of the X_i.

The second will allow us to use Corollary 9.5 to ensure that the matchings M_i exist. Setting $G_i = G - \cup_{j<i} M_j$, we require:

(P2) If $d_{G_i}(v) \geq (\Delta - i) - 3\log^8 \Delta$ then $|N_{G_i}(v) \cap X_i| \geq 2\log^8 \Delta$.

We will actually only choose $X_1, \ldots, X_{i_0}$ where $i_0 = \Delta - \log^{10} \Delta$, because for higher values of i, $\Delta(G_i)$ becomes too small to be manageable. It won't harm us to stop here, because this change will only increase by $\log^{10} \Delta + 1$ the number of colours required to finish off the edges not covered by the sequence of matchings, and so we will obtain a $\Delta + 2\log^{10} \Delta + 2$ total colouring.

There is a fairly straightforward manner in which to choose each X_i randomly in order to try to apply the Local Lemma to show that a sequence of sets satisfying P1, P2 exists. Unfortunately, this method fails for a very subtle reason.

False Proof. For each $1 \leq i \leq i_0$ and for each vertex $v \notin S_i$, we place v into X_i with probability $8 \log^8 \Delta/(\Delta - i)$, where of course these random choices are all independent. $A_{v,i}$ is the event that v, i violate property P2, and B_v is the event that v violates property P1.

Since $\Delta - i \geq \log^{10} \Delta$, and since v has at most $\log^8 \Delta$ neighbours in S_i, if $d_{G_i}(v) > (\Delta - i) - 3\log^8 \Delta$ then v has at least $\frac{1}{2}(\Delta - i)$ neighbours in $G_i - S_i$. Therefore, by the Chernoff Bound,

$$\mathbf{Pr}(A_{v,i}) \leq \mathbf{Pr}\left(BIN\left(\frac{1}{2}(\Delta - i), \frac{8\log^8 \Delta}{\Delta - i}\right) < 2\log^8 \Delta\right) < 2\mathrm{e}^{-\log^8 \Delta/3}.$$

For any vertex v, let $X(v)$ denote the number of sets X_i that v is in. By Linearity of Expectation and the fact that $\sum_{i=1}^{j} \frac{1}{j} \leq 1 + \log j$, we have:

$$\mathbf{E}(X(v)) \leq \sum_{i=1}^{i_0} \frac{8\log^8 \Delta}{\Delta - i} < 8\log^8 \Delta \times \sum_{i=1}^{\Delta} \frac{1}{i} < 16\log^9 \Delta.$$

Because the probability that v is in X_i varies with i, $X(v)$ is not a binomial variable and so we cannot apply the Chernoff Bound . However, very similar bounds, such as the Simple Concentration Bound presented in Chap. 10, will yield

$$\mathbf{Pr}(B_v) \leq \mathbf{Pr}(|X(v) - \mathbf{E}(X(v))| > \log^9 \Delta) < 2\mathrm{e}^{-\log^8 \Delta}.$$

(This bound really holds. This is not the mistake in the proof.)

Furthermore, $A_{v,i}$ is defined only by the random choices of whether u goes into X_i for each neighbour u of v, and B_v is determined only by the random choices of whether v goes into X_i for each i. Therefore, by the Mutual Independence Principle, any event $A_{v,i}$ or B_v is mutually independent of all events $A_{u,i}$ and B_u such that u and v are at distance at least 3 in G. Thus, each event is mutually independent of all but at most Δ^3 other events. Since $\Delta^3 \times 2\mathrm{e}^{-\log^8 \Delta/3} < \frac{1}{4}$ for Δ sufficiently large, the Local Lemma implies the existence of sets $X_1, \ldots, X_{i_0}$ satisfying P1, P2. □

The reader who found this proof convincing should try to find the subtle error before reading further.

Error in Proof. The reader who is reading ahead without trying to find the error himself, should feel shame and turn back. For the rest of you:

The error in the proof is our use of the innocuous phrase: *for each neighbour u of v.* Here neighbour refers to neighbour in G_i. Thus, the event $A_{v,i}$ is determined not just by the choices of whether the neighbours of v in G go into X_i. It is also determined by any random choices which help to determine which members of $N_G(v)$ lie in $N_{G_i}(v)$. In other words, it is determined by all random choices that determine which edges incident to v lie in the matchings $M_1, \ldots, M_{i-1}$. Since these matchings are selected deterministically according to Corollary 9.5, for any vertex u anywhere in G, changing whether u is in X_j

could have sweeping affects on M_j and could conceivably affect which, if any, edge incident to v lies in M_j. Therefore, $A_{v,i}$ is dependent on virtually every random choice concerning X_j for every $j < i$, and so the Local Lemma does not apply.

In order to overcome this problem, we must carry out several applications of the Local Lemma – one for each $1 \leq i \leq i_0$. More specifically, for each i, in sequence, we pick X_i at random and prove that with positive probability P2 holds for X_i. This will ensure that M_i exists, and so we set $G_{i+1} = G_i - M_i$, choose X_{i+1} at random, and use another application of the Local Lemma to show that with positive probability P2 holds for X_{i+1}. Since we are applying a separate application of the Local Lemma for each i, we don't have to worry about dependency on the choices concerning X_j for any $j < i$. The only drawback is that we cannot enforce property P1 through these independent applications of the Local Lemma. So we must enforce P1 in a different manner, as follows.

For each i, we will have a set F_i of vertices which are forbidden to go into X_i. We wish to place each vertex v into enough sets F_i that v cannot appear in more than $\log^{10} \Delta$ sets X_i, thus enforcing P1. The most natural way to do this would be to simply wait until v has been placed into $\log^{10} \Delta$ of the X_i and then put v into every subsequent F_j. It turns out, however, that this is not the best approach (see Exercise 9.2), and it is better to instead choose the F_j so that the sets X_i to which v belongs are spread out over the entire sequence of sets. Of the various ways to do so, the best is to use the following rule:

9.6 *If $v \in X_i$ then $v \in F_j$ for all $i < j \leq i + \frac{\Delta - i}{\log^9 \Delta}$.*

That is, after v is placed in some X_i, the number of subsequent sets for which v is forbidden is directly proportional to the number of sets remaining. The reader can verify (we will do so in a minute), that this ensures that each vertex is in at most $\log^{10} \Delta$ of the X_i, i.e. that P1 holds.

Of course, we will run into problems if F_i ever grows too big. For example, if through some unfortunate stroke of luck, every vertex goes into X_1, then for the next several values of i we will have $F_i = V(G)$ and so X_i will be empty. To ensure that this does not happen, we will enforce the following property:

(P3) For every vertex v and each i, $|N_{G_i}(v) \cap X_i| \leq 20 \log^8 \Delta$.

Our aim is to show that the maximum degree of G_i decreases sufficiently quickly as i increases. To do this, for each i we must bound the number of sets X_j, with $j < i$, in which a vertex can appear. By focusing only on the restriction imposed by (9.6), we get such a bound as follows.

Set $a_1 = 1$ and recursively define $a_{l+1} = a_l + \frac{\Delta - a_l}{\log^9 \Delta}$. Set $k_1 = 0$ and for each $2 \leq i \leq i_0 + 1$, set k_i to be the largest value of k with $a_k < i$. Clearly, the maximum number of sets X_j with $j < i$ that a vertex can lie in, subject to (9.6), is k_i.

For each $1 \le i \le i_0 + 1$, set

$$D_i = \Delta - i + k_i + 2.$$

D_i is the best bound that we can hope for on the maximum degree of G_i, since a vertex v of degree Δ in G might lie in up to k_i previous sets X_j as well as one previous set S_j, and thus would be unmatched in at least $k_i + 1$ of the matchings $M_1, \ldots, M_{i-1}$. As we shall see, our construction will in fact achieve this bound.

Note that $\Delta - a_k = (\Delta - 1)(1 - \frac{1}{\log^9 \Delta})^{k-1}$ and so $k_{i_0+1} < \log^{10} \Delta - 2$. Thus we have $D_i < \Delta - i + \log^{10} \Delta$, and in particular know that G_{i_0} has maximum degree at most $2 \log^{10} \Delta$.

Our main step is to make use of the Local Lemma to obtain an inductive proof of the following:

Lemma 9.7 *For each $1 \le i \le i_0$, there is a set $X_i \subset V(G) - F_i - S_i$ such that*

(a) P2 holds,
(b) P3 holds,
(c) G_i has a matching M_i which misses S_i and hits every vertex in $V(G) - S_i - X_i$ which has degree at least $D_i - 2$ in G_i.

G_i and F_i are, of course, defined recursively as discussed earlier.

With Lemma 9.7 in hand, the proof of the main theorem follows via some routine but tedious calculations, which we dispose of first.

Proof of Theorem 9.2. The main step is to prove that $\Delta(G_i) \le D_i$, as mentioned earlier. This is quite simple, we just have to be a little careful.

Consider any vertex v and any $1 \le i \le i_0$. We wish to show that $d_{G_i}(v) \le D_i$. Let j be the smallest index such that for all $j \le r \le i$, $d_{G_r}(v) \ge D_r - 2$. (If no such j exists then clearly $d_{G_i}(v) \le D_i$.) Note that if $j = 1$ then $d_{G_j}(v) \le D_j - 1 = \Delta$, and if $j > 1$ then

$$d_{G_j}(v) \le d_{G_{j-1}}(v) \le D_{j-1} - 3 \le D_j - 2.$$

It is easily verified that v lies in X_t for at most $k_i - k_j + 1$ values of t with $j \le t < i$ if $j > 1$, and for at most $k_i - k_j$ such values when $j = 1$. Also, v might lie in S_t for some $j \le t < i$. Furthermore, our choice of j ensures that by condition (c) of Lemma 9.7, v is not missed by M_t for any other $j \le t < i$. Therefore, for $j > 1$ we have:

$$\begin{aligned} d_{G_i}(v) &\le (D_j - 2) - ((i - j) - (k_i - k_j + 2)) \\ &= D_j - (k_j - j) + (k_i - i) \\ &= D_i \end{aligned}$$

If $j = 1$, we have:

$$d_{G_i}(v) \leq (D_j - 1) - ((i - j) - (k_i - k_j + 1)) = D_i.$$

Now it is straightforward to find our $(\Delta + 2\log^{10}\Delta + 2)$-total colouring of G which extends $\{S_1, \ldots, S_{\Delta+1}\}$. For $1 \leq i \leq i_0$, we assign colour i to $M_i \cup S_i$, and for $i_0 < i \leq \Delta + 1$, we assign colour i to S_i. All that remains is to colour the edges of G_{i_0+1}. Since $\Delta(G_{i_0+1}) \leq D_{i_0+1} < 2\log^{10}\Delta$, Vizing's Theorem implies that these edges can be coloured with $2\log^{10}\Delta + 1$ new colours. □

It only remains to prove Lemma 9.7, which we do in the next section.

9.3 The Main Proof

Proof of Lemma 9.7. We assume that the lemma holds for each $j < i$ and proceed by induction. Observe that since the lemma holds for all $j < i$, it follows as in the proof of Theorem 9.2 that $\Delta(G_i) \leq D_i$.

Our first step will be to apply the Local Lemma to show that we can choose X_i such that P2 and P3 hold. We will then show that P2 implies condition (c).

As in our false proof, we place each vertex $v \notin F_i \cup S_i$ into X_i with probability $8\log^8\Delta/(\Delta - i)$, where each of these random choices is made independently of the others. For any vertex v, let A_v denote the event that v violates P2 and let B_v be the event that v violates $P3$.
As mentioned earlier, $k_i < \log^{10}\Delta - 2$ and so $D_i < \Delta - i + \log^{10}\Delta < 2(\Delta - i)$. Since $d_{G_i}(v) \leq D_i$, the Chernoff Bound yields

$$\mathbf{Pr}(B_v) \leq \mathbf{Pr}(BIN\left(2(\Delta - i), \frac{8\log^8\Delta}{\Delta - i}\right) > 20\log^8\Delta) < 2e^{-\log^8\Delta/3}.$$

A similar argument yields a bound on $\mathbf{Pr}(A_v)$, the first step is to bound $|N_{G_i}(v) - F_i|$.

Any vertex $u \in F_i$ is there because $u \in X_j$ for some j with $j < i \leq j + \frac{\Delta - j}{\log^9\Delta}$. For such a j, we have $\Delta - i > (1 - \frac{1}{\log^9\Delta})(\Delta - j)$. Combining these two facts, we see that for such a j, we have: $j \geq i - \frac{\Delta - i}{2\log^9\Delta}$.

In the following, the first line comes from this fact along with the fact that our inductive hypothesis P3 holds for each $j < i$.

$$\begin{aligned} |N_{G_i}(v) \cap F_i| &\leq 20\log^8\Delta \times 2\frac{\Delta - i}{\log^9\Delta} \\ &= \frac{40(\Delta - i)}{\log\Delta}. \end{aligned}$$

Since $i \le i_0$, $\Delta - i \ge \log^{10}\Delta$. Therefore, if $d_{G_i}(v) \ge (\Delta - i) - 3\log^8\Delta$ then since v has at most $\log^8\Delta$ neighbours in S_i, $|N_{G_i}(v) - F_i - S_i| \ge \frac{1}{2}(\Delta - i)$. It now follows as in the (correct) calculation in our false proof that by the Chernoff Bound, $\mathbf{Pr}(A_v) \le 2e^{-\log^8\Delta/3}$.

Each event A_v or B_v is mutually independent of all events A_u and B_u where u, v are of distance at least 3, i.e. of all but at most $2\Delta^2$ other events. Since $2\Delta^2 \times 2e^{-\log^8\Delta/3} < \frac{1}{4}$ for Δ sufficiently large, it follows from the Local Lemma that with positive probability none of these events occur, and so it is possible to choose X_i such that P2 and P3 both hold.

Now we show that since P3 holds, we can find our desired matching M_i. We will apply Corollary 9.5 to $G_i - S_i$, setting $\ell = 2\log^8\Delta - 1$ and letting R be the set of vertices of degree at least $D_i - 2$ in $G_i - S_i - X_i$.

Since $\Delta(G_i - S_i) \le \Delta(G_i) \le D_i$, and since by the frugality of our vertex colouring, each vertex has at most $\log^8\Delta$ neighbours in S_i, if $v \in R$ then

$$d_{G_i - S_i}(v) \ge D_i - 2 - \log^8\Delta \ge \Delta(G_i - S_i) - 2\log^8\Delta + 1,$$

and so condition (i) of Corollary 9.5 holds.

Furthermore, $\Delta(G_i - S_i) \ge \Delta(G_i) - \log^8\Delta \ge (\Delta - i) - \log^8\Delta$. Thus, property P2 implies condition (ii) of Corollary 9.5. Therefore, $G - S_i$ has a matching hitting R as required. □

Exercises

Exercise 9.1 Prove Corollary 9.5 as follows. Show that for any subset $T \subseteq V(G)$, we have the following:

(i) every odd component of $(G - T)$ which is contained entirely within R has at least $\Delta(G) - \ell$ neighbours in T;
(ii) every vertex in T has at most $\Delta(G) - \ell$ neighbours in $R - T$.

Now argue that Lemma 9.4 implies Corollary 9.5.

Exercise 9.2 Consider the following two possible ways to construct F_i. In each case, explain why it would cause difficulties in our proof.

1. If v lies in at least $\log^{10}\Delta$ sets X_i, then we place v into F_j for every subsequent index j.
2. We divide the indices $1, \ldots, i_0$ into intervals of equal length. In particular, we set $a_k = k \times \frac{\Delta}{\log^{10}\Delta}$ and set $I_k = \{a_{k-1} + 1, \ldots, a_k\}$. If $v \in X_i$ for some $i \in I_k$, then we place v into F_j for every subsequent index j in I_k.

Part IV

A Naive Colouring Procedure

The proofs presented in the next two chapters use a simple but surprisingly powerful technique. As we will see, this technique is the main idea behind many of the strongest results in graph colouring over the past decade or so. We suggest that the primary goal of the reader of this book should be to learn how to use this method.

The idea is to generate a random partial colouring of a graph in perhaps the simplest way possible. Assign to each vertex a colour chosen uniformly at random. Of course, with very high probability this will not be a partial colouring, so we fix it by uncolouring any vertex which receives the same colour as one of its neighbours. What remains must be a proper partial colouring of the graph. We then extend this partial colouring (perhaps greedily) to obtain a colouring of G.

If there are many repeated colours in each neighbourhood, then using our greedy procedure we can finish off the colouring with significantly fewer than Δ colours (see for example Exercise 1.1). This is the general approach taken here. Of course, if $N(v)$ is a clique then there will be no repeated colours in $N(v)$ under any partial colouring. Thus, the procedure works best on graphs in which each neighbourhood spans only a very few edges. Our first application will be to triangle-free graphs, i.e. those in which the neighbourhoods span stable sets.

The key to analyzing this procedure is to note that changing the colour assigned to a vertex v cannot have an extensive impact on the colouring we obtain. It can only affect the colouring on v and its neighbourhood. This permits us to to apply the Local Lemma to obtain a colouring in which every vertex has many repeated colours in its neighbourhood provided that for each vertex v the probability that $N(v)$ has too few repeated colours is very small.

Now, we perform this local analysis in two steps. We first show that the expected number of colours which appear twice on $N(v)$ is large. We then show that this random variable is concentrated around its expected value. In order to do so, we need to introduce a new tool for proving concentration results, as neither Markov's Inequality nor the Chernoff Bound is appropriate. In fact, we will introduce two new tools for proving concentration results, one in each of the next two chapters.

As we shall see, even this naive procedure yields surprisingly strong results. We answer a question of Erdős and Nešetřil posed in 1985 and prove a conjecture of Beutelspacher and Hering. Later in the book, we will use more sophisticated variants of the same approach to obtain even more impressive results.

10. Talagrand's Inequality and Colouring Sparse Graphs

10.1 Talagrand's Inequality

In Chap. 5 we saw the Chernoff Bound, our first example of a concentration bound. Typically, this bound is used to show that a random variable is very close to its expected value with high probability. Such tools are extremely valuable to users of the probabilistic method as they allow us to show that with high probability, a random experiment behaves approximately as we "expect" it to.

The Chernoff Bound applies to a very limited type of random variable, essentially the number of heads in a sequence of tosses of the same weighted coin. While this limited tool can apply in a surprisingly large number of situations, it is often not enough. In this chapter, we will discuss Talagrand's Inequality, one of the most powerful concentration bounds commonly used in the probabilistic method. We will present another powerful bound, Azuma's Inequality, in the next chapter. These two bounds are very similar in nature, and both can be thought of as generalizations of the following:

Simple Concentration Bound *Let X be a random variable determined by n independent trials $T_1, \dots, T_n$, and satisfying*

$$\textit{changing the outcome of any one trial can affect X by at most c,} \tag{10.1}$$

then

$$\mathbf{Pr}(|X - \mathbf{E}(X)| > t) \le 2e^{-\frac{t^2}{2c^2 n}} .$$

Typically, we take c to be a small constant.

To motivate condition (10.1), we consider the following random variable which is *not* strongly concentrated around its expected value:

$$A = \begin{cases} n, \text{ with probability } \frac{1}{2} \\ 0, \text{ with probability } \frac{1}{2} \end{cases} .$$

To make A fit the type of random variable discussed in the Simple Concentration Bound, we can define $T_1, \dots, T_n$ to be binomial random variables, each equal to 0 with probability $\frac{1}{2}$ and 1 with probability $\frac{1}{2}$, and set $A = 0$

if $T_n = 0$ and $A = n$ if $T_n = 1$. Here, $\mathbf{E}(A) = \frac{n}{2}$ but with probability 1, $|A - \mathbf{E}(A)| \geq \frac{n}{2}$. Contrast this with the random variable $B = \sum_{i=1}^{n} T_i$ (i.e. the number of 1's amongst $T_1, \ldots, T_n$). The expected value of B is also $\frac{n}{2}$, but by the Chernoff Bound, the probability that $|B - \mathbf{E}(B)| \geq \alpha n$ is at most $2e^{-\frac{4\alpha^2 n}{3}}$. The difference is that B satisfies condition (10.1) with $c = 1$, while A clearly does not satisfy this condition unless we take c to be far too large to be useful. In essence, the outcomes of each of $T_1, \ldots, T_n$ combine equally to determine B, while A is determined by a single "all-or-nothing" trial. In the language of the stock markets, a diversified portfolio is less risky than a single investment.

It is straightforward to verify that the Simple Concentration Bound implies that $\mathbf{Pr}(|BIN(n, \frac{1}{2}) - \frac{n}{2}| > t) \leq 2e^{-\frac{t^2}{2n}}$, which is nearly as tight as the Chernoff Bound. (This is very good, considering that the Simple Concentration Bound is so much more widely applicable than the Chernoff Bound!) In the same way, for any constant p, the Simple Concentration Bound yields as good a bound as one can hope for on $\mathbf{Pr}(|BIN(n,p) - np| > t)$, up to the constant term in the exponent. However, when $p = o(1)$, the Simple Concentration Bound performs rather poorly on $BIN(n,p)$. For example, if $p = n^{-\frac{1}{2}}$, then it yields $\mathbf{Pr}\left(|BIN(n,p) - np| > \frac{1}{2}np\right) \leq 2e^{-\frac{1}{16}}$ which is far worse than the bound of $2e^{-\frac{\sqrt{n}}{12}}$ provided by the Chernoff Bound.

In general, we would often like to show that for any constant $\alpha > 0$ there exists a $\beta > 0$ such that $\mathbf{Pr}(|X - \mathbf{E}(X)| > \alpha\mathbf{E}(X)) \leq e^{-\beta\mathbf{E}(X)}$. The Simple Concentration Bound can only do this if $\mathbf{E}(X)$ is at least a constant fraction of n. Fortunately, when this property does not hold, Talagrand's Inequality will often do the trick.

Talagrand's Inequality adds a single condition to the Simple Concentration Bound to yield strong concentration even in the case that $\mathbf{E}(X) = o(n)$. In its simplest form, it yields concentration around the *median* of X, $\mathbf{Med}(X)$ rather than $\mathbf{E}(X)$. Fortunately, as we will see, if X is strongly concentrated around its median, then its expected value must be very close to its median, and so it is also strongly concentrated around its expected value, a fact that is usually much more useful as medians can be difficult to compute.

Talagrand's Inequality I *Let X be a non-negative random variable, not identically 0, which is determined by n independent trials $T_1, \ldots, T_n$, and satisfying the following for some $c, r > 0$:*

1. *changing the outcome of any one trial can affect X by at most c, and*
2. *for any s, if $X \geq s$ then there is a set of at most rs trials whose outcomes* certify *that $X \geq s$ (we make this precise below),*

then for any $0 \leq t \leq \mathbf{Med}(X)$,

$$\mathbf{Pr}(|X - \mathbf{Med}(X)| > t) \leq 4e^{-\frac{t^2}{8c^2 r \mathbf{Med}(X)}}.$$

As with the Simple Concentration Bound, in a typical application c and r are small constants.

More precisely, condition 2 says that there is a set of trials $T_{i_1}, \ldots, T_{i_t}$ for some $t \leq rs$ such that changing the outcomes of all the other trials cannot cause X to be less than s, and so in order to "prove" to someone that $X \geq s$ it is enough to show her just the outcomes of $T_{i_1}, \ldots, T_{i_t}$. For example, if each T_i is a binomial variable equal to 1 with probability p and 0 with probability $1-p$, then if $X \geq s$ we could take $T_{i_1}, \ldots, T_{i_t}$ to be s of the trials which came up "1".

The fact that Talagrand's Inequality proves concentration around the median rather than the expected value is not a serious problem, as in the situation where Talagrand's Inequality applies, those two values are very close together, and so concentration around one implies concentration around the other:

Fact 10.1 *Under the conditions of Talagrand's Inequality I,* $|\mathbf{E}(X) - \mathbf{Med}(X)| \leq 40c\sqrt{r\mathbf{E}(X)}$.

This fact, which we prove in Chap. 20, now allows us to reformulate Talagrand's Inequality in terms of $\mathbf{E}(X)$.

Talagrand's Inequality II *Let X be a non-negative random variable, not identically 0, which is determined by n independent trials $T_1, \ldots, T_n$, and satisfying the following for some $c, r > 0$:*

1. *changing the outcome of any one trial can affect X by at most c, and*
2. *for any s, if $X \geq s$ then there is a set of at most rs trials whose outcomes* certify *that $X \geq s$,*

then for any $0 \leq t \leq \mathbf{E}(X)$,

$$\mathbf{Pr}(|X - \mathbf{E}(X)| > t + 60c\sqrt{r\mathbf{E}(X)}) \leq 4e^{-\frac{t^2}{8c^2 r\mathbf{E}(X)}}.$$

Remarks

1. The reason that the "40" from Fact 10.1 becomes a "60" here is that there is some loss in replacing $\mathbf{Med}(X)$ with $\mathbf{E}(X)$ in the RHS of the inequality.
2. In almost every application, c and r are small constants and we take t to be asymptotically much larger than $\sqrt{\mathbf{E}(X)}$ and so the $60c\sqrt{r\mathbf{E}(X)}$ term is negligible. In particular, if the asymptotic order of t is greater than $\sqrt{\mathbf{E}(X)}$, then for any $\beta < \frac{1}{8c^2 r}$ and $\mathbf{E}(X)$ sufficiently high, we have:

$$\mathbf{Pr}(|X - \mathbf{E}(X)| > t) \leq 2e^{-\frac{\beta t^2}{\mathbf{E}(X)}}.$$

3. This formulation is probably the simplest useful version of Talagrand's Inequality, but does not express its full power. In fact, this version *does not* imply the Simple Concentration Bound (as the interested reader may verify). In Chap. 20, we will present other more powerful versions of Talagrand's Inequality, including some from which the Simple Concentration Bound is easy to obtain. The version presented here is along the lines of a version developed by S. Janson, E. Shamir, M. Steele and J. Spencer (see [145]) shortly after the appearance of Talagrand's paper [148].

The reader should now verify that Talagrand's Inequality yields a bound on the concentration of $BIN(n,p)$ nearly as good as that obtained from the Chernoff Bound for all values of p.

To illustrate that Talagrand's Inequality is more powerful than the Chernoff Bound and the Simple Concentration Bound, we will consider a situation in which the latter two do not apply.

Consider a graph G. We will choose a random subgraph $H \subseteq G$ by placing each edge in $E(H)$ with probability p, where the choices for the edges are all independent. We define X to be the number of vertices which are endpoints of at least one edge in H.

Here, each random trial clearly affects X by at most 2, and so X satisfies condition (10.1). The problem is that if v is the number of vertices in G, then clearly $\mathbf{E}(X) \leq v$, while it is entirely possible that the number of edges in G, and hence the number of random trials, is of order v^2. Thus, the Simple Concentration Bound does not give a good bound here (and the Chernoff Bound clearly doesn't apply). However, if $X \geq s$ then it is easy to find s trials which certify that X is at least s, namely a set of s edges which appear in H and which between them touch at least s vertices. Thus Talagrand's Inequality suffices to show that X is strongly concentrated.

We will present one final illustration, perhaps the most important of the simple applications of Talagrand's Inequality.

Let $\sigma = x_1, \ldots, x_n$ be a uniformly random permutation of $1, \ldots, n$, and let X be the length of the longest increasing subsequence of σ[1]. A well-known theorem of Erdős and Szekeres [47] states that any permutation of $1, \ldots, n$ contains either a monotone increasing subsequence of length $\lceil \sqrt{n} \rceil$ or a monotone decreasing subsequence of length $\lceil \sqrt{n} \rceil$. It turns out that the expected value of X is approximately $2\sqrt{n}$, i.e. twice the length of a monotone subsequence guaranteed by the Erdős-Szekeres Theorem (see [104, 153]). A natural question is whether X is highly concentrated. Prior to the onset of Talagrand's Inequality, the best result in this direction was due to Frieze [61] who showed that with high probability, X is within a distance of roughly $\mathbf{E}(X)^{2/3}$ of its mean, somewhat weaker than our usual target of $\mathbf{E}(X)^{1/2}$.

At first glance, it is not clear whether Talagrand's Inequality applies here, since we are not dealing with a sequence of independent random trials.

[1] In other words, a subsequence $x_{i_1} < x_{i_2} < \ldots < x_{i_k}$ where, of course, $i_1 < \ldots < i_k$.

Thus, we need to choose our random permutation in a non-straightforward manner. We choose n uniformly random real numbers, $y_1, \ldots, y_n$, from the interval $[0, 1]$. Now arranging $y_1, \ldots, y_n$ in increasing order induces a permutation σ of $1, \ldots, n$ in the obvious manner[2].

If $X \geq s$, i.e. if there is an increasing subsequence of length s, then the s corresponding random reals clearly certify the existence of that increasing subsequence, and so certify that $X \geq s$. It follows that changing the value of any one y_i can affect X by at most one. So, Talagrand's Inequality implies:

$$\mathbf{Pr}(|X - \mathbf{E}(X)| > t + 60\sqrt{\mathbf{E}(X)}) < 4e^{-\frac{t^2}{8\mathbf{E}(X)}}.$$

This was one of the original applications of Talagrand's Inequality in [148]. More recently, Baik, Deift and Johansson [15] have shown that a similar result holds when we replace $\sqrt{\mathbf{E}(X)}$ by $\mathbf{E}(X)^{\frac{1}{3}}$, using different techniques (see also [96]).

We will find Talagrand's Inequality very useful when analyzing the Naive Colouring Procedure discussed in the introduction to this part of the book. To illustrate a typical situation, suppose that we apply the procedure using $\beta\Delta$ colours for some fixed $\beta > 0$, and consider what happens to the neighbourhood of a particular vertex v. Let A denote the number of colours assigned to the vertices in N_v, and let R denote the number of colours retained by the vertices in N_v after we uncolour vertices involved in conflicts.

If $\deg(v) = \Delta$, then it turns out that $\mathbf{E}(A)$ and $\mathbf{E}(R)$ are both of the same asymptotic order as Δ. A is determined solely by the colours assigned to the Δ vertices in N_v, and so a straightforward application of the Simple Concentration Bound proves that A is highly concentrated. R, on the other hand, is determined by the colours assigned to the up to Δ^2 vertices of distance at most two from v, and so the Simple Concentration Bound is insufficient here. However, as we will see, by applying Talagrand's Inequality we can show that R is also highly concentrated.

10.2 Colouring Triangle-Free Graphs

Now that we have Talagrand's Inequality in hand, we are ready to carry out our analysis of the naive random procedure we presented in the introduction to this part of the book. In this section we consider the special case of triangle-free graphs. We shall show:

Theorem 10.2 *There is a Δ_0 such that if G is a triangle-free graph with $\Delta(G) \geq \Delta_0$ then $\chi(G) \leq (1 - \frac{1}{2e^6})\Delta$.*

[2] Because these are uniformly random real numbers, it turns out that with probability 1, they are all distinct.

We shall improve significantly on this theorem in Chap. 13 where with the same hypotheses we obtain: $\chi(G) \le O(\frac{\Delta}{\log \Delta})$.

Proof of Theorem 10.2. We will not specify Δ_0, rather we simply insist that it is large enough so that certain inequalities implicit below hold. We can assume G is Δ-regular because our procedure from Sect. 1.5 for embedding graphs of maximum degree Δ in Δ-regular graphs maintains the property that G is triangle free.

We prove the theorem by finding a partial colouring of G using fewer than $\Delta - \frac{\Delta}{2e^6}$ colours such that in every neighbourhood there are at least $\frac{\Delta}{2e^6} + 1$ colours which appear more than once. As discussed in the introduction to this part of the book, we can then complete the desired colouring using a greedy colouring procedure (see Exercise 1.1). We find the required partial colouring by analyzing our naive random colouring procedure.

So, we set $C = \lfloor \frac{\Delta}{2} \rfloor$ and consider running our random procedure using C colours. That is, for each vertex w we *assign* to w a uniformly random colour from $\{1, \ldots, C\}$. If w is assigned the same colour as a neighbour we uncolour it, otherwise we say w *retains* its colour.

We are interested, for each vertex v of G, in the number of colours which are assigned to at least two neighbours of v and retained on at least two of these vertices. In order to simplify our analysis, we consider the random variable X_v which counts the number of colours which are assigned to at least two neighbours of v and are retained by *all* of these vertices.

For each vertex v, we let A_v be the event that X_v is less than $\frac{\Delta}{2e^6} + 1$. We let $\mathcal{E} = \{A_v | v \in V(G)\}$. To prove the desired partial colouring exists, we need only show that with positive probability, none of these bad events occurs. We will apply the Local Lemma.

To begin, note that A_v depends only on the colour of vertices which are joined to v by a path of length at most 2. Thus, setting

$$S_v = \{A_w | \text{v and w are joined by a path of length at most 4}\},$$

we see that A_v is mutually independent of $\mathcal{E} - S_v$. But $|S_v| < \Delta^4$. So, as long as no A_v has probability greater than $\frac{1}{4\Delta^4}$, we are done.

We compute a bound on $\mathbf{Pr}(A_v)$ using a two step process which will be a standard technique throughout this book. First we will bound the expected value of X_v, and then we show that X_v is highly concentrated around its expected value:

Lemma 10.3 $\mathbf{E}(X_v) \ge \frac{\Delta}{e^6} - 1$.

Lemma 10.4 $\mathbf{Pr}(|X_v - \mathbf{E}(X_v)| > \log \Delta \times \sqrt{\mathbf{E}(X_v)}) < \frac{1}{4\Delta^5}$.

These two lemmas will complete our proof, because

$$\frac{\Delta}{e^6} - 1 - \log \Delta \times \sqrt{\mathbf{E}(X_v)} \ge \frac{\Delta}{e^6} - 1 - \log \Delta \sqrt{\Delta} > \frac{\Delta}{2e^6} + 1,$$

for Δ sufficiently large, and so these lemmas imply that $\mathbf{Pr}(A_v) < \frac{1}{4\Delta^5}$ as required.

Proof of Lemma 10.3. For each vertex v we define $X_v^{'}$ to be the number of colours which are assigned to exactly two vertices in $N(v)$ and are retained by both those vertices. Note that $X_v \geq X_v^{'}$.

A pair of vertices $u, w \in N(v)$ will both retain the same colour α which is assigned to no other neighbour of v, iff α is assigned to both u and w and to no vertex in $S = N(v) \cup N(u) \cup N(w) - u - v$. Because $|S| \leq 3\Delta - 3 \leq 6C$, for any colour α, the probability that this occurs is at least $\left(\frac{1}{C}\right)^2 \times \left(1 - \frac{1}{C}\right)^{6C}$. There are C choices for α and $\binom{\Delta}{2}$ choices for $\{u, w\}$. Using Linearity of Expectation and the fact that $\mathrm{e}^{-1/C} < 1 - \frac{1}{C} + \frac{1}{2C^2}$, we have:

$$\mathbf{E}(X_v^{'}) \geq C\binom{\Delta}{2} \times \left(\frac{1}{C}\right)^2 \times \left(1 - \frac{1}{C}\right)^{6C} \geq \frac{\Delta - 1}{\mathrm{e}^6}\left(1 - \frac{4}{C}\right) > \frac{\Delta}{\mathrm{e}^6} - 1,$$

for C sufficiently large. □

Proof of Lemma 10.4. Instead of proving the concentration of X_v directly, we will focus on two related variables. The first of these is AT_v (assigned twice) which counts the number of colours assigned to at least two neighbours of v. The second is Del_v (deleted) which counts the number of colours assigned to at least two neighbours of v but removed from at least one of them. We note that $X_v = AT_v - \mathrm{Del}_v$, and so to prove Lemma 10.4, it will suffice to prove the following concentration bounds on these two related variables, which hold for any $t \geq \sqrt{\Delta \log \Delta}$.

Claim 1: $\mathbf{Pr}(|AT_v - \mathbf{E}(AT_v)| > t) < 2e^{-\frac{t^2}{8\Delta}}$.

Claim 2: $\mathbf{Pr}(|\mathrm{Del}_v - \mathbf{E}(\mathrm{Del}_v)| > t) < 4e^{-\frac{t^2}{100\Delta}}$.

To see that these two claims imply Lemma 10.4, we observe that by Linearity of Expectation, $\mathbf{E}(X_v) = \mathbf{E}(AT_v) - \mathbf{E}(\mathrm{Del}_v)$. Therefore, if $|X_v - \mathbf{E}(X_v)| > \log \Delta \sqrt{\mathbf{E}(X_v)}$, then setting $t = \frac{1}{2} \log \Delta \sqrt{\mathbf{E}(X_v)}$, we must have either $|AT_v - \mathbf{E}(AT_v)| > t$ or $|\mathrm{Del}_v - \mathbf{E}(\mathrm{Del}_v)| > t$. Applying our claims, along with the Subadditivity of Probabilities, the probability of this happening is at most

$$2\mathrm{e}^{-\frac{t^2}{8\Delta}} + 4\mathrm{e}^{-\frac{t^2}{100\Delta}} < \frac{1}{4\Delta^5}.$$

It only remains to prove our claims.

Proof of Claim 1. The value of AT_v depends only on the Δ colour assignments made on the neighbours of v. Furthermore, changing any one of these assignments can affect AT_v by at most 2, as this change can only affect whether the old colour and whether the new colour are counted by AT_v. Therefore, the result follows from the Simple Concentration Bound with $c = 2$. □

Proof of Claim 2. The value of Del_v depends on the up to nearly Δ^2 colour assignments made to the vertices of distance at most 2 from v. Because $\mathbf{E}(\mathrm{Del}_v) \le \Delta = o(\Delta^2)$ (since in fact Del_v is always at most Δ), the Simple Concentration Bound will not apply here. So we will use Talagrand's Inequality.

As with AT_v, changing any one colour assignment can affect Del_v by at most 2. Furthermore, if $\mathrm{Del}_v \ge s$ then there is a set of at most $3s$ colour assignments which certify that $\mathrm{Del}_v \ge s$. Namely, for each of s colours counted by Del_v, we take 2 vertices of that colour in N_v and one of their neighbours which also received that colour. Therefore, we can apply Talagrand's Inequality with $c = 2$ and $r = 3$ to obtain:

$$\mathbf{Pr}(|\mathrm{Del}_v - \mathbf{E}(\mathrm{Del}_v)| > t) < 4\mathrm{e}^{-\frac{\left(t-120\sqrt{3\mathbf{E}(\mathrm{Del}_v)}\right)^2}{96\mathbf{E}(\mathrm{Del}_v)}} < 4\mathrm{e}^{-\frac{t^2}{100\Delta}},$$

since $t \ge \sqrt{\Delta \log \Delta}$ and $\mathbf{E}(\mathrm{Del}_v) \le \Delta$. □

So, we have proven our two main lemmas and hence the theorem. □

10.3 Colouring Sparse Graphs

In this section, we generalize from graphs in which each neighbourhood contains no edges to graphs in which each neighbourhood contains a reasonable number of non-edges. In other words, we consider graphs G which have maximum degree Δ and such that for each vertex v there are at most $\binom{\Delta}{2} - B$ edges in the subgraph induced by $N(v)$ for some reasonably large B.

We note that to ensure that G has a $\Delta - 1$ colouring we will need to insist that B is at least $\Delta - 1$ as can be seen by considering the graph obtained from a clique of size Δ by adding a vertex adjacent to one element of the clique. More generally, we cannot expect to colour with fewer than $\Delta - \lfloor \frac{B}{\Delta} \rfloor$ colours for values of B which are smaller than $\Delta^{\frac{3}{2}} - \Delta$ (see Exercise 10.1). We shall show that if B is not too small, then we can get by with nearly this small a number of colours.

Theorem 10.5 *There exists a Δ_0 such that if G has maximal degree $\Delta > \Delta_0$ and $B \ge \Delta(\log\ \Delta)^3$, and no $N(v)$ contains more than $\binom{\Delta}{2} - B$ edges then $\chi(G) \le \Delta + 1 - \frac{B}{e^6\Delta}$.*

Proof The proof of this theorem mirrors that of Theorem 10.2. Once again, we will consider X_v, the number of colours assigned to at least two *non-adjacent* neighbours of v and retained on all the neighbours of v to which it is assigned. Again, the proof reduces to the following two lemmas, whose (omitted) proofs are nearly identical to those of the corresponding lemmas in the previous section.

Lemma 10.6 $\mathbf{E}(X_v) \geq \frac{2B}{e^6 \Delta}$.

Lemma 10.7 $\mathbf{Pr}\left(|X_v - \mathbf{E}(X_v)| > \log \Delta \sqrt{\mathbf{E}(X_v)}\right) < \frac{1}{4\Delta^5}$.

It is easy to see that provided B is at least $\Delta(\log \Delta)^3$ then we can combine these two facts to obtain the desired result. □

Remarks

1. Using much more complicated techniques, such as those introduced in Part 6, we can show that Theorem 10.5 holds for every B.
2. As you will see in Exercise 10.2, a slight modification to this proof yields the same bound on the list chromatic number of G.

Using our usual argument, we can obtain a version of Theorem 10.5 which holds for every Δ, by weakening our constant e^6:

Corollary 10.8 *There exists a constant $\delta > 0$ such that if G has maximal degree $\Delta > 0$ and $B \geq \Delta(\log\ \Delta)^3$, and no $N(v)$ contains more than $\binom{\Delta}{2} - B$ edges then $\chi(G) \leq \Delta + 1 - \delta\frac{B}{\Delta}$.*

Proof of Corollary 10.8. Set $\delta = \min\{\frac{1}{\Delta_0}, \frac{1}{e^6}\}$. Now, if $\Delta > \Delta_0$ then the result holds by Theorem 10.5. Otherwise it holds because $B < \frac{\Delta}{\delta}$ so we are simply claiming that G is Δ colourable, which is true by Brooks' Theorem . □

10.4 Strong Edge Colourings

We close this chapter by describing an application of Theorem 10.5 to strong edge colourings, which appears in [118] and which motivated Theorem 10.5. The notion of a strong edge colouring was first introduced by Erdős and Nešetřil (see [52]), and is unrelated to the strong vertex colourings which were discussed in Chap. 8.

A *strong edge-colouring* of a graph, G, is a proper edge-colouring of G with the added restriction that no edge is adjacent to two edges of the same colour, i.e. a 2-frugal colouring of $L(G)$ (note that in any proper edge-colouring of G, no edge is adjacent to three edges of the same colour and so the corresponding colouring of $L(G)$ is 3-frugal). Equivalently, it is a proper vertex-colouring of $L(G)^2$, the square of the line graph of G which has the same vertex set as $L(G)$ and in which two vertices are adjacent precisely if they are joined by a path of length at most two in $L(G)$. The *strong chromatic index* of G, $s\chi_e(G)$ is the least integer k such that G has a strong edge-colouring using k colours. We note that each colour class in a strong edge colouring is an *induced matching*, that is a matching M such that no edge of $G - M$ joins endpoints of two distinct elements of M.

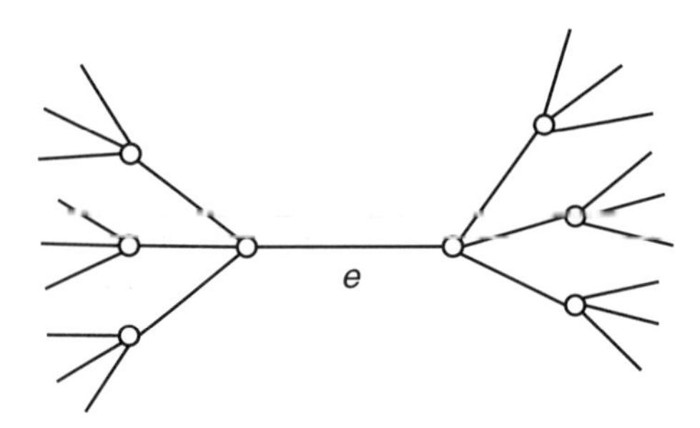

Fig. 10.1. An edge and its neighbourhood in $L(G)^2$

If G has maximum degree Δ, then trivially $s\chi_e(G) \leq 2\Delta^2 - 2\Delta + 1$, as $L(G)^2$ has maximum degree at most $2\Delta^2 - 2\Delta$. In 1985, Erdős and Nešetřil pointed out that the graph G_k obtained from a cycle of length five by duplicating each vertex k times (see Fig. 10.2) contains no induced matching of size 2. Therefore, in any strong colouring of G_k, every edge must get a different colour, and so

$$s\chi_e(G_k) = |E(G_k)| = 5k^2 .$$

We note that $\Delta(G_k) = 2k$, and so $s\chi_e(G_k) = \frac{5}{4}\Delta(G_k)^2$.

Erdős and Nešetřil conjectured that these graphs are extremal in the following sense:

Conjecture *For any graph G, $s\chi_e(G) \leq \frac{5}{4}\Delta(G)^2$.*

They also asked if it could even be proved that for some fixed $\epsilon > 0$ every graph G satisfies $s\chi_e(G) \leq (2 - \epsilon)\Delta(G)^2$. In this section we briefly point out

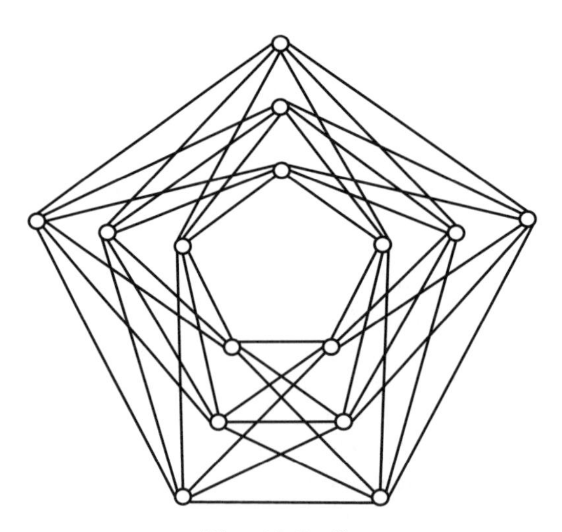

Fig. 10.2. G_3

how Theorem 10.5 can be used to answer this question in the affirmative. For other work on this and related problems, see [11], [33], [53], [83] and [84].

Theorem 10.9 *There is a* Δ_0 *such that if* G *has maximum degree* $\Delta \geq \Delta_0$, *then* $s\chi_e(G) \leq 1.99995\Delta^2$.

Corollary 10.10 *There exists a constant* $\epsilon > 0$ *such that for every graph* G, $s\chi_e(G) \leq (2-\epsilon)\Delta^2$.

Proof of Corollary 10.10. Set $\epsilon = min\{.00005, \frac{1}{\Delta_0}\}$. If $\Delta(G) \geq \Delta_0$ then the result follows from the theorem. Otherwise, we simply need to use the fact that $\Delta(L(G)^2) \leq 2\Delta(G)^2 - 2\Delta(G)$ and apply Brook's Theorem.

To prove Theorem 10.9, the main step is to show the following, whose tedious but routine proof can be found in [118]:

Lemma 10.11 *If* G *has maximum degree* Δ *sufficiently large then, for each* $e \in V(L(G)^2)$, $N_{L(G)^2}(e)$ *has at most* $(1 - \frac{1}{36})\binom{2\Delta^2}{2}$ *edges.*

Using this lemma, it is a straightforward matter to apply Theorem 10.5 to yield Theorem 10.9.

Exercises

Exercise 10.1 Show that for every Δ and $B \leq \Delta^{\frac{3}{2}} - \Delta$ there exists a graph with $\Delta + 1$ vertices in which the neighbourhood of each vertex contains at most $\binom{\Delta}{2} - B$ edges, whose chromatic number is $\lfloor \Delta - \frac{B}{\Delta} \rfloor$.

Exercise 10.2 Complete the proof of Theorem 10.5 and then modify it to get the same bound on the list chromatic number of G.

11. Azuma's Inequality and a Strengthening of Brooks' Theorem

11.1 Azuma's Inequality

In this chapter, we introduce a new tool for proving bounds on concentration. It differs from the tools we have mentioned so far, in that it can be applied to a sequence of dependent trials. To see a concrete example of such a situation, imagine that we are colouring the vertices of a graph one by one, assigning to each vertex a colour chosen uniformly *from those not yet assigned to any of its coloured neighbours.* This ensures that the colouring obtained is indeed a proper colouring, and analyzing such a random process may yield good bounds on the minimum number of colours required to obtain vertex colourings with certain properties. However, our choices at each vertex are now no longer independent of those made at the other vertices.

In this chapter, we introduce a new concentration inequality which can be used to handle such situations, Azuma's Inequality. It applies to a random variable R which is determined by a sequence $X_1, \ldots, X_n$ of random trials. Our new tool exploits the ordering on the random trials; it obtains a bound on the concentration of R using bounds on the maximum amount by which we *expect* each trial to affect R when it is performed. This approach bears fruit even when we are considering a set of independent trials. To illustrate this point, we consider the following simple game.

A player rolls a fair six-sided die $n+1$ times, with outcomes $r_0, r_1, \ldots, r_n$. Roll 0 establishes a target. The players winnings are equal to X, the number of rolls $i \geq 1$ such that $r_i = r_0$.

It should be intuitively clear that X is highly concentrated. However, changing the outcome of r_0 can have a dramatic effect on X. For example, if our sequence is $1, 1, \ldots, 1$ then any change to r_0 will change X from n to 0. Thus, we cannot directly apply the Simple Concentration Bound or Talagrand's Inequality to this problem. It turns out that we can apply Azuma's Inequality because the conditional expected value of X after the first roll is $\frac{n}{6}$ regardless of the outcome of this trial. Thus, the first trial has no effect whatsoever on the conditional expected value of X.

Remark Of course, we do not need a bound as powerful as Azuma's Inequality, to prove that X is highly concentrated – we could prove this by first rolling the die to determine r_0 and then simply applying the Chernoff Bound to the sequence $r_1, \ldots, r_n$. However, the reader can easily imagine that she could contrive a similar but more complicated scenario where it is not so easy to apply our other bounds. For one such example, see Exercise 11.1.

Like Talagrand's Inequality, Azuma's Inequality can be viewed as a strengthening of the Simple Concentration Bound. There are three main differences between Azuma's Inequality and the Simple Concentration Bound.

(i) We must compute a bound on the amount by which the outcome of each trial can affect the *conditional* expected value of X,
(ii) Azuma's Inequality can be applied to sequences of dependent trials and is therefore much more widely applicable,
(iii) The concentration bound given by the Simple Concentration Bound is in terms of an upper bound c on the maximum amount by which changing the outcome of a trial can affect the value of X. To apply Azuma's Inequality we obtain distinct values $c_1, \ldots, c_n$ where c_i bounds the amount by which changing the outcome of T_i can affect the conditional expected value of X. We then express our concentration bound in terms of $c_1, \ldots, c_n$. This more refined approach often yields stronger results.

Azuma's Inequality [14] *Let X be a random variable determined by n trials $T_1, \ldots, T_n$, such that for each i, and any two possible sequences of outcomes $t_1, \ldots, t_i$ and $t_1, \ldots, t_{i-1}, t_i'$:*

$$|\,\mathbf{Exp}(X\,|\,T_1 = t_1, \ldots, T_i = t_i) - \mathbf{Exp}(X\,|\,T_1 = t_1, \ldots, T_i = t_i')\,| \leq c_i \quad (11.1)$$

then

$$\mathbf{Pr}(|X - \mathbf{E}(X)| > t) \leq 2\mathrm{e}^{-t^2/(2\sum c_i^2)}.$$

Condition (11.1) corresponds to condition (10.1) in the Simple Concentration Bound, however the two inequalities are very different. The following discussion underscores the difference between them. Suppose we have an adversary who is trying to make X as large as he can, and a second adversary who is trying to make X as small as she can. Either adversary is allowed to change the outcome of exactly one trial T_i. Condition (10.1) says that if the adversaries wait until all trials have been carried out, and then change the outcome of T_i, then their power is always limited. Condition (11.1) says that if they must make their changes as soon as T_i is carried out, without waiting for the outcomes of all future trials, then their power is limited.

The above discussion suggests that condition (10.1) is more restrictive than condition (11.1), and thus that Azuma's Inequality implies the Simple Concentration Bound. It is, in fact, straightforward to verify this implication (this is Exercise 11.2). As we will see, Azuma's Inequality is actually much more powerful than the Simple Concentration Bound. For example, in the game discussed earlier, we satisfy condition (11.1) with $c_0 = 0$ and $c_i = 1$ for $i > 0$ (by Linearity of Expectation) and so Azuma's Inequality implies that X is highly concentrated.

Azuma's Inequality is an example of a Martingale inequality. For further discussion of Martingale inequalities, we refer the reader to [10], [112] or [114].

If we apply Azuma's Inequality to a set of independent trials with each c_i equal to a small constant c, then the resulting bound is $e^{-\epsilon t^2/n}$ for a positive constant ϵ rather than the often more desirable $e^{-\epsilon t^2/\mathbf{E}(X)}$ which is typically obtained from Talagrand's Inequality. While the bound given by applying Azuma's inequality using such c_i is usually sufficient when $\mathbf{E}(X) = \beta n$ for some constant $\beta > 0$, it is often not strong enough when $\mathbf{E}(X) = o(n)$. However, in such situations, by taking at least some of the c_i to be very small, we can often apply Azuma's Inequality to get the desired bound of $e^{-(\epsilon t^2/\mathbf{E}(X))}$. We will see an example of this approach in the proof of Lemma 11.8 later in this chapter.

We now consider variables determined by sequences of dependent trials, where the change in the conditional expectation caused by each trial is bounded. Our discussion focuses on one commonly occurring situation. Suppose X is a random variable determined by a uniformly random permutation P of $\{1, \dots, n\}$, with the property that interchanging any two values $P(i), P(j)$ can never affect X by more than c. Then, as we discuss below, we can apply Azuma's Inequality to show that X is concentrated.

For each $1 \leq i \leq n$, we let T_i be a uniformly random element of $\{1, \dots, n\} - \{T_1, \dots, T_{i-1}\}$. It is easy to see that $T_1, \dots, T_n$ forms a uniformly random permutation. Furthermore, we will show that this experiment satisfies condition (11.1).

Consider any sequence of outcomes $T_1 = t_1, \dots, T_{i-1} = t_{i-1}$, along with two possibilities for T_i, t_i, t_i'. For any permutation P satisfying $P(1) = t_1, \dots, P(i) = t_i$ and $P(j) = t_i'$ for some $j > i$, we let P' be the permutation obtained by interchanging $P(i)$ and $P(j)$. Our hypotheses yield $|X(P) - X(P')| \leq c$. Furthermore, it is easy to see that

$$\mathbf{Pr}(P|T_1 = t_1, \dots, T_i = t_i) = \mathbf{Pr}(P'|T_1 = t_1, \dots, T_i = t_i') = \frac{1}{(n-i)!}.$$

It is straightforward to verify that these two facts ensure that condition (11.1) holds, and so we can apply Azuma's Inequality to show that X is highly concentrated. (Note that it is important that Azuma's Inequality does not require our random trails to be independent.) Of course, Azuma's Inequality also applies in a similar manner when X is determined by a sequence of several random permutations.

Remark As discussed in the last chapter, we can generate a uniformly random permutation by generating n independent random reals between 0 and 1. We applied Talagrand's Inequality to this model to prove that the length of the longest increasing subsequence is concentrated. We cannot use Talagrand's inequality in the same way to prove the above result as the condition that swapping two values $P(i)$ and $P(j)$ can affect X by at most c

does not guarantee that changing the value of one of the random reals affects the value of X by a bounded amount.

For a long time, Azuma's Inequality (or, more generally, the use of Martingale inequalities) was the best way to prove many of the difficult concentration bounds arising in probabilistic combinatorics. However, the conditions of Talagrand's inequality are often much easier to verify. Thus in situations where they both apply, Talagrand's Inequality has begun to establish itself as the "tool of choice".

It is worth noting, in this vein, that Talagrand showed that his inequality can also be applied to a single uniformly random permutation (see Theorem 5.1 of [148]). More recently, McDiarmid obtained a more general version which applies to sequences of several permutations, as we will discuss in Chap. 16. Thus, we can now prove concentration for variables which depend on such a set of random trials, using a Talagrand-like inequality rather than struggling with Azuma. To see the extent to which this simplifies our task, compare some of the lengthy concentration proofs in [132] and [119] (which predated McDiarmid's extension of Talagrand's Inequality) with the corresponding proofs in Chaps. 16 and 18 of this book.

Nevertheless, there are still many sequences of dependent trials to which Talagrand cannot be applied but Azuma's Inequality can (see for example [116]).

11.2 A Strengthening of Brooks' Theorem

Brooks' Theorem characterizes graphs for which $\chi \leq \Delta$. For Δ at least 3, they are those that contain no $\Delta + 1$ clique. Characterizing which graphs have $\chi \leq \Delta - 1$ seems to be more difficult, Maffray and Preissmann [110] have shown it is NP-complete to determine if a 4-regular graph has chromatic number at most three (if you do not know what *NP-complete* means, replace it by *hard*). However, Borodin and Kostochka [29] conjectured that if $\Delta(G) \geq 9$ then an analogue of Brooks' Theorem holds; i.e. $\chi(G) \leq \Delta(G) - 1$ precisely if $\omega(G) \leq \Delta - 1$ (this is Problem 4.8 in [85] to which we refer readers for more details). To see that 9 is best possible here, consider the graph G, depicted in Fig. 11.1, obtained from five disjoint triangles $T_1, \ldots, T_5$ by adding all edges between T_i and T_j if $|i - j| \equiv 1 \bmod 5$. It is easy to verify that $\Delta(G) = 8$, $\omega(G) = 6$, and $\chi(G) = 8$. Beutelspacher and Hering [22] independently posed the weaker conjecture that this analogue of Brooks' Theorem holds for sufficiently large Δ. We prove their conjecture. That is, we show:

Theorem 11.1 *There is a Δ_2 such that if $\Delta(G) \geq \Delta_2$ and $\omega(G) \leq \Delta(G) - 1$ then $\chi(G) \leq \Delta(G) - 1$.*

It would be natural to conjecture that Theorem 11.1 could be generalized as follows:

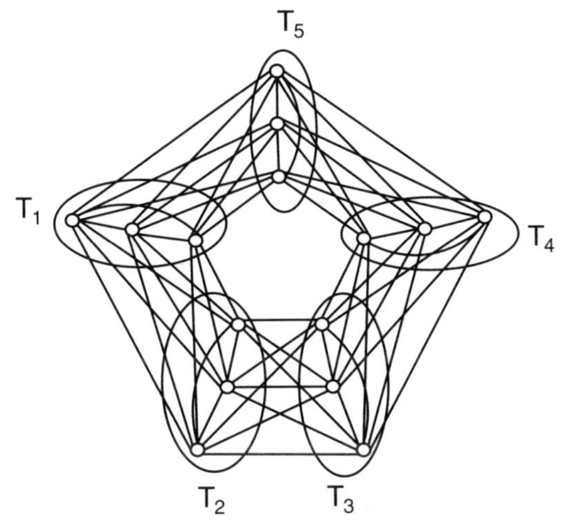

Fig. 11.1. G

For all k, there is a Δ_k such that if $\Delta(G) \geq \Delta_k$ and $\omega(G) \leq \Delta(G)+1-k$ then $\chi(G) \leq \Delta(G)+1-k$.

However this conjecture turns out to be false even for $k = 3$, as the following example shows: For $\Delta \geq 5$, let G_Δ be a graph obtained from a clique $K_{\Delta-4}$ with $\Delta - 4$ vertices and a chordless cycle C with 5 vertices by adding all edges between C and $K_{\Delta-4}$ (see Fig. 11.2). It is easy to verify that G_Δ has maximum degree Δ, clique number $\Delta - 2$, and chromatic number $\Delta - 1$.

Nevertheless, we can generalize Theorem 11.1 in two ways. Firstly, we can bound how quickly χ must decrease as ω moves away from $\Delta + 1$.

Theorem 11.2 *For all k, there is a Δ_k such that if $\Delta(G) \geq \Delta_k$ and $\omega(G) \leq \Delta(G)+1-2k$ then $\chi(G) \leq \Delta(G)+1-k$.*

This result is a corollary of Theorem 16.4 discussed in Chap. 16. As pointed out in that chapter, the theorem is essentially best possible for large k.

Secondly, we can show that if χ is sufficiently near Δ then although we may not be able to determine χ precisely simply by considering the sizes of the cliques in G, we can determine it by considering only the chromatic numbers of a set of subgraphs of G which are very similar to cliques. For example, we have:

There is a Δ_0 such that for any $\Delta \geq \Delta_0$ and $k > \Delta - \sqrt{\Delta} + 2$,

> there is a collection of graphs $H_1, \ldots, H_t$, which are similar to k-cliques in that $\chi(H_i) = k$, $|V(H_i)| \leq \Delta + 1$ and $\delta(H_i) \geq k - 1$, such that the following holds:
> For any graph G with maximum degree Δ, $\chi(G) \geq k$ iff G contains at least one H_i as a subgraph.

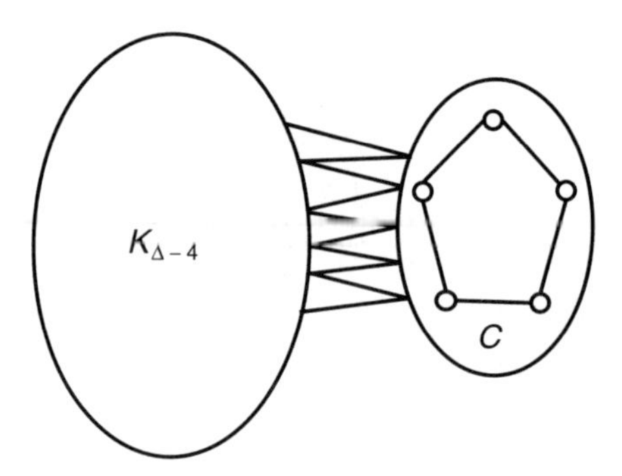

Fig. 11.2. G_Δ

We discuss a number of results of this type in Chap. 15. The proofs of these generalizations of Theorem 11.1 all use ideas introduced in its proof.

Proof of Theorem 11.1. We fix a Δ_2 which satisfies a number of implicit inequalities scattered throughout the proof and prove the theorem for this value of Δ_2. To this end, we assume the theorem is false and let G be a counter-example to it with the fewest number of vertices. Thus, G has maximum degree $\Delta \geq \Delta_2$, $\omega(G) \leq \Delta - 1$, and $\chi(G) = \Delta$.

Before presenting the key ideas of the proof, we make the following easy observations.

11.3 *Every subgraph H of G with $H \neq G$ has a $\Delta - 1$ colouring.*

Proof If $\Delta(H) = \Delta$ then the result follows by the minimality of G. Otherwise, the result follows from Brooks' Theorem because $\omega(H) \leq \omega(G) \leq \Delta - 1$. □

11.4 *Every vertex of G has degree at least $\Delta - 1$.*

Proof For any vertex v of G, by (11.3), $G - v$ has a $\Delta - 1$ colouring. If v had fewer than $\Delta - 1$ neighbours then we could extend this to a $\Delta - 1$ colouring of G. □

We have already seen in Sect. 10.3 that for Δ sufficiently large, if no vertex in G has more than $\binom{\Delta}{2} - (\log\ \Delta)^3\Delta$ edges in its neighbourhood, then $\chi(G) \leq \Delta - \frac{(\log\ \Delta)^3}{e^6} \leq \Delta - 1$. Thus, the crux of the proof will be to deal with vertices which have such dense neighbourhoods. This motivates the following:

Definitions A vertex v of G is *dense* if $N(v)$ contains fewer than $\Delta(\log\ \Delta)^3$ non-adjacent pairs of vertices. Otherwise, it is *sparse*.

We need to investigate the structure surrounding such dense vertices.

Definition Set $r = (\log\ \Delta)^4$. We say a clique is *big* if $|C| \geq \Delta - r$.

We shall prove:

Lemma 11.5 *Every dense vertex is contained in a big clique.*

Lemma 11.6 *We can partition $V(G)$ into $D_1, \ldots, D_l, S$ so that*

(i) *each D_i contains a big clique C_i Furthermore, either $D_i = C_i$ or $D_i = C_i + v_i$ for some vertex v_i which sees at least $\frac{3\Delta}{4}$ but not all of the vertices of C_i;*
(ii) *no vertex of $V - D_i$ sees more than $\frac{3\Delta}{4}$ vertices of D_i;*
(iii) *each vertex of S is sparse; and*
(iv) *each vertex v of C_i has at most one neighbour outside C_i which see more than $r+4$ vertices of C_i, furthermore if $|C_i| = \Delta - 1$ then v has no such neighbours.*

It is this decomposition of G into l dense sets and a set of sparse vertices which allows us to prove Theorem 11.1. For, having proved Lemma 11.6, to deal with the dense vertices we need only colour each D_i. This will be relatively easy, for these sets are disjoint and there are very few edges from D_i to $V - D_i$.

For ease of exposition, we consider the Δ-regular graph G' obtained from G by taking two copies of G and adding an edge between the two copies of each vertex of degree $\Delta - 1$. We note that applying Lemma 11.6 to both copies of G yields:

Corollary 11.7 *There is a decomposition of G' satisfying conditions (i)–(iv) of Lemma 11.6.*

Now, by taking advantage of this corollary, we can extend the proof technique of the last chapter to prove a useful lemma.

Definition Consider a decomposition as in Corollary 11.7. For $1 \leq i \leq l$, if D_i is the clique C_i set $K_i = C_i$ else set $K_i = C_i \cap N(v_i)$.

Lemma 11.8 *There is a partial $\Delta - 1$ colouring of G' satisfying the following two conditions.*

(a) *for every vertex $v \in S$ there are at least 2 colours appearing twice in the neighbourhood of v,*
(b) *every K_i contains two uncoloured vertices w_i and x_i whose neighbourhoods contain two repeated colours.*

To complete a partial $\Delta - 1$ colouring satisfying (a) and (b) to a $\Delta - 1$ colouring of G' and thereby obtain a $\Delta - 1$ colouring of G, we proceed as follows.

We let U_2 be the set of uncoloured vertices whose neighbourhoods contain at least two repeated colours and we let U_1 be the remaining uncoloured vertices. We complete the colouring greedily by colouring the uncoloured vertices one at a time. The only trick is to colour all the vertices of U_1 before colouring any vertex of U_2. Consider a vertex v of U_1. Since every sparse vertex is in U_2 by (a), v is in some D_i. By (b), the vertices x_i and w_i are in U_2 and hence are uncoloured when we come to colour v. Since v has these two uncoloured neighbours, it follows that there is a colour available with which to colour v. Thus, we can extend our partial colouring of G' to a $\Delta-1$ colouring of $G'-U_2$. We can then complete the colouring because for each vertex u in U_2 there are two repeated colours in $N(u)$. □

Remark We note that in proving Theorem 11.1 from Lemma 11.8, we used a slight refinement of our greedy colouring procedure. We carefully chose the order in which we would complete the colouring, and then coloured each vertex greedily when we came to it. This idea, which we first encountered in the proof of Brooks' Theorem in Chap. 1, will prove crucial to many of the proofs to follow both in this chapter and future ones.

We have yet to prove Lemmas 11.5, 11.6, and 11.8. We prove the last of these in the next section, and prove the first two in the third and final section of the chapter.

11.3 The Probabilistic Analysis

Proof of Lemma 11.8. We find a partial $\Delta-1$ colouring satisfying conditions (a) and (b) of Lemma 11.8 by analyzing our naive colouring procedure. In doing so, we take advantage of the partition given by Corollary 11.7. Once again, we will use the Local Lemma.

To do so, we need to define two kinds of events. For each $v \in S$, we let A_v be the event that (a) fails to hold for v, i.e., that there are fewer than 2 repeated colours on $N(v)$. For each D_i, we let A_i be the event that (b) fails to hold for D_i, i.e. that there do not exist two uncoloured vertices of K_i each of which has two repeated colours in its neighbourhood. We note that if none of the events in the set $\mathcal{E} = (\cup A_v) \cup (\cup A_i)$ hold then the random colouring satisfies (a) and (b) of Lemma 11.8. To finish the proof we use the Local Lemma to show that this occurs with positive probability.

We note that A_v depends only on the colours within distance two of v. Also each A_i depends only on the colour of the vertices in D_i or within distance two of D_i. It follows that each event in $\mathcal{E}$ is mutually independent of a set of all but at most Δ^5 other events. So, we need only show that each event in $\mathcal{E}$ holds with probability at most Δ^{-6}.

11.9 *Each A_v has probability at most Δ^{-6}.*

To prove this result we consider (as in the last chapter) the variable X_v which counts the number of colours assigned to at least two neighbours of v and retained by all such neighbours. We first obtain a bound on the expected value of X_v and then bound its concentration using Talagrand's Inequality. As the proof is almost identical to one in the last chapter, we omit the details.

11.10 *Each A_i has probability at most Δ^{-6}.*

To prove this result, we need the following simple corollary of Corollary 11.7.

Lemma 11.11 *For any D_i, there are at least $\frac{\Delta}{4r}$ disjoint triples each of which consists of a vertex v of K_i and two neighbours of v outside of K_i both of which have at most $r+4$ neighbours in K_i.*

Proof of Lemma 11.11. Consider any D_i, and corresponding K_i. By definition, $|K_i| \geq \frac{3\Delta}{4}$. Take a maximal set of disjoint triples each of which consists of a vertex in K_i and two neighbours of this vertex outside of D_i, each of which has at most $r+4$ neighbours in K_i. Suppose there are k triples in this set and let W be the $2k$ vertices in these triples which are not in D_i. By (iv) of Corollary 11.7 and the maximality of our set of triples, every vertex in K_i is a neighbour of some vertex in W. Hence, $(r+4)|W| \geq |K_i|$, which yields $k \geq \frac{\Delta}{4r}$ as required. □

To compute the probability bound on A_i, we consider the set $\mathcal{T}_i$ of $\frac{\Delta}{4r}$ disjoint triples guaranteed to exist by Lemma 11.11. We let T_i be the union of the vertex sets of these triples. We let M_i be the number of these triples for which (i) the vertex in K_i is uncoloured, (ii) both the other vertices are coloured with a colour which is also used to colour a vertex of K_i, and (iii) no vertex of the triple is assigned a colour assigned to any other vertex in T_i. This last condition is present to ensure that changing the colour of a vertex can only affect the value of M_i by two.

To begin, we compute the expected value of M_i. We note that M_i counts the number of triples (a,b,c) in $\mathcal{T}_i$ with $c \in K_i$ such that there are colours j,k,l and vertices x,y,z with $x \in K_i - T_i - N(a)$, $y \in K_i - T_i - N(b)$, $z \in N(c) - T_i$, such that

1. j is assigned to a and x but to none of the rest of $T_i \cup N(a) \cup N(x)$,
2. k is assigned to b and y but to none of the rest of $T_i \cup N(b) \cup N(y)$,
3. l is assigned to z and c but on none of the rest of T_i.

To begin, we fix a triple $\{a,b,c\}$ in $\mathcal{T}_i$. We let $A_{j,k,l,x,y,z}$ be the event that (1), (2), and (3) hold. Since $|T_i| \leq \frac{\Delta}{4}$, the probability of $A_{j,k,l,x,y,z}$ is at least $(\Delta-1)^{-6}\frac{1}{e^5}$. Furthermore, two such events with different sets of indices are disjoint. Now, there are at least $\frac{2\Delta}{3}$ choices for both x and y. There are at least $\frac{9\Delta}{10}$ choices for z and $(\Delta-1)(\Delta-2)(\Delta-3)$ choices for distinct j,k,l. So, a straightforward calculation shows that the probability that (1), (2), and (3) hold for some choice of $\{j,k,l,x,y,z\}$ is at least $(\Delta-1)^{-6}(\Delta-1)$ $(\Delta-2)(\Delta-3)\frac{2\Delta}{3}^2\frac{9\Delta}{10}\frac{1}{e^5} \geq \frac{1}{e^6}$. Since, there are $\frac{\Delta}{4r}$ triples in $\mathcal{T}_i$, the expected value of M_i is at least $\frac{\Delta}{4r}\frac{1}{e^6} \geq \frac{\Delta}{r^2}$.

We now prove that M_i is concentrated around its mean, and hence at least two with high probability, by applying Azuma's Inequality. To apply Azuma's Inequality, we must be careful about the order in which we assign the random colours to $V(G)$. We will colour the vertices of $V - T_i - K_i$ first and then the vertices of $T_i \cup K_i$. So, we order the vertices of G as $w_1, \ldots, w_n$ where for some s, we have $\{w_1, \ldots, w_s\} = V - T_i - K_i$ and $\{w_{s+1} =, \ldots, w_n\} = T_i \cup K_i$. We then choose the random colour assignments for the vertices in the given order.

For each of these choices we now obtain our bound c_j on the effect of the choice on the conditional expected value of $\mathbf{E}(M_i)$. We note that changing the colour of any vertex can affect the conditional expected value of M_i by at most 2 since it affects the value of M_i by at most 2 for any given assignment of colours to the remaining vertices. So, $\sum_{n-s}^{n} c_j^2 \le 2^2|T_i \cup K_i| \le 5\Delta$. Furthermore, changing the colour assigned to a vertex w_j of $V - T_i - K_i$ from α to β will only affect M_i if some neighbour of w_j in $T_i \cup K_i$ receives either α or β. This occurs with probability at most $\frac{2d_j}{\Delta}$ where d_j is the number of neighbours of w_j in $T_i \cup K_i$. Hence by colouring w_j we can change the conditional expected value of M_i by at most $c_j = \frac{4d_j}{\Delta}$. Since the d_j sum to at most Δ^2, $\sum_{i=1}^{n-s} c_j$ is at most 4Δ. As, each c_j is at most 4, we see that $\sum_{i=1}^{n-s} c_j^2 \le 16\Delta$. Thus, the sum of all the c_j^2 is at most 21Δ. Applying Azuma's Inequality with $t = \frac{\Delta}{r^2} - 2$ yields $\mathbf{Pr}(A_i) < \Delta^{-6}$, as desired. □

11.4 Constructing the Decomposition

In this section we prove our two lemmas on the local structure surrounding dense vertices, i.e. Lemmas 11.5 and 11.6. The proofs of these lemmas are not probabilistic. We include them for completeness. In these proofs, we repeatedly apply the refinement of the greedy colouring procedure discussed above. That is, we repeatedly find some partial $\Delta - 1$ colouring and complete it to a $\Delta - 1$ colouring by greedily colouring the uncoloured vertices in an appropriate order. Crucial to the proofs is the following:

Observation 11.12 *If H is a subgraph of G with at most $\Delta + 1$ vertices such that every vertex of H has at least $\frac{9\Delta}{10}$ neighbours in H then H is either a clique or is a clique and a vertex.*

Proof Let $X_1, \ldots, X_l$ be a maximum size family of disjoint stable sets of size two in H. If $l = 0$ then H is a clique and we are done, so we assume l is at least 1. Clearly $l \le \frac{\Delta+1}{2}$. We let $S = \cup_{i=1}^{l} X_i$. We can $\Delta - 1$ colour $G - H$ since it is a proper subgraph of G. We claim that if H is not a clique and a vertex then we can extend any such colouring to a $\Delta - 1$ colouring of G.

We will first extend our colouring to the vertices of S so that for each i, the vertices of X_i get the same colour. To do so, we colour the two vertices of X_i

at the same time. Between them, these vertices have at most $\frac{2\Delta}{10}$ neighbours outside of H. Our colouring procedure ensures that there are at this point at most l colours used on H. Thus, there are at least $\frac{3\Delta}{10} - 1$ colours which can be assigned to both vertices of X_i. So we can indeed extend the colouring to S so that each X_i is monochromatic.

Case 1: $l \geq \frac{\Delta}{10} + 2$.

By our degree condition on H, each vertex of $H - S$ misses at most $\frac{\Delta}{10}$ vertices of S and hence has two repeated colours in its neighbourhood. So we can complete our $\Delta - 1$ colouring greedily.

Case 2: $l < \frac{\Delta}{10} + 2$.

Note that $C = H - S$ is a clique with at least $\frac{7\Delta}{10} - 3$ vertices. By our degree and size conditions on H, there are at most $\frac{2\Delta}{10}$ vertices of C which miss a vertex in X_1. Thus, we can find vertices u and v of C both of which see both vertices of X_1. In fact, a similar argument allows us to insist that if $l \geq 2$ then u and v both see all of $X_1 \cup X_2$.

Now, we claim that if $l \geq 2$ then we can complete our colouring greedily provided we colour u and v last. When we colour a vertex of $C - u - v$ it has two uncoloured vertices (u and v). Both u and v have two repeated colours in their neighbourhoods. The claim follows.

Finally, if $l = 1$ then we let $X_1 = \{x, y\}$. Since H is not a clique and a vertex, both x and y miss a vertex of C. Since there are no two disjoint stable sets of size two in H, they both miss some vertex z of C and see all of $C - z$. In this case, we insist that when extending our colouring of $G - H$ to $G - (H - X_1)$ we actually extend it to a colouring of $G - (H - X_1 - z)$ by using one of the $\frac{7\Delta}{10} - 1$ colours which do not appear on $N(x) \cup N(y) \cup N(z) - H$ to colour $\{x, y, z\}$. Now we can complete the colouring greedily, as all the vertices of $C - z$ see the three vertices x, y, z on which we have used one colour. □

Proof of Lemma 11.5. Consider a dense vertex v. Define $S(v) = v + \{w | w \in N(v),\ |N(w) \cap N(v)| \geq \frac{9\Delta}{10} + r\}$. Since $|N(v)| \geq \Delta - 1$ and $|E(\overline{N(v)})| \leq \Delta(\log \Delta)^3$, it follows that $|N(v) - S(v)| < r$. This implies that $\Delta + 1 > |S(v)| > \Delta - r + 1$, and that every vertex of S_v has more than $\frac{9\Delta}{10}$ neighbours in S_v. By Observation 11.12, either S_v is a clique C_v or there is a vertex w of S_v such that $S_v - w$ is a clique C_v. In either case, C_v is the desired big clique containing v. □

A bit more work is required to prove Lemma 11.6.

Proof of Lemma 11.6. We have just proven that every dense vertex is in a big clique. We now examine these cliques more closely.

11.13 *If two maximal big cliques C_1 and C_2 with $|C_1| \leq |C_2|$ intersect, then $|C_1 - C_2| \leq 1$.*

Proof By considering a vertex in the intersection of the two cliques, we see that their union contains at most $\Delta + 1$ vertices. Hence, we can apply Observation 11.12 to the graph obtained from their union. □

We then obtain:

11.14 *No maximal big clique C intersects two other maximal big cliques.*

Proof By (11.13), the union S of these three big cliques would contain at most $|C| + 2 \leq \Delta + 1$ vertices. Applying Lemma 11.12 to S yields that S is a clique or a clique and a vertex. This contradicts our assumption that S contains three maximal big cliques. □

Now, (11.13), (11.14), and Lemma 11.5 imply that we can partition G up into sets $E_1, \ldots, E_l$ and T such that T is the set of vertices in no big clique and hence contains no dense vertices, and each E_i is either a maximal big clique C_i or consists of a maximal big clique C_i and a vertex v_i seeing at least $\Delta - r - 1$ but not all of the vertices of C_i.

To complete the proof of Lemma 11.6, we need the following result.

Observation 11.15 *For each vertex v of C_i, there is at most one neighbour of v in $G - C_i$ which sees more than $r + 4$ vertices of C_i. Furthermore, if $|C_i| = \Delta - 1$ then there is no such vertex.*

Proof For $|C_i| < \Delta - 1$, this proof is similar to the case $l = 2$ of Observation 11.12 and is left as an exercise. For $|C_i| = \Delta - 1$, the proof has the same flavour but is slightly more complicated. The reader may work through the details by solving Exercises 11.3 -11.5 □

Corollary 11.16 *For each C_i, there is at most one vertex in $G - C_i$ which sees at least $\frac{3\Delta}{4}$ vertices of C_i.*

This corollary ensures that we can obtain $D_1, \ldots, D_l$ and S satisfying (i),(ii), and (iii) of Lemma 11.6 by

1. Setting $D_i = E_i$ if E_i is not a clique,
2. Setting $D_i = E_i$ if E_i is C_i but no vertex v of $G - C_i$ satisfies $|N(v) \cap C_i| \geq \frac{3\Delta}{4}$,
3. Setting $D_i = E_i + v_i$ for the unique vertex v_i of $G - C_i$ satisfying $|N(v_i) \cap C_i| \geq \frac{3\Delta}{4}$ otherwise.

Now, Observation 11.15 implies (iv) of Lemma 11.6 holds as well, and the proof is complete. □

Exercises

Exercise 11.1 Consider the experiment in which we toss a fair coin once and then one of two biased coins $n-1$ times. If the ith flip came up heads then the coin we use for the $(i+1)$st flip will yield heads with probability $\frac{2}{3}$. If the ith flip came up tails then the coin we use for the $(i+1)$st flip will yield tails with probability $\frac{2}{3}$. Let X be the total number of flips which come up heads. Prove that each coin flip changes the conditional expected value of X by at most 3. Use Azuma's Inequality to prove that X is concentrated around its expected value. Can you apply Talagrand's Inequality to obtain this result?

Exercise 11.2 Show that Azuma's Inequality implies the Simple Concentration Bound.

In the following exercises, C is a clique of size $\Delta - 1$ in our minimal counterexample G, z is a vertex outside of C which sees at least $r+5$ vertices of C, and v is a neighbour of z in C.

Exercise 11.3 Mimic the proof of the case $l \leq 2$ of Observation 11.12 to show

(a) v has degree Δ, and
(b) the other external neighbour y of v has at most two other neighbours in C.

Exercise 11.4 Assume every neighbour of z in C has degree Δ and no vertex outside $C+z$ has more than three neighbours in $C \cap N(z)$. Let y be the other neighbour of v outside C. Show

(a) there is a vertex w in $N(z) \cap C - N(y)$ such that adding an edge between y and the neighbour of w in $V - C - z$ does not create a clique of size Δ,
(b) for any such w there is a colouring of $G - (C - w) - z$ in which y and w receive the same colour, and
(c) for any such w and vertex x in $C - N(z)$ there is a colouring of $G - (C - w - x)$ in which y and w receive the same colour and z and x receive the same colour.

Exercise 11.5 Combine the last two results to show that z does not exist.

Part V

An Iterative Approach

In this part of the book, we will see how to prove some very strong results by applying several iterations of the Naive Colouring Procedure introduced in the previous two chapters. In a typical application, we begin by constructing a partial colouring just as in Part IV. Instead of completing it greedily, we produce a series of refinements, each time using the Naive Colouring Procedure again.

In the first two chapters, we see how applying several iterations yields a significant improvement on Theorem 10.2. First we present Kim's proof [95] that the chromatic number of any graph without triangles or 4-cycles is at most $O(\Delta/\log\Delta)$. Then we present a more sophisticated proof by Johansson [86] who shows that the same bound applies to all triangle-free graphs. In the third chapter, we present Kahn's proof [89] that the List Colouring Conjecture is asymptotically true. This last result is the one for which Kahn first introduced the Naive Colouring Procedure.

This approach is a special case of the following general technique. We construct an object X (for example, a colouring or a stable set of a given graph) via a series of partial objects $X_1, X_2, ..., X_t = X$. At each step, we prove the existence of an extension of X_i to a suitable X_{i+1} by considering a random choice for that extension and applying the probabilistic method. We gain a surprising amount of power by applying the probabilistic method to several incremental random choices rather than to a single random choice of X. The technique is first used in this book in the proof of Lemma 3.6.

This technique is often referred to as the "semi-random method", the "pseudo-random method" or the "Rödl Nibble". The latter name comes from a series of well known papers beginning with Rödl's [138] and continuing through [60, 129, 88, 89] and further. However, this is a misnomer since pseudo-random arguments appeared much earlier, for example in [7].

12. Graphs with Girth at Least Five

12.1 Introduction

In Chap. 10, we saw that the chromatic number of a triangle-free graph with maximum degree Δ (sufficiently large) is at most $(1 - \frac{1}{2e^6})\Delta$. The main step was to show that after a single iteration of the Naive Colouring Procedure, with positive probability every vertex has more than $\frac{1}{2e^6}\Delta$ colours appearing at least twice in its neighbourhood, and so the colouring can be completed greedily. As we remarked, the proof left much room for improvement. For example, our lower bound on the difference between the number of vertices coloured in each neighbourhood and the number of colours used on those vertices was very loose. More importantly, rather than completing the colouring greedily, we could have used further iterations of the Naive Colouring Procedure on the uncoloured vertices.

In this chapter and the next we will tighten up the argument and obtain the best result possible, up to a constant multiple, namely that the chromatic number of such a graph is at most $O(\frac{\Delta}{\ln \Delta})$, as shown by Johansson [86]. On the other hand, as we see in Exercise 12.7, for every g, Δ there are graphs with girth at least g and with maximum degree Δ whose chromatic number is at least $\frac{\Delta}{2 \ln \Delta}(1 + o(1))$ (where the asymptotics are in terms of Δ).

Of course, the fact that G is triangle-free is important here, as there are many graphs, for example cliques, with chromatic number close to Δ. The point is that all of them contain many triangles. Vizing [155] seems to have been the first to ask whether the bound in Brooks' Theorem can be improved significantly for triangle-free graphs. The first non-trivial result was due independently to Borodin and Kostochka [29], Catlin [31] and Lawrence [102], who showed that if G has no triangles (and in fact even if G merely has no K_4) then $\chi(G) \le \frac{3}{4}(\Delta + 2)$. Later, Kostochka (see [85]) showed that for G triangle-free, $\chi(G) \le \frac{2}{3}\Delta + 2$. This was the best bound known for triangle-free graphs until Johansson's result.

As we will see, 4-cycles are somewhat of a nuisance when applying the Naive Colouring Procedure, so in this chapter we will focus on the case where our graph has girth at least 5, and present the following result of Kim [95]:

Theorem 12.1 *If G has girth at least 5 and maximum degree Δ, then* $\chi(G) \leq \frac{\Delta}{\ln \Delta}(1 + o(1))$.

Note that this is within a factor of 2 of the best possible bound.

As with most applications of the Naive Colouring Procedure, the proof applies equally well to list colourings, and so Kim in fact obtains the same bound for the list chromatic number of G (as will the reader in Exercise 12.8).

To prove Theorem 12.1, we apply the Naive Colouring Procedure repeatedly, in fact poly($\ln \Delta$) times. For each vertex v we will keep track of a list, L_v, of colours which do not yet appear on the neighbourhood of v. Our hope is that the sizes of these lists will not decrease too quickly, and in particular that none of the lists will shrink to size 0 before the colouring is complete.

We will close this section by discussing intuitively why the lack of 4-cycles helps us here, and why this procedure should succeed only when there are at least roughly $\frac{\Delta}{\ln \Delta}$ available colours.

Suppose that we have carried out a number of iterations of the Naive Colouring Procedure, using the colours $\{1, \ldots, C\}$. Consider any two vertices v_1, v_2. Their respective lists, L_{v_1}, L_{v_2} are two random subsets of $\{1, \ldots, C\}$. Are they independent? The answer is certainly NO if they have many common neighbours, because every common neighbour which is coloured will cause the same colour to be deleted from both lists, and so many such neighbours will cause the two lists to coincide much more than two independently chosen lists would. But G has no 4-cycles and so v_1 and v_2 have at most one common neighbour! Therefore, the dependency caused by common neighbours is very small. Of course, dependency arises in other ways (see Exercise 12.1), but it turns out that such dependency is relatively minor. Thus, after several iterations of the Naive Colouring Procedure, the set of lists L_v look very similar to a collection of independently chosen subsets of $\{1, \ldots, C\}$.

Consider any vertex v which eventually gets coloured. By symmetry, its colour is uniformly chosen from $\{1, \ldots, C\}$. Now consider two vertices, v_1, v_2 which both get coloured. Are their two colours independent? The answer is certainly NO if v_1 and v_2 are adjacent as then the colours could not be identical. But what if they are not adjacent? Well again the answer would be NO if L_{v_1}, L_{v_2} were highly dependent, because at each step v_1 and v_2 are assigned colours drawn at random from these lists. For example, if L_{v_1} and L_{v_2} tended to be very similar then the colours eventually retained by v_1 and v_2 would be the same with probability much higher than $\frac{1}{C}$. But as we discussed in the preceding paragraph, the dependency between L_{v_1} and L_{v_2} is very small, and thus so is the dependency between the two colours. In fact, given any independent set of vertices, the colours (if any) which these vertices eventually retain are very similar to a collection of colours chosen uniformly and independently from $\{1, \ldots, C\}$.

Let's see now why the foregoing discussion implies, at least intuitively, that every vertex will be coloured provided C is just slightly bigger than $\frac{\Delta}{\ln \Delta}$. The only reason that a vertex would remain uncoloured is that all C colours eventually appear on its neighbourhood. So consider the colours assigned to the neighbourhood of a vertex v. As G is triangle-free, $N(v)$ is an independent set and so, as we proceed, the colours which are retained on $N(v)$ are nearly independent. We need to determine how big C must be in order to ensure that at least one colour never appears on $N(v)$. This is closely related to a special case of what is known as the Coupon Collector's Problem [141]. Suppose that you collect a sequence of coupons, each with a uniformly random colour from $\{1, \dots, C\}$ on it. How many coupons do you have to collect before you have at least one of each colour? It turns out that with high probability (see Exercise 12.2) the number of coupons will be very close to $C \ln C$. In other words, if $d(v)$ is much higher than $C \ln C$, then with high probability the size of L_v will eventually shrink to 0, while if $d(v)$ is much lower than $C \ln C$ then with high probability the size of L_v will always be fairly large. Thus, at least intuitively, the breaking point is roughly $\Delta \approx C \ln C$, i.e. $C \approx \frac{\Delta}{\ln \Delta}$.

12.2 A Wasteful Colouring Procedure

12.2.1 The Heart of The Procedure

Most of the work in finding our colouring will be achieved through several iterations of a variation of our Naive Colouring Procedure. To obtain this variation we will tweak our Naive Colouring Procedure in three ways. The first tweak is necessary. The other two tweaks are not necessary, but they simplify the proof significantly.

Modification 1: *At each iteration, we only assign colours to a few of the vertices.*

Recall that in our proof of Theorem 10.2, we carried out a single iteration of the Naive Colouring Procedure using $\frac{\Delta}{2}$ colours. For each vertex v, the probability that v retained its colour was $(1 - \frac{1}{\Delta/2})^{\Delta}$ which is approximately e^{-2}. In fact, for any $\epsilon > 0$ the reader can easily verify that if we were to use $\epsilon\Delta$ colours, then the probability that v retained its colour would be approximately δ for some positive δ defined in terms of ϵ. However, in this application we are using $o(\Delta)$ colours and so we need to be more careful.

To see this, suppose that we carry out the first iteration of our standard Naive Colouring Procedure using $\frac{\Delta}{\ln \Delta}$ colours. The probability that v retains its colour is $(1 - \frac{1}{\Delta/\ln \Delta})^{\Delta} \approx \frac{1}{\Delta}$. Thus, the expected number of neighbours of a vertex which retain their colours is approximately 1, which is far too small for our purposes. For one thing, the probability that this number is zero is approximately $\frac{1}{\mathrm{e}}$ which is much too high for us to be able to apply the Local Lemma to show that there is a partial colouring in which every vertex has even one neighbour which retains its colour.

Remark The astute reader may notice that if we start with $(1+\epsilon)\frac{\Delta}{\ln \Delta}$ colours, $\epsilon > 0$ (as in fact we will do in this chapter), then the expected number of neighbours of v which retain their colour is $\Delta^{\frac{\epsilon}{1+\epsilon}}$ which is large enough for us to apply the Local Lemma . However, after several iterations of our procedure, the expected value of this number drops to 1, and even lower, and so we run into problems.

To overcome this problem, at each iteration we will activate a small number of the vertices, and assign colours only to these activated vertices. More precisely, a vertex becomes activated with probability $\frac{K}{\ln \Delta}$, for some small constant $K < \frac{1}{2}$ to be specified later. Now during the first iteration, the expected number of neighbours of v which are activated is $\Delta \times \frac{K}{\ln \Delta}$, and the probability that an activated vertex retains its colour is $(1 - \frac{K}{\ln \Delta}\frac{1}{\Delta/\ln \Delta})^{\Delta} \approx \mathrm{e}^{-K}$. Thus, the expected number of neighbours of a vertex which retain their colours is approximately $\frac{K}{\mathrm{e}^K}\frac{\Delta}{\ln \Delta}$ which is high enough for our purposes.

Modification 2: *We remove more colours from L_v than we need to.*

The purpose of the lists L_v is to avoid conflict between colours assigned during different iterations. At the end of each iteration, we remove every colour that is retained by one of the neighbours of a vertex v from its list, L_v. This ensures that those colours will not subsequently be assigned to v. To simplify the analysis, we will actually remove a colour from L_v if it is ever *assigned* to a neighbour of v. So in fact, L_v may very well be missing several colours which do not appear in $N(v)$ at all, i.e. those which were assigned to but not retained by neighbours of v.

At first glance, it seems that this is far too wasteful to have any chance of success. However, as discussed above, we activate so few vertices in each iteration that the probability of a colour assigned to a vertex being retained is bounded below by e^{-K} for some small constant K. Hence the number of colours removed from L_v should be no more than e^K times the number of colours retained on $N(v)$. Since K is small, e^K is very close to 1, and so this wastefulness does not actually hurt us very much. Furthermore, the benefit of wastefulness to the ease of the proof will be substantial.

Modification 3: *At the end of each iteration, for each vertex v, and colour c still in L_v, we perform a coin flip which provides one last chance to remove c from L_v.*

If we did not have this modification, the probability that a particular vertex keeps a particular colour in its list would vary somewhat over different vertices and colours. This final coin flip will serve to make these probabilities equal, and thus simplify our computations significantly. (See Exercise 12.3.)

Remark The use of equalizing coin flips is one of two popular ways to equalize these probabilities. The other is to add artificial vertices to the graph in such a way that these probabilities all become equal. This technique can be seen, for example, in [95, 89].

In summary, we will initially set all lists to $\{1, \ldots, C\}$ where C is the number of colours that we are using, and then apply several iterations of the following procedure:

Wasteful Colouring Procedure

1. For each uncoloured vertex v, *activate* v with probability $\frac{K}{\ln \Delta}$, where K is a small constant to be named later.
2. For each activated vertex v, assign to v a colour chosen uniformly at random from L_v.
3. For each activated vertex v, remove the colour assigned to v from L_u for each $u \in N(v)$.
4. Uncolour every vertex which receives the same colour as a neighbour.
5. Conduct an "equalizing" coin flip (to be specified later) for each vertex v and colour $c \in L_v$, removing c from L_v if it loses the coin flip.

It is worth noting that if a colour c is removed from a vertex v in Step 4, then c was assigned to a neighbour of v and thus must have been removed from L_v in Step 3.

12.2.2 The Finishing Blow

We will show that with positive probability, after carrying out enough iterations of our procedure, the situation will be such that it is quite simple to complete our colouring.

The most natural goal would be to show that we can complete the colouring greedily, i.e. to show that for some value L, every list has size at least L, while every vertex has fewer than L uncoloured neighbours. Unfortunately, we are unable to prove this using our approach (see Exercise 12.6). Instead, we will aim to complete the colouring using Theorem 4.3. To do so, we only have to achieve the easier goal that for some L, every list has size at least L, while for every vertex v and colour c, v has fewer than $L/8$ uncoloured neighbours u such that $c \in L_u$.

Remark Actually, by using a more complicated proof, it is possible to get to a situation where we can complete the colouring greedily. See, for example, [95].

12.3 The Main Steps of the Proof

We now present a detailed proof of Theorem 12.1. In this section, we provide an outline of the proof, listing all the main lemmas. Their proofs will appear in the next two sections.

As usual, we assume G to be Δ-regular. A construction similar to that in Sect. 1.5 shows that the general theorem can be reduced to this case (see Exercise 12.4). We will use the Wasteful Colouring Procedure from the previous section to show that for any $0 < \epsilon < \frac{1}{100}$ and Δ sufficiently large as a function of ϵ, G has a proper $(1+\epsilon)\frac{\Delta}{\ln \Delta}$ colouring. Note that this implies Theorem 12.1.

We have been discussing the intuition behind this procedure as though it were a random rather than a pseudo-random one. The first step in our analysis is to decide what conditions we need to impose on the output of each iteration. These conditions must ensure that the colouring on each vertex's neighbourhood looks sufficiently like that of a typical neighbourhood in the random colouring to allow us to complete the analysis.

It turns out that we only need to focus on two parameters. Specifically, for each vertex v, colour $c \in L_v$ and iteration i, we define the following to be the respective values at the beginning of iteration i of the Wasteful Colouring Procedure:

$$\ell_i(v) \text{ -- the size of } L_v$$

$$t_i(v,c) \text{ -- the number of uncoloured neighbours } u \text{ of } v \text{ with } c \in L_u$$

Of course, we define $\ell_1(v) = (1+\epsilon)\frac{\Delta}{\ln \Delta}$ and $t_1(v,c) = \Delta$.

The importance of these parameters is self-evident, we must know their values if we wish to eventually apply Theorem 4.3 to the uncoloured vertices. We insist that these parameters behave at each vertex as they would at a typical vertex under the random process. As we see below, it turns out that to do this we do not need to keep track of any other parameters.

Rather than keeping track of the values of $\ell_i(v)$ and $t_i(v,c)$ for every v, c, we focus on their extreme values. In particular, we will recursively define appropriate L_i and T_i such that we can show that with positive probability, for each i the following property holds at the beginning of iteration i:

Property P(i): For each uncoloured vertex v and each colour $c \in L_v$,

$$\ell_i(v) \geq L_i$$

$$t_i(v,c) \leq T_i$$

In order to motivate our definition of L_i and T_i, we will briefly consider the idealized and simplified situation where at the beginning of iteration i, we have that for every v, c, $\ell_i(v) = L_i$ and $t_i(v,c) = T_i$, and where we do not carry out any equalizing coin flips.

In this case, for each colour $c \in L_v$, the probability that L_v keeps c, i.e. that c is not assigned to any neighbour of v is

$$\mathrm{Keep}_i = \left(1 - \frac{K}{\ln \Delta} \times \frac{1}{L_i}\right)^{T_i}.$$

So the expected value of $\ell_{i+1}(v)$ is $\ell_i(v) \times \mathrm{Keep}_i$. Note that Keep_i is also the probability that an activated vertex retains the colour it is assigned. So a similar, but slightly more complicated, calculation shows that the expected value of $t_{i+1}(v,c)$ is roughly $t_i(v,c) \times (1 - \frac{K}{\ln \Delta}\mathrm{Keep}_i) \times \mathrm{Keep}_i$.

Now of course, it is too much to hope that we can show that with positive probability for *every* v, c we have $\ell_{i+1}(v) \geq \mathbf{E}(\ell_{i+1}(v))$ and $t_{i+1}(v,c) \leq \mathbf{E}(t_{i+1}(v,c))$. However, by using our concentration tools, we can show that all these parameters are within a small error term of their expected values. Thus we use the following definition:

We set $L_1 = (1+\epsilon)\frac{\Delta}{\ln \Delta}, T_1 = \Delta$, and we recursively define

$$L_{i+1} = L_i \times \mathrm{Keep}_i - L_i^{2/3};$$

$$T_{i+1} = T_i\left(1 - \frac{K}{\ln \Delta}\mathrm{Keep}_i\right) \times \mathrm{Keep}_i + T_i^{2/3}.$$

Initially, $T_1 >> L_1$, but as we can see, T_i drops more quickly than L_i, so we can be hopeful that eventually $T_i < \frac{1}{8}L_i$ at which point we can apply Theorem 4.3. More specifically, at each iteration, the ratio T_i/L_i decreases by roughly a factor of $1 - \frac{K}{\ln \Delta}\mathrm{Keep}_i$ which is at most approximately $1 - \frac{K}{\ln \Delta}\mathrm{e}^{-K}$. Thus, after $O(\ln \Delta \ln \ln \Delta)$ iterations, T_i will drop below $\frac{1}{8}L_i$.

Our only concern is to make sure that L_i does not get too small before this happens. For example, if L_i gets as small as $\ln \Delta$, then applying the Local Lemma becomes problematic and we will no longer be able to ensure that property $P(i)$ holds. Initially, $\mathrm{Keep}_i \approx \mathrm{e}^{-K}$ and so L_i drops by a linear factor at each step. If it were to continue to decrease this quickly, then L_i would get too small within $O(\ln \Delta)$ iterations which is too soon. Fortunately, as the ratio T_i/L_i decreases, Keep_i increases, and so the rate at which L_i decreases slows down. In fact, we will see that it slows down enough for us to be able to continue for the required number of iterations.

In the previous discussion, we assumed that every $\ell_i(v)$ is exactly L_i and every $t_i(v,c)$ is exactly T_i. Of course, we are typically not in this idealized situation. That is why we must use our equalizing coin flips, which we are now ready to define precisely:

For any vertex v and colour c, we define $\mathrm{Keep}_i(v,c)$ to be the probability that no neighbour of v is assigned c. $\mathrm{Keep}_i(v,c)$ is the product over all uncoloured $u \in N(v)$ with $c \in L_u$ of $1 - \frac{K}{\ln \Delta} \times \frac{1}{\ell_i(u)}$. If $P(i)$ holds, then this is at least $\mathrm{Keep}_i = (1 - \frac{K}{\ln \Delta} \times \frac{1}{L_i})^{T_i}$. Thus, by defining

$$\mathrm{Eq}_i(v,c) = 1 - \mathrm{Keep}_i/\mathrm{Keep}_i(v,c),$$

and performing the equalizing coin flip described in Modification 3 by removing colour c from L_v with probability $\mathrm{Eq}_i(v,c)$ then for every vertex v and colour $c \in L_v$, we ensure that the probability of c remaining in L_v is precisely Keep_i.

A subtle problem arises when dealing with $t_{i+1}(v,c)$. It turns out that this parameter is not strongly concentrated, as it can be affected by a large amount if c is assigned to v (in particular it drops to zero). To overcome this problem, we focus instead on a closely related variable: $t'_{i+1}(v,c)$ – the number of vertices $u \in N(v)$ which were counted by $t_i(v,c)$ and such that during iteration i, u did not retain a colour, c was not assigned to any vertex of $N(u) - v$, and c was not removed from L_u because of an equalizing coin flip. Because assigning c to v has no effect on $t'_{i+1}(v,c)$, we overcome our problem and we can show that $t'_{i+1}(v,c)$ is indeed strongly concentrated. To justify focusing on this derived variable, we note that if c is not assigned to v then $t_{i+1}(v,c) = t'_{i+1}(v,c)$. On the other hand, if c is assigned to v then $t_{i+1}(v,c) = 0$ which is fine since we are only interested in an upper bound on $t_{i+1}(v,c)$.

The rest of the proof is quite straightforward, following a series of standard steps, despite the fact that some of the mathematical expressions may be a little daunting. We will list the main lemmas here, saving their proofs for later sections.

In what follows, all probabilistic computations are in terms of the random choices made during a particular iteration of our Wasteful Colouring Procedure. That is, we have already carried out iterations 1 to $i-1$, and we are about to carry out iteration i. The first step is to compute the expected values of our parameters at the end of iteration i:

Lemma 12.2 *If $P(i)$ holds then for every uncoloured vertex v and colour $c \in L_v$*

(a) $\mathbf{E}(\ell_{i+1}(v)) = \ell_i(v) \times \mathrm{Keep}_i$;
(b) $\mathbf{E}(t'_{i+1}(v,c)) \leq t_i(v,c)\left(1 - \frac{K}{\ln \Delta} \times \mathrm{Keep}_i\right) \times \mathrm{Keep}_i + O(\frac{T_i}{L_i})$.

As usual, the next step is to prove that these variables are strongly concentrated.

Lemma 12.3 *If $P(i)$ holds, and if $T_i, L_i \geq \ln^7 \Delta$, then for any uncoloured vertex v and colour $c \in L_v$,*

(a) $\mathbf{Pr}\left(|\ell_{i+1}(v) - \mathbf{E}(\ell_{i+1}(v))| > L_i^{2/3}\right) < \Delta^{-\ln \Delta}$;
(b) $\mathbf{Pr}\left(|t'_{i+1}(v,c) - \mathbf{E}(t'_{i+1}(v,c))| > \frac{1}{2}T_i^{2/3}\right) < \Delta^{-\ln \Delta}$.

Thus, by applying the Local Lemma, it is straightforward to prove:

Lemma 12.4 *With positive probability, $P(i)$ holds for every i such that for all $1 \leq j < i$: $L_j, T_j \geq \ln^7 \Delta$ and $T_j \geq \frac{1}{8}L_j$.*

As is common, the computations would be much easier if the recursive formulae for L_{i+1} and T_{i+1} did not have the $L_i^{2/3}, T_i^{2/3}$ terms. So our next step will be to show that these "error terms" do not accumulate significantly, using the same straightforward type of argument as in the proof of Theorem 8.1.

Lemma 12.5 *Define* $L_1^{'} = (1+\epsilon)\frac{\Delta}{\ln \Delta}, T_1^{'} = \Delta$, *and recursively define*

$$L_{i+1}^{'} = L_i^{'} \times \mathrm{Keep}_i;$$

$$T_{i+1}^{'} = T_i^{'}\left(1 - \frac{K}{\ln \Delta}\mathrm{Keep}_i\right) \times \mathrm{Keep}_i.$$

If for all $1 \le j < i$ *we have* $L_j, T_j \ge \ln^7 \Delta$ *and* $T_j \ge \frac{1}{8}L_j$, *then*

(a) $|L_i - L_i^{'}| \le (L_i^{'})^{5/6} = \mathrm{o}(L_i^{'})$;
(b) $|T_i - T_i^{'}| \le (T_i^{'})^{5/6} = \mathrm{o}(T_i^{'})$.

(Note that in the preceding definition, Keep_i is still defined in terms of T_i, L_i, *not* $T_i^{'}, L_i^{'}$.)

With this lemma in hand, it is easy to show the lists never get too small, i.e,

Lemma 12.6 *There exists* i^* *such that*

(a) For all $i \le i^*$, $T_i > \ln^8 \Delta, L_i > \Delta^{\frac{\epsilon}{3}}$, *and* $T_i \ge \frac{1}{8}L_i$;
(b) $T_{i^*+1} \le \frac{1}{8}L_{i^*+1}$.

And finally, Lemmas 12.4 and 12.6 yield our main proof:

Proof of Theorem 12.1. We carry out our Wasteful Colouring Procedure for up to i^* iterations. If $P(i)$ fails to hold for any iteration i, then we halt. By Lemmas 12.4 and 12.6, with positive probability $P(i)$ holds for each iteration and so we do in fact perform i^* iterations. After iteration i^* every list has size at least 8 times the maximum over all uncoloured v and $c \in L_v$ of $t_{i^*+1}(v,c)$. Therefore, by Theorem 4.3 we can complete the colouring.

We have proven that for any $\epsilon > 0$, if Δ is sufficiently large then $\chi(G) \le (1+\epsilon)\frac{\Delta}{\ln \Delta}$. Therefore $\chi(G) \le \frac{\Delta}{\ln \Delta}(1 + \mathrm{o}(1))$. □

In the next two sections, we will fill in the proofs of these lemmas.

12.4 Most of the Details

In this section, we present all but the concentration details. That is, we give the proofs of all the lemmas but Lemma 12.3, which we leave for the next section.

Proof of Lemma 12.2.

(a) For any colour $c \in L_v$, the probability that c remains in L_v is precisely Keep_i. The rest follows from Linearity of Expectation.

(b) Unless v is assigned c in iteration i, we have $t'_{i+1}(v,c) = t_{i+1}(v,c)$. Therefore, the probability that these two variables differ is at most $\frac{1}{L_i}$. Since $t'_{i+1}(v,c)$ cannot exceed T_i, this implies that $\mathbf{E}(t'_{i+1}(v,c)) \le \mathbf{E}(t_{i+1}(v,c)) + \frac{T_i}{L_i}$. This fact allows us to focus on $\mathbf{E}(t_{i+1}(v,c))$.

Consider any uncoloured vertex $u \in N(v)$ such that $c \in L_u$. We will show that the probability that u does not retain a colour and that c remains in L_u is at most $\left(1 - \frac{K}{\ln \Delta} \times \text{Keep}_i\right) \times \text{Keep}_i + \mathrm{o}(\frac{1}{L_i})$. By Linearity of Expectation, this will suffice to prove our lemma.

If u is not activated, then u will not be assigned a colour, and the probability that c remains in L_u is Keep_i.

Suppose u is activated and is not assigned c. For each $\gamma \in L_u - c$, we compute the probability, conditional on u being assigned γ, that (1) c remains in L_u and (2) at least one neighbour of u is activated and assigned γ.

Note that conditioning on u being assigned γ has no effect on the colours assigned on $N(u)$. So the probability that (1) holds is simply Keep_i. For any particular vertex $w \in N(u)$ with $\gamma \in L_w$, we consider the probability that w is activated and assigned γ, conditional on (1) holding. Since the colour activations and assignments are independent over different vertices, this is the same as the probability that w is activated and assigned γ, conditional on the event D_w that w is not assigned c, which by definition of conditional probabilities, is equal to

$$\begin{aligned}
&\mathbf{Pr}((w \text{ is activated and assigned } \gamma) \cap D_w)/\mathbf{Pr}(D_w) \\
&= \mathbf{Pr}(w \text{ is activated and assigned } \gamma)/\mathbf{Pr}(D_w) \\
&\le \left(\frac{K}{\ln \Delta} \times \frac{1}{L_i}\right) \Big/ \left(1 - \frac{K}{\ln \Delta} \times \frac{1}{L_i}\right) \\
&= \frac{K}{\ln \Delta} \times \frac{1}{L_i} + O\left(\frac{1}{L_i^2 \ln^2 \Delta}\right).
\end{aligned}$$

Therefore, given that u is assigned γ, the probability of (2) conditional on (1) holding, is at most

$$1 - \left(1 - \frac{K}{\ln \Delta}\frac{1}{L_i} + O\left(\frac{1}{L_i^2 \ln^2 \Delta}\right)\right)^{t_i(u,\gamma)} \le 1 - \text{Keep}_i + \mathrm{o}\left(\frac{1}{L_i}\right).$$

If u is activated and receives c, then the probability that c remains in L_u and u does not retain c is zero.

Therefore, the probability that u does not retain a colour and that c remains in L_u is at most:

$$\left(1\frac{K}{\ln \Delta}\right) \times \text{Keep}_i + \frac{K}{\ln \Delta} \times (L_i - 1) \times \frac{1}{L_i} \times \text{Keep}_i \left(1 - \text{Keep}_i + \text{o}\left(\frac{1}{L_i}\right)\right)$$

$$= \left(1 - \frac{K}{\ln \Delta} \times \text{Keep}_i\right) \times \text{Keep}_i + \text{o}\left(\frac{1}{L_i}\right)$$

as required. □

We postpone the proof of Lemma 12.3 until the next section.

Proof of Lemma 12.4. We will prove this lemma by induction on i. Property $P(1)$ clearly holds. For $i \geq 1$, we assume that $P(i)$ holds and we prove that with positive probability $P(i+1)$ holds, by analyzing iteration i.

For every v and $c \in L_v$ we define A_v to be the event that $\ell_{i+1}(v) < L_{i+1}$ and $B_{v,c}$ to be the event that $t_{i+1}(v, c) > T_{i+1}$. If none of these bad events hold, then $P(i+1)$ holds.

$T_1/L_1 = \text{O}(\ln \Delta)$ and it is straightforward to verify that $T_i/L_i = \text{O}(\ln \Delta)$ for every relevant i. (In fact this follows immediately from Lemma 12.5 whose proof does not rely on any previous lemma.) Furthermore, recall that for any vertex v and colour c such that at the beginning of iteration $i+1$, v is uncoloured and $c \in L_v$, we have $t_{i+1}(v, c) = t'_{i+1}(v, c)$. Therefore by Lemma 12.2, if $B_{v,c}$ holds then $t'_{i+1}(v, c)$ differs from its expected value by at least $\frac{1}{2}T_i^{2/3}$.

By Lemma 12.3, the probability of any one of our events is at most $\Delta^{-\ln \Delta}$. Furthermore, each event is determined by the colours assigned to and equalizing coin flips for vertices of distance at most 2 from the vertex by which the event is indexed. Therefore, by the Mutual Independence Principle, each event E is mutually independent of all events involving vertices of distance greater than 4 from the vertex by which E is indexed, i.e. of all but fewer than $\Delta^4 \times (1+\epsilon)\frac{\Delta}{\ln \Delta} < \Delta^5$ other events. For Δ sufficiently large, $\Delta^{-\ln \Delta}\Delta^5 < \frac{1}{4}$ and so the result follows from the Local Lemma. □

In the proofs of Lemmas 12.5 and 12.6, it will be useful to have upper and lower bounds on Keep_i. Clearly Keep_i is closely related to the ratio T_i/L_i, so we will obtain our lower bound by showing that this ratio is decreasing.

Lemma 12.7 *If for all $j < i, L_j, T_j \geq \ln^7 \Delta$ and $T_j \geq \frac{1}{8}L_j$ then $T_i/L_i < T_{i-1}/L_{i-1}$.*

Proof The proof is by induction. We assume that it is true for all values up to i and consider $i+1$. Note that this assumption implies $T_i/L_i \leq T_1/L_1 < \ln \Delta$ and so $\text{Keep}_i \geq \mathrm{e}^{-KT_1/(L_1 \ln \Delta)} + o(1) = \mathrm{e}^{-K(1+\epsilon)} + o(1)$. Now,

$$L_{i+1} = L_i\left(\text{Keep}_i - L_i^{-1/3}\right)$$

and, using the facts that $T_i \geq \ln^7 \Delta$ and $\mathrm{Keep}_i = \Omega(1)$, we have

$$\begin{aligned}
T_{i+1} &= T_i \times \mathrm{Keep}_i - T_i \times \frac{K \times \mathrm{Keep}_i^2}{\ln \Delta} + T_i^{2/3} \\
&\leq T_i \times \mathrm{Keep}_i - \left(K \times \mathrm{Keep}_i^2\right) T_i^{6/7} + T_i^{2/3} \\
&< T_i \times \mathrm{Keep}_i - T_i^{5/6} \\
&< T_i \left(\mathrm{Keep}_i - L_i^{-1/3}\right),
\end{aligned}$$

which implies the lemma. □

Corollary 12.8 *If for all* $j \leq i, L_j, T_j \geq \ln^7 \Delta$ *and* $T_j \geq \frac{1}{8} L_j$ *then* $\mathrm{e}^{-K(1+\epsilon)} + o(1) \leq \mathrm{Keep}_i \leq 1 - \frac{K}{10 \ln \Delta}$.

Proof The lower bound is from the third sentence of the previous proof. The upper bound follows from $T_i \geq \frac{1}{8} L_i$ since this implies $\mathrm{Keep}_i \leq \mathrm{e}^{-K/8 \ln \Delta} < 1 - \frac{K}{10 \ln \Delta}$. □

Proof of Lemma 12.5. This proof is a straightforward induction along the same lines as the similar part of the proof of Theorem 8.1. Clearly $L_i^{'} > L_i$, so for part (a) we just have to prove inductively that $L_i^{'} \leq L_i + L_i^{'5/6}$. By applying Corollary 12.8 along with the fact that for K sufficiently small, the function $x^{5/6} - x$ is decreasing on the interval $[\mathrm{e}^{-K(1+\epsilon)} + o(1), 1]$, and by examining the Taylor Series for $(1-y)^{5/6}$ around $y = 0$ we obtain that for sufficiently large Δ:

$$\mathrm{Keep}_i^{5/6} - \mathrm{Keep}_i \geq \left(1 - \frac{5}{6} \times \frac{K}{10 \ln \Delta}\right) - \left(1 - \frac{K}{10 \ln \Delta}\right) = \frac{K}{60 \ln \Delta}.$$

Now, we proceed with our induction, using the facts that $L_i^{'} \approx L_i$ and $L_i \geq \ln^7 \Delta$.

$$\begin{aligned}
L_{i+1}^{'} &= \mathrm{Keep}_i L_i^{'} \\
&\leq \mathrm{Keep}_i (L_i + L_i^{'5/6}) \\
&= L_{i+1} + L_i^{2/3} + \mathrm{Keep}_i L_i^{'5/6} \\
&\leq L_{i+1} + \mathrm{Keep}_i^{5/6} L_i^{'5/6} + L_i^{2/3} - \frac{K}{60 \ln \Delta} L_i^{'5/6} \\
&< L_{i+1} + L_{i+1}^{'5/6}.
\end{aligned}$$

For part (b), a virtually identical argument shows that $T_i^{'} \geq T_i - T_i^{5/6}$. □

Our next step is to prove Lemma 12.6. Before doing so, it is interesting to recall our intuitive discussion from the end of Sect. 12.1. There, we considered what would happen to L_v if each of the neighbours of v were independently assigned a random colour from $\{1,\dots,C\}$, before v was coloured. If $C = (1+\epsilon)\frac{\Delta}{\ln\Delta}$ then the expected number of colours remaining in L_v would be $C \times (1-\frac{1}{C})^{\Delta} = (1+\epsilon)\Delta \times \Delta^{-1/(1+\epsilon)}/\ln\Delta = (1+\epsilon)\Delta^{\epsilon/(1+\epsilon)}/\ln\Delta$. We will show that we can ensure L_v is never much smaller than this - we will obtain $L_i \geq \Delta^{\epsilon/10}$.

Proof of Lemma 12.6. As we discussed earlier, our main goal is to show that L_i does not decrease too quickly. Of course, it is simpler to focus on L_i' instead, and to use the fact that L_i and L_i' are very close, by Lemma 12.5.

The rate at which L_i' decreases is equal to Keep_i. Our first step will be to obtain a lower bound on Keep_i which is stronger than that given by Corollary 12.8. Again, we will focus on the ratio $r_i = T_i/L_i \approx r_i' = T_i'/L_i'$.

$$r_i' = r_1' \prod_{j=1}^{i-1}\left(1 - \frac{K}{\ln\Delta}\text{Keep}_j\right).$$

By Lemma 12.7, r_i is decreasing. Furthermore, for Δ sufficiently large, we have $(1-\frac{K}{L_i\ln\Delta}) > \exp(-\frac{K}{(1-\epsilon/4)L_i\ln\Delta})$, and so

$$\text{Keep}_i > \exp\left(-\frac{K}{(1-\epsilon/4)\ln\Delta}r_i\right) > \exp\left(-\frac{K}{(1-\epsilon/4)\ln\Delta}r_1\right).$$

Therefore, since $\epsilon < \frac{1}{100}$:

$$\begin{aligned} r_i' &\leq r_1'\left(1 - \frac{K}{\ln\Delta}\exp\left(-\frac{K}{(1-\epsilon/4)\ln\Delta}r_1\right)\right)^{i-1} \\ &\leq \frac{\ln\Delta}{1+\epsilon}\left(1 - \frac{K}{\ln\Delta}\exp\left(-\frac{K}{(1+\epsilon/2)}\right)\right)^{i-1}. \end{aligned}$$

Applying Lemma 12.5, we get nearly the same bound on r_i:

$$\begin{aligned} r_i &\leq r_i' \times \frac{1+\frac{1}{T_i'^{1/4}}}{1-\frac{1}{L_i'^{1/4}}} \\ &= r_i'\left(1 + O\left(\ln^{-7/4}\Delta\right)\right) \\ &< \frac{\ln\Delta}{1+\epsilon/2}\left(1 - \frac{K}{\ln\Delta}\exp\left(-\frac{K}{(1+\epsilon/2)}\right)\right)^{i-1}. \end{aligned}$$

This yields a better lower bound on Keep_i.

$$\text{Keep}_i > \exp\left(-\frac{K}{(1-\epsilon/4)\ln\Delta}r_i\right)$$

$$\geq \exp\left(-\frac{K}{(1-\epsilon/4)\ln \Delta} \times \frac{\ln \Delta}{1+\epsilon/2}\left(1-\frac{K}{\ln \Delta}\exp\left(-\frac{K}{1+\epsilon/2}\right)\right)^{i-1}\right)$$

$$> \exp\left(-\frac{K}{1+\epsilon/8}\left(1-\frac{K}{\ln \Delta}\exp\left(-\frac{K}{1+\epsilon/2}\right)\right)^{i-1}\right).$$

Therefore we have:

$$L_i' = L_1 \times \prod_{j=1}^{i-1} \text{Keep}_j$$

$$\geq (1+\epsilon)\frac{\Delta}{\ln \Delta} \exp\left(-\frac{K}{1+\epsilon/8}\sum_{j\geq 1}\left(1-\frac{K}{\ln \Delta}\exp\left(-\frac{K}{1+\epsilon/2}\right)\right)^{j-1}\right)$$

$$= (1+\epsilon)\frac{\Delta}{\ln \Delta} \exp\left(-\frac{\ln \Delta}{1+\epsilon/8}\exp\left(\frac{K}{1+\epsilon/2}\right)\right)$$

$$> (1+\epsilon)\frac{\Delta}{\ln \Delta} \times \Delta^{-\mathrm{e}^{\frac{K}{1+\epsilon/2}}/(1+\frac{\epsilon}{8})}.$$

We take $K = (1+\frac{\epsilon}{2})\ln\left(1+\frac{\epsilon}{100}\right) \approx \frac{\epsilon}{100}$. Thus, since $\epsilon < \frac{1}{100}$ we obtain:

$$\frac{\mathrm{e}^{K/(1+\epsilon/2)}}{1+\epsilon/8} = \frac{1+\epsilon/100}{1+\epsilon/8} < 1-\frac{\epsilon}{9},$$

and so

$$L_i' > (1+\epsilon)\Delta^{\epsilon/9}/\ln \Delta > \Delta^{\epsilon/10}.$$

Therefore, L_i' never gets too small for our purposes. Thus, neither does L_i.

The next step is to show that T_i eventually gets much smaller than L_i. As we discussed earlier, this is straightforward since we have shown $r_i = T_i'/L_i' \leq (1-\frac{K}{\ln \Delta}\text{Keep}_1)^{i-1}$ which tends to 0. In particular, since Keep_1 is bounded from below by a positive constant (see Corollary 12.8), we have $T_i' < \frac{1}{10}L_i'$ for large enough $i = O(\ln \Delta/\ln\ln \Delta)$. By Lemma 12.5, $T_i < \frac{1}{8}L_i$ for the same values of i. This implies the existence of i^* as required. □

The only step that remains is to prove Lemma 12.3 which we will do in the next section.

12.5 The Concentration Details

Now we will complete the proof of Theorem 12.1 by proving Lemma 12.3, i.e. that for any $v, c \in L_v$, $\ell_{i+1}(v)$ and $t'_{i+1}(v,c)$ are highly concentrated. We will use Talagrand's Inequality, and so to prove that a variable X is concentrated, it will suffice to verify the following two conditions for our set of random choices.

1. Changing the outcome of a single random choice can affect X by at most 1.
2. For any s, if $X \geq s$ then there is a set of at most $3s$ random choices whose outcomes certify that $X \geq s$.

By Talagrand's Inequality, if these conditions hold then $\mathbf{Pr}(|X - \mathbf{E}(X)| > t \leq e^{-\beta t^2/\mathbf{E}(X)}$ for some constant $\beta > 0$ and for any t with $\sqrt{\mathbf{E}(X)} << t \leq \mathbf{E}(X)$. For each of the X, t that we will consider here this bound is less than $\Delta^{-\ln \Delta}$.

Recall that the random choices for an iteration are as follows. First, for each vertex, we choose whether or not to activate it. Next, for each activated vertex, we choose a random colour to assign it. Finally, after uncolouring some vertices and removing some colours from lists, we carry out an equalizing coin flip for each v and colour c which is still in L_v.

In order to apply Talagrand's Inequality, these choices must be independent. But, strictly speaking, they are not since whether we carry out a colour choice for a vertex depends on the outcome of a previous activation choice. Similarly, whether we carry out an equalizing coin flip for a pair v, c depends on the outcomes of the earlier choices which determine whether c was removed from L_v.

To overcome this problem, we add a set of dummy choices, where we choose a colour for every unactivated vertex and we conduct an equalizing coin flip for every v, c such that c was removed from L_v during the iteration. Of course, we do not assign the chosen colour to an unactivated vertex and if c is no longer in L_v then we do not need to remove c if it loses the equalizing coin flip. So these dummy choices have no effect on the performance of the iteration. Their only effect is that now the entire set of choices is independent and so we can apply Talagrand's Inequality.

The fact that we need to add dummy choices may seem like an overly pedantic point to make, but it was just this kind of dependency which led to the subtle error in the false proof from Chap. 9.

Proof of Lemma 12.3. (a) Here it is easier to show that the number of colours which are removed from L_v during iteration $i+1$, $\overline{\ell}$, is highly concentrated.

Changing the assignment to any vertex $u \in N(v)$ can change $\overline{\ell}$ by at most 1, and changing the assignment to any other vertex cannot affect $\overline{\ell}$ at all. Furthermore, changing the outcome of any equalizing coin flip can affect $\overline{\ell}$ by at most 1.

If $\overline{\ell} \geq s$ then for some $s_1 + s_2 = s$, there are s_1 neighbours of v who were each assigned a different colour from L_v, and s_2 equalizing coin flips which each resulted in a colour being removed from L_v, and so the outcomes of these $s_1 + s_2$ trials certify that $\overline{\ell} \geq s$.

Therefore Talagrand's Inequality implies that

$$\mathbf{Pr}\left(|\overline{\ell} - \mathbf{E}(\overline{\ell})| > L_i^{2/3}\right) < \Delta^{-\ln \Delta}.$$

By Linearity of Expectation, $\mathbf{E}(\ell_{i+1}(v)) = \ell_i(v) - \mathbf{E}(\bar{\ell})$, and so

$$\mathbf{Pr}\left(|\ell_{i+1}(v) - \mathbf{E}(\ell_{i+1}(v))| > L_i^{2/3}\right) = \mathbf{Pr}\left(|\bar{\ell} - \mathbf{E}(\bar{\ell})| > L_i^{2/3}\right) < \Delta^{-\ln \Delta}.$$

(b) Denote by T the set of uncoloured neighbours of v whose lists contain c at the beginning of the iteration. We define X to be the number of vertices in T which remain uncoloured after the iteration, and Y to be the number of these vertices which remain uncoloured but which either have a neighbour other than v that is assigned c, or have c removed from their lists as the result of an equalizing coin flip.

Clearly $t'_{i+1}(v,c) = X - Y$, and so by Linearity of Expectation, it will suffice to show that X and Y are both sufficiently concentrated.

First we focus on X. We have to be a little careful here because changing the colour of v can affect X by a great deal, possibly dropping it to 0 if every activated vertex in T receives the same colour and the colour of v is changed to that colour. Thus we do not satisfy condition (1) above. To get around this problem, we isolate the effect that the colour assigned to v can have as follows.

Let X_1 denote the number of vertices $u \in T$ which are assigned either the same colour as a neighbour other than v or which are assigned no colour (i.e. are not activated) Let X_2 denote the number of vertices in T which are assigned the same colour as v. As we will see, it is straightforward to apply Talagrand's Inequality to X_1 since the colour assigned to v has no effect on X_1. Furthermore, we will see that X_2 is a binomial variable with expected value O(1), and so it is easy to show that with very high probability X_2 is quite small.

First, we consider X_1. There are no edges within T, and so changing a choice for some $u \in T$ can only affect whether or not u remains uncoloured, and thus affects X_1 by at most 1. Furthermore, no vertex outside of $v \cup T$ has more than one neighbour in T, and so changing one of its choices can affect X_1 by at most 1.

If a vertex $u \in T$ does not receive a colour, then the activation choice for u certifies this fact. If u receives a colour but is uncoloured because a neighbour also receives that colour, then the activation choices and assignments for u and that neighbour certify this fact. Therefore, if the number of neighbours of v which do not receive and retain a colour is at least s, then there is a set of at most $4s$ choices which certify this fact.

As $X_1 \le T_i$, we have $\mathbf{E}(X_1) \le T_i$ and so Talagrand's Inequality now implies that $\mathbf{Pr}\left(|X_1 - \mathbf{E}(X_1)| > \frac{1}{4}T_i^{2/3}\right) < \frac{1}{4}\Delta^{-\ln \Delta}$.

X_2 is bounded from above in distribution by the binomial variable $BIN(T_i, \frac{K}{\ln \Delta} \times \frac{1}{L_i})$. As discussed in the proof of Lemma 12.6, $T_i/L_i = O(\ln \Delta)$ and so $\frac{K}{\ln \Delta} \times \frac{1}{L_i} < \frac{\alpha}{T_i}$ for some absolute constant α. Since $T_i \ge \ln^7 \Delta$, it follows from Exercise 2.12 that $\mathbf{Pr}\left(X_2 > \frac{1}{4}T_i^{2/3}\right) < \frac{1}{4}\Delta^{-\ln \Delta}$.

Since $X_1 \leq X \leq X_1 + X_2$, these two bounds together imply that

$$\mathbf{Pr}\left(|X - \mathbf{E}(X)| \geq \frac{1}{2}T_i^{2/3}\right) < \frac{1}{2}\Delta^{-\ln \Delta}.$$

By applying a similar argument, in conjunction with an argument similar to that from part (a), we get that

$$\mathbf{Pr}\left(|Y - \mathbf{E}(Y)| \geq \frac{1}{2}T_i^{2/3}\right) < \frac{1}{2}\Delta^{-\ln \Delta},$$

and it immediately follows that t'_{i+1} is as concentrated as required. We leave the details to the reader (Exercise 12.5), remarking only that a certificate of size up to $6s$ is required. □

Remark Note that it is only in proving these concentration bounds that we used the fact that G has no 4-cycles.

Exercises

Exercise 12.1 Suppose that u and v are at distance 10 in G. Explain why after several iterations of the Wasteful Colouring Procedure the random sets L_u and L_v are not independent.

Exercise 12.2 Suppose that you collect coupons, one at a time, where each one contains a random colour from $\{1, \dots, c\}$. Prove that for any constant $\epsilon > 0$ and c sufficiently large, after collecting $(1+\epsilon)c \ln c$ coupons, with high probability you will have at least one of each colour. (Hint: use the First Moment Method).

Exercise 12.3 Another possibility for an equalizing coin flip would be to flip a coin which provides one last chance to uncolour a vertex, thus ensuring that the probability that an uncoloured vertex gets coloured is the same for every such vertex. Explain why in this particular application, such an equalizing flip would not be as useful as the one that we chose to use.

Exercise 12.4 Use a construction similar to that given in Sect. 1.5 to prove that every graph with girth 5 and maximum degree Δ can be embedded in a Δ-regular graph with girth 5. The problem with the construction in Sect. 1.5 is that it may create some 4-cycles.

Exercise 12.5 Complete the proof of Lemma 12.3(b).

Exercise 12.6 Introduce a new variable, D_i, such that at the beginning of iteration i, each vertex has at most D_i uncoloured neighbours. Obtain a recursive formula for D_i analogous to those for L_i and T_i. Analyze the rate at which D_i decreases, comparing it to L_i. In particular, show that it is much easier to reach a point where $T_i < \frac{1}{8}L_i$ than to reach a point where $D_i < L_i$.

Exercise 12.7 (Hard) See [24] pages 47–53 for a description of how to generate a random Δ-regular graph for any constant Δ, as well as a proof that for any constant g, such a graph has girth at least g with probability tending to a positive constant (which, of course, is a function of Δ and g) as n, the number of vertices, tends to infinity.

Show that for some $k = \frac{\Delta}{2\ln\Delta}(1 + o(1))$, the expected number of k-colourings of such a graph tends to zero as n tends to infinity, and then use the First Moment Method to argue that for Δ, g arbitrarily large there exist Δ-regular graphs with girth at least g and chromatic number at least $\frac{\Delta}{2\ln\Delta}(1 + o(1))$.

Exercise 12.8 Modify the proof of Theorem 12.1 to show that if G has maximum degree Δ and girth at least 5, then $\chi_\ell(G) \leq \frac{\Delta}{\ln\Delta}(1 + o(1))$.

13. Triangle-Free Graphs

In Chap. 12 we proved that graphs with girth at least 5, i.e. graphs with no triangles or 4-cycles, have chromatic number at most $O(\frac{\Delta}{\ln \Delta})$. In this chapter, we will present Johansson's stronger result [86] that the same bound holds even for triangle-free graphs.

Theorem 13.1 *There exists Δ_0 such that every triangle-free graph G with maximum degree $\Delta \geq \Delta_0$, has $\chi(G) \leq \frac{160\Delta}{\ln \Delta}$.*

Remarks

1. The constant term "160" is not optimal – in fact, Johansson's proof replaces it by "9". However, reducing it to a "1" to match the result in Chap. 12 looks more difficult, and might require a different approach.
2. In a subsequent paper [87], Johansson proves that for any fixed t, if G is K_t-free and has maximum degree Δ, then $\chi(G) \leq O(\frac{\Delta \ln \ln \Delta}{\ln \Delta})$. It would be very interesting to remove the extra $\ln \ln \Delta$ term from this bound and so extend Theorem 13.1 to K_t-free graphs for $t > 3$.
3. The bounds in Theorem 13.1 and Remark 2 also hold for $\chi_\ell(G)$.
4. Alon, Krivelevich and Sudakov [7] and Vu [157, 158] provide extensions of Theorem 13.1 to sparse graphs.
5. In contrast to Theorem 13.1, for every k there exist triangle-free graphs with colouring and chromatic number k, see Exercise 13.1.

At first glance, it may seem like a minor improvement to reduce the girth requirement from 5 to 4. However, it is not as simple as one might expect, since the absence of 4-cycles played a crucial role in our proof of Theorem 12.1, and allowing for their presence requires some significant modifications in our approach.

Recall our intuitive explanation for the success of our iterative procedure in the previous chapter. For any vertex v, because our graph had girth at least 5, the lists on the neighbours of v evolved virtually independently of each other. This suggested that the colours appearing on $N(v)$ would be very similar to a collection of Δ colours, each drawn at random independently and with replacement from the set $\{1, \dots, C\}$. As long as C is a bit larger than $\frac{\Delta}{\ln \Delta}$, such a collection will, with high probability, not include every colour. Thus, with high probability, we will never run out of colours for v.

How does the presence of 4-cycles affect this intuition? If two vertices $u_1, u_2 \in N(v)$ share several common neighbours, then their lists do not evolve independently. Colours which are assigned to their common neighbours are removed from both L_{u_1} and L_{u_2} and so those two lists will tend to be similar. But that should only help us, since intuitively, this should increase the probability that u_1 and u_2 get the same colour! If v lies in many 4-cycles, then the set of colours appearing on $N(v)$ should tend to have more repetitions than an independently drawn set, so it should tend to be smaller and thus the probability of always having a colour available for v should be even higher.

So it seems likely that the procedure we used in the previous chapter should still work. However, because of the 4-cycles, it is more difficult to analyze the procedure, and the methods we used to analyze it before will not work here. In particular, without 4-cycles, we were able to control the rates at which each $|T_{v,c}|$ and $|L_v|$ decreased. With the presence of 4-cycles, we cannot do this.

Consider, for example, $T_{v,c}$. In an extreme case, if G is a complete bipartite graph, then every vertex in $N(v)$ will have the same neighbourhood, and so will have the same list. Therefore, after several iterations (ignoring the effect of the equalizing coin flips), the size of $T_{v,c}$ will either have decreased all the way to zero, or will only have decreased to the number of uncoloured neighbours of v, and both of these extremes will typically have been achieved by several colours. Furthermore, the probability of a colour c remaining in a list L_v depends heavily on the size of $T_{v,c}$, and so with the presence of 4-cycles, the size of L_v becomes very difficult to analyze.

So we will not focus on these parameters. Instead, we will focus on the probability that two adjacent vertices are assigned the same colour. At first glance, this might not seem like the most natural parameter to consider. However, as we shall see, it allows us to bound the probability that an activated vertex is involved in a conflict which was, in fact, the most important reason for considering the two parameters in the previous chapter[1]. Furthermore, we will see that an elegant modification of our procedure makes this parameter very simple to analyze, thus allowing us to complete our proof.

13.1 An Outline

13.1.1 A Modified Procedure

We will modify the procedure used in Chap. 12, and apply it to colour G with the colours $\{1, \dots, C\}$ where $C = \frac{160\Delta}{\ln \Delta}$. We begin by presenting this procedure in a very general form.

[1] We used $1 - \text{Keep}_i \leq 1 - (1 - \frac{K}{\ln \Delta L_i})^{T_i}$

The essence of the procedure is that in each iteration,

(i) we make a random choice as to whether to activate each uncoloured vertex, where these choices are made independently;
(ii) for each activated vertex v, we choose a random colour from L_v to assign to v;
(iii) we uncolour every vertex assigned the same colour as a neighbour.

To specify an instance of this paradigm, we must do two things. First, in step (i) we must specify the activation probability. Usually we choose a very low probability, so that very few vertices are activated. This is so the probability of a vertex receiving the same colour as a neighbour is very small – almost (but not quite) negligible.

Secondly, in step (ii), we must specify the probability distribution from which we choose the colour for v. In Chap. 12, we used the most natural distribution on L_v – the uniform one. In this chapter, we will see that it can be advantageous to use a non-uniform distribution for this step, and so some colours will be more likely to be assigned than others.

Our goal is to reach a final stage where so much of the graph is coloured, that with positive probability, we can complete our colouring with an argument very similar to that in Theorem 4.3.

In order to show that we will colour a lot of our graph, we want to show that in each iteration, the probability that an uncoloured vertex v retains a colour is high. To do that, we need to show the following:

(a) the probability that v is assigned a colour is high; and
(b) the probability that a neighbour of v gets the same colour is low.

Now, (a) is easy to deal with – it is essentially just the probability that v is activated. So we will focus principally on (b). In particular, the key to our analysis is ensuring that the following property holds:

13.2 *For every pair of adjacent uncoloured vertices u and v, the probability that u and v are assigned the same colour given that they are both activated is at most $\frac{2}{C}$.*

Recall that in analyzing the last iteration of the procedure from the previous chapter, we showed that because the probability of a conflict between adjacent activated vertices was sufficiently small, we could apply Theorem 4.3 to prove that if we assigned a colour to every vertex, then with positive probability there would be no conflicts. I.e., an analogue of (13.2) was crucial in the analysis of this final iteration. In this chapter, (13.2) will be the key, not only to the analysis of the evolution of the iterative procedure, but also to administering the final blow. For, given that (13.2) holds, to show that we can complete the colouring using one final iteration in which all the vertices are activated, we only need a sufficiently good upper bound on the number of uncoloured neighbours of each vertex.

As we have already stated, the main new ingredient in our procedure is that we no longer choose the colour to assign to an activated vertex v uniformly from L_v. Rather, some colours in L_v will have a higher probability of being assigned than others. We will choose the probability distribution for the uncoloured vertices in the ith iteration by modifying that used in the previous iteration in such a way that the probability that two adjacent vertices are involved in a conflict does not change significantly. This is what allows us to ensure that (13.2) continues to hold.

This ingenious variation on our Naive Colouring Procedure is the fundamental change that is required to modify the proof of Theorem 12.1 to yield a proof of Theorem 13.1.

13.1.2 Fluctuating Probabilities

To describe precisely how we iteratively define the probability distributions we use to assign colours, we need some definitions.

In every iteration, each vertex is activated to receive a colour with probability α. For each vertex v and colour c, we use $p_i(v,c)$, or simply $p(v,c)$, to denote the probability that c is assigned to v during iteration i if v is activated in iteration i. Thus, c is assigned to v with probability $\alpha\ p_i(v,c)$. We use $p_i(v)$, or simply $p(v)$ to denote the vector $(p_i(v,1),\dots,p_i(v,C))$. We indicate that c is not in L_v simply by assigning $p_i(v,c) = 0$, and so, for example, we can express $\sum_{c\in L_v} p_i(v,c)$ as $\sum_c p_i(v,c)$. For any colour $c \in L_v$, we let $\text{Keep}_i(v,c)$, as usual, be the probability that c is not assigned to any neighbour of v during iteration i, i.e. that L_v keeps c.

Now, initially, we do choose the colours assigned uniformly. That is, for each vertex v and colour c, we have $p_1(v,c) = 1/C$.

Essentially, we will define:

$$p_{i+1}(v,c) = \begin{cases} p_i(v,c)/\text{Keep}_i(v,c), & \text{if } L_v \text{ keeps } c \\ 0, & \text{otherwise}. \end{cases}$$

Observing that $\mathbf{E}(p_{i+1}(v,c)) = (p_i(v,c)/\text{Keep}_i(v,c)) \times \mathbf{Pr}(L_v \text{ keeps } c)$, this immediately yields the following, which is one of the key elements of our analysis:

13.3 $\mathbf{E}(p_{i+1}(v,c)) = p_i(v,c)$.

One of the important benefits of (13.3) is:

13.4 $\mathbf{E}\Big(\sum_c p_{i+1}(v,c)\Big) = \sum_c p_i(v,c)$.

In particular, $E\Big(\sum_c p_2(v,c)\Big) = 1$, clearly a very desirable fact. However, a moment's thought should convince the reader that, although this sum may often be close to its expected value, it is unlikely to be equal to it. So it seems that this method of iteratively defining the probability distributions is problematic as we cannot hope to maintain the property that the sum of the probabilities involved is exactly one.

One way to avoid this problem would be to scale these probabilities so that they sum to one, using (13.4) to show that this does not change them significantly. However, this would introduce a lot of clutter to our analysis and so we take a different approach, described in Sect. 13.1.3 which allows us to ignore the requirement that the p_i sum to one. It will again be important for our approach that the probabilities sum to approximately one.

Leaving aside this technical complication for the moment, let's see why defining the p_i iteratively in this fashion allows us to keep the probability of conflict low.

Our preliminary discussion suggests that we will be interested in the following variable, which is the probability bounded by (13.2):

Definition For each edge uv, let $H_i(u,v) = \sum_c p_i(u,c)p_i(v,c)$.

The key to our analysis is the following:

Fact 13.5 $\mathbf{E}(H_{i+1}(u,v)) = H_i(u,v)$.

This crucial fact is precisely the point at which the triangle-freeness of G is required in our proof. It is also the most important benefit of the way in which we allow our assignment probabilities to vary.

Proof The key observation is that since u and v have no common neighbours and we are using the Wasteful Colouring Procedure, we have for every $c \in L_u \cap L_v$:

$$\mathbf{Pr}(c \text{ remains in } L_u \cap L_v) = \mathbf{Pr}(c \text{ remains in } L_u) \times \mathbf{Pr}(c \text{ remains in } L_v).$$

Thus,

$$\begin{aligned}
\mathbf{E}(H_{i+1}(u,v)) &= \sum_{c \in L_u \cap L_v} \mathbf{Pr}(c \text{ remains in } L_u \cap L_v) \\
&\qquad \times (p_i(u,c)/\mathrm{Keep}_i(u,c)) \times (p_i(v,c)/\mathrm{Keep}_i(v,c)) \\
&= \sum_{c \in L_u \cap L_v} \mathbf{Pr}(c \text{ remains in } L_u) \times \mathbf{Pr}(c \text{ remains in } L_v) \\
&\qquad \times (p_i(u,c)/\mathrm{Keep}_i(u,c)) \times (p_i(v,c)/\mathrm{Keep}_i(v,c)) \\
&= \sum_{c \in L_u \cap L_v} \mathrm{Keep}_i(u,c) \times \mathrm{Keep}_i(v,c) \\
&\qquad \times (p_i(u,c)/\mathrm{Keep}_i(u,c)) \times (p_i(v,c)/\mathrm{Keep}_i(v,c)) \\
&= \sum_{c \in L_u \cap L_v} p_i(u,c)p_i(v,c) \\
&= H_i(u,v),
\end{aligned}$$

□

Fact 13.5, along with our upcoming proof that $H_{i+1}(u,v)$ is strongly concentrated, allows us to maintain that for every iteration i,

$$H_i(u,v) \approx H_1(u,v) = \frac{1}{C}.$$

In particular, we can ensure that no $H_i(u,v)$ exceeds $\frac{2}{C}$, i.e. that (13.2) holds. As mentioned earlier, this is the crux of the proof.

Having described the key ideas needed in our proof, we are almost ready to begin the formal analysis. First however, we must deal with the technical difficulty mentioned earlier and one other slight complication. We do so in the next two sections. The formal analysis begins in Sect. 13.2.

13.1.3 A Technical Fiddle

At first glance, it seems crucial that we maintain $\sum_c p_i(v,c) = 1$ for every v, i. In order to free ourselves from this requirement, we will change the way in which we randomly assign colours:

Instead of making a single random choice as to which colour, if any, is assigned to v, we are going to make a separate choice for each colour. That is, for each colour $c \in L_v$, we assign c to v with probability $\alpha \times p_i(v,c)$, where this choice is made independently of the corresponding choices for the other colours in L_v. Note that this means that we are not performing a single activation flip for v. In effect, we are performing a separate activation flip for each colour c, and if v is activated for c, then we assign c to v with probability $p_i(v,c)$.

At first, this may seem like a foolish thing to do, because it is quite possible that several colours will be assigned to v. However, it turns out that this possibility does not pose a serious problem. If more than one colour appears on v, then we simply remove all of them from v – regardless of whether they are assigned to any neighbours of v. The probability that v receives a second colour, conditional on v receiving at least one colour, will be roughly α which we will choose to be quite small. Thus, the probability that v is uncoloured because of a multiple colour assignment is negligible and will not have a serious effect on our analysis. Furthermore, we will see that there are significant benefits to making the colour assignments independently in this way – for example, we will be able to obtain very simple proofs of some of our concentration results using the Simple Concentration Bound.

Remark Here it is important that for a vertex v, we implicitly carry out a different activation flip for each colour. Suppose that instead we were to activate v with probability α and then if v is activated, assign each colour $c \in L_v$ to v independently with probability $p_i(v,c)$. Then the possibility of v receiving multiple colours, given that it is activated, would be $\Omega(1)$ which is large enough to affect our analysis.

13.1.4 A Complication

The way in which we define our assignment probabilities creates a complication which we have overlooked thus far:

Complication: *Some of the probabilities might become too large.*

As it stands, there is nothing preventing the value of $p_i(v,c)$ from exceeding one, as it might if $p_{i-1}(v,c) > \text{Keep}_{i-1}(v,c)$. Obviously, such a situation would be problematic. In fact, it turns out that problems arise in proving some of our concentration bounds even if $p_i(v,c)$ rises up to around $\frac{1}{\sqrt{\Delta}}$.

To avoid these problems, we will introduce a restriction which prevents any $p_i(v,c)$ from exceeding $\hat{p} = \Delta^{-11/12}$. If $p_i(v,c)$ reaches $\hat{p}$ then c will become problematic for v and we will no longer allow c to be assigned to v (although for technical reasons, it will be convenient to keep $p_i(v,c) = \hat{p}$ rather than to set $p_i(v,c) = 0$). We will provide more details of this in the next section.

Most of the work in the proof of Theorem 13.1 will be to show that for each vertex v, very few assignment probabilities reach $p_i(v,c) = \hat{p}$.

In the next section we will give a more thorough description of our procedure, including the details of how we keep the assignment probabilities from exceeding $\hat{p}$. We then show how to use this procedure to prove Theorem 13.1.

13.2 The Procedure

13.2.1 Dealing with Large Probabilities

We start by formally describing how we keep the assignment probabilities from exceeding $\hat{p} = \Delta^{-11/12}$.

If $p(v,c)$ is ever increased above $\hat{p}$, then we set $p(v,c) = \hat{p}$, and we will keep $p(v,c) = \hat{p}$ for every subsequent iteration.

This creates a problem in trying to maintain property (13.3). If $p_i(v,c) = \hat{p}$ then $p_{i+1}(v,c)$ cannot exceed $p_i(v,c)$. Thus, if there is a chance that c will be removed from L_v during iteration i, i.e. that $p_{i+1}(v,c) = 0$, then we have $\mathbf{E}(p_{i+1}(v,c)) < p_i(v,c)$ which violates (13.3).

To deal with this problem, we insist that $p_{i+1}(v,c)$ cannot be set to 0, i.e., $p_j(v,c)$ will be fixed at $\hat{p}$ for every iteration $j \geq i$. Of course, it is possible that c might be assigned to, and even retained by, a neighbour of v at some point in the future, and yet this fact will never be reflected by setting $p(v,c) = 0$. To avoid conflicts with such neighbours, we will never allow c to be assigned to v in any future iteration. However, it will be very convenient to keep $p(v,c) = \hat{p}$ in order to continue to satisfy (13.3).

Thus, at iteration i, for each vertex v we define the set of *big colours*

$$B_i(v) = \{c : p_i(v,c) = \hat{p}\},$$

and we do not allow any colour in $B_i(v)$ to be assigned to v.

We must make one more slight adjustment in the way that we increase $p_i(v,c)$ in order to maintain (13.3). If $p_i(v,c) < \hat{p} < p_i(v,c)/\text{Keep}_i(v,c)$, then we define

$$\text{Eq}_i(v,c) = \frac{\frac{p_i(v,c)}{\hat{p}} - \text{Keep}_i(v,c)}{1 - \text{Keep}_i(v,c)}.$$

If c is not assigned to any neighbour of v during iteration i, we set $p_{i+1}(v,c) = \hat{p}$, as usual. If c is assigned to a neighbour of v then we set $p_{i+1}(v,c) = \hat{p}$ with probability $\text{Eq}_i(v,c)$ and we set $p_{i+1}(v,c) = 0$ otherwise. Thus, the probability that $p_i(v,c)$ is set to $\hat{p}$ is $\text{Keep}_i(v,c)+(1-\text{Keep}_i(v,c))\times\text{Eq}_i(v,c) = p_i(v,c)/\hat{p}$, and so (13.3) holds.

13.2.2 The Main Procedure

We set $K = 1$, i.e. we set the activation probability to be $\alpha = \frac{1}{\ln \Delta}$. (If we were interested in optimizing the constant term in our main theorem, as we were in Chap. 12, then we would need to be more careful in our choice of K.)

Other than the occasional extra coin flip when $p_i(v,c)$ is particularly close to $\hat{p}$, there are no equalizing coin flips in our procedure.

In summary, each iteration i of our procedure is as follows:

1. For each uncoloured vertex v, and colour $c \notin B_i(v)$, we assign c to v with probability $\frac{1}{\ln \Delta} p_i(v,c)$.
2. For each vertex v and colour $c \notin B_i(v)$, if c is assigned to at least one neighbour of v, then
   - if $p_i(v,c)/\text{Keep}_i(v,c) \le \hat{p}$ set $p_{i+1}(v,c) = 0$,
   - else set $p_{i+1}(v,c) = \hat{p}$ with probability $\text{Eq}_i(v,c)$ and set $p_{i+1}(v,c) = 0$ otherwise.
3. If c is assigned to v, and either (i) c is assigned to a neighbour $u \in N(v)$, or (ii) another colour is also assigned to v, then we remove c from v. Otherwise v retains c.
4. If $p_i(v,c) > 0$ and if $p_{i+1}(v,c)$ was not updated in step 2, then
   - if $p_i(v,c)/\text{Keep}_i(v,c) \le \hat{p}$ then set $p_{i+1}(v,c) = p_i(v,c)/\text{Keep}_i(v,c)$,
   - else set $p_{i+1}(v,c) = \hat{p}$.

13.2.3 The Final Step

Initially, $p_1(v,c) = \frac{1}{C}$ for all v, c and so $H_1(u,v) = C \times \left(\frac{1}{C}\right)^2 = \frac{1}{C}$ for every edge uv. Since we have $\mathbf{E}(H_{i+1}(u,v)) = H_i(u,v)$ for every iteration i and this variable turns out to be strongly concentrated, we will be able to ensure that $H_i(u,v)$ is never greater than $\frac{2}{C}$. We will also ensure that the size of $B_i(v)$ never exceeds $\frac{3}{10\hat{p}}(1+\text{o}(1))$.

We will continue our procedure until the maximum degree of the subgraph induced by the uncoloured vertices is at most $\frac{C}{64}$, at which point we will be able to complete the colouring as follows:

For each uncoloured vertex v, we will remove $B(v)$ from L_v and scale $p(v,c)$ for every other $c \in L_v$ so that the assignment probabilities sum to 1. That is, for each $c \in L_v - B(v)$ we set

$$p^*(v,c) = \frac{p(v,c)}{\sum_{c \in L_v - B(v)} p(v,c)}.$$

We then assign to v a single colour chosen from this normalized distribution, i.e. v receives c with probability $p^*(v,c)$.

Since $B_i(v) \le \frac{3}{10\hat{p}}(1+\text{o}(1))$ and $\sum_c p(v,c) \approx 1$, we have $\sum_{c \in L_v - B(v)} p(v,c) \ge \frac{1}{2}$ and so $p^*(v,c) \le 2p(v,c)$.

For each edge uv we define $E_{u,v}$ to be the event that u and v are both assigned the same colour.

$$\mathbf{Pr}(E_{u,v}) = \sum_c p^*(u,c)p^*(v,c) \le 4 \sum_c p(u,c)p(v,c) = 4H(u,v) < \frac{8}{C}.$$

Since the maximum degree of the subgraph induced by the uncoloured vertices is at most $\frac{C}{64}$, each of our events is mutually independent of all but at most $\frac{C}{32}$ other events (namely those corresponding to the edges which share an endpoint with uv). Therefore a straightforward application of the Local Lemma implies that with positive probability none of these events hold, i.e. that we have successfully completed our colouring.

13.2.4 The Parameters

We will keep track of four parameters. We have already mentioned that the first three are important:

$C_i(v)$ – defined for each vertex v to be $\sum_c p_i(v,c)$

$H_i(u,v)$ – defined for each edge uv to be $\sum_c p_i(u,c)p_i(v,c)$

$D_i(v)$ – defined for each vertex v to be the number of uncoloured neighbours of v at the beginning of iteration i

The final parameter is required to bound the size of $B_i(v)$. There are a number of ways that we might do this, but the simplest way is to focus on what is known as the *entropy* of the assignment probabilities:

$$Q_i(v) = -\sum_c p_i(v,c) \ln p_i(v,c),$$

where we take $p_i(v,c) \ln p_i(v,c) = 0$ if $p_i(v,c) = 0$. Note that $Q_i(v)$ is non-negative since $\ln p_i(v,c)$ is never positive.

Roughly speaking, $Q_i(v)$ measures how widely the assignment probabilities vary – the more they vary, the smaller their entropy will be. For example, on one extreme, when there is only one non-zero term and it is equal to 1, $Q = 0$, and on the other extreme, when all C terms are equal to $\frac{1}{C}$, $Q = \ln C$. So intuitively, if $Q_i(v)$ is not too small, then the values of $p_i(v, c)$ don't vary too much, and so not very many of these values are as high as $\hat{p}$. Later on, we will look more precisely at how a bound on $Q_i(v)$ yields a bound on the size of $B_i(v)$.

As usual, our main steps are to bound the expected values of these parameters after each iteration, and then to show that they are highly concentrated. In doing so, we will show that with positive probability, at the beginning of each iteration we satisfy the following property:

Property P: For each uncoloured vertex v and neighbour $u \in N(v)$,

$$\begin{aligned} C_i(v) &= 1 + \mathrm{O}\left(\frac{1}{\Delta^{1/10}}\right) \\ H_i(u,v) &\le \frac{1}{C} + \frac{1}{C\Delta^{1/10}} \\ D_i(v) &\le \Delta \times \left(1 - \frac{3}{5\ln\Delta}\right)^{i-1} \\ Q_i(v) &\ge Q_1(v) - \frac{1}{10}\ln\Delta. \end{aligned}$$

Our two main lemmas are:

Lemma 13.6 *If at the beginning of iteration i we satisfy Property P, then for every uncoloured vertex v and edge uv, we have*

(a) $\mathbf{E}(C_{i+1}(v)) = C_i(v)$;
(b) $\mathbf{E}(H_{i+1}(u,v)) = H_i(u,v)$.
(c) $\mathbf{E}(D_{i+1}(v)) \le D_i(v)(1 - \frac{2}{3\ln\Delta})$;
(d) $\mathbf{E}(Q_{i+1}(v)) \ge Q_i(v) - \frac{2}{\ln\Delta} \times \frac{D_i}{C}$.

Lemma 13.7 *For each iteration i, uncoloured vertex v and edge uv we have*

(a) $\mathbf{Pr}\left(|C_{i+1}(v) - \mathbf{E}\left(C_{i+1}(v)\right)| > \frac{1}{\Delta^{1/6}}\right) \le \Delta^{-5}$;
(b) $\mathbf{Pr}\left(|H_{i+1}(u,v) - \mathbf{E}(H_{i+1}(u,v))| > \frac{1}{C\Delta^{1/3}}\right) \le \Delta^{-5}$;
(c) $\mathbf{Pr}\Big(|D_{i+1}(v) - \mathbf{E}(D_{i+1}(v))| > \Delta^{2/3}\Big) \le \Delta^{-5}$;
(d) $\mathbf{Pr}\left(|Q_{i+1}(v) - \mathbf{E}(Q_{i+1}(v))| > \frac{\ln\Delta}{\Delta^{1/6}}\right) \le \Delta^{-5}$.

These two lemmas yield our main Theorem:

Proof of Theorem 13.1. We continue our procedure for i^* iterations, where i^* is the minimum integer for which $\Delta\left(1 - \frac{3}{5\ln\Delta}\right)^{i^*} < C/64$.

Hence $i^* = \mathrm{O}(\ln \varDelta \ln\ln \varDelta)$. Via iterative applications of the Local Lemma, mimicking the proof used in Chap. 12, Lemmas 13.6 and 13.7 yield that we can indeed preserve property P for this many iterations, and that more strongly, with positive probability, for every vertex v, edge uv, and iteration $1 \le i \le i^*$,

$$\begin{aligned}
C_i(v) &= 1 + \mathrm{O}\left(\frac{i}{\varDelta^{1/6}}\right) \\
H_i(u,v) &\le \frac{1}{C} + \frac{2i}{C\varDelta^{1/3}} < \frac{1}{C} + \frac{1}{C\varDelta^{1/10}} \\
D_i(v) &\le \left(1 - \frac{2}{3\ln \varDelta}\right)^{i-1} \varDelta + i\varDelta^{2/3} < \left(1 - \frac{3}{5\ln\varDelta}\right)^{i-1} \varDelta \\
Q_i(v) &\ge Q_1 - \left(\sum_{j=1}^{i-1} \frac{2D_j}{C\ln\varDelta}\right) - \frac{i\ln\varDelta}{\varDelta^{\frac{1}{6}}} \\
&> Q_1 - \frac{1}{80}\sum_{j=1}^{i-1}\left(1 - \frac{3}{5\ln\varDelta}\right)^{j-1} - \frac{i\ln\varDelta}{\varDelta^{\frac{1}{6}}} \\
&> Q_1 - \frac{1}{40}\ln\varDelta.
\end{aligned}$$

Note that our bound on D_i and our choice of i^* ensures that $D_{i^*}(v) \le \frac{C}{64}$ for every v.

We will now use our bound on $Q_i(v)$ to obtain a bound on $|B_i(v)|$. Our strategy will be to show that each colour in $B_i(v)$ contributes roughly $\hat{p} \times \frac{1}{12}\ln\varDelta$ to the total change in entropy $Q_1(v) - Q_i(v)$. Since this change is at most $\frac{1}{40}\ln\varDelta$, it follows that $|B_i(v)|$ is at most roughly $\frac{3}{10} \times \frac{1}{\hat{p}}$, or more importantly, that the total weight of the colours in $B_i(v)$ is at most roughly $\frac{3}{10}$.

The most straightforward way to express $Q_1(v) - Q_i(v)$ is as the sum: $\sum_c p_1(v,c)\ln(p_1(v,c)) - p_i(v,c)\ln(p_i(v,c))$. However, the terms of this sum turn out to be somewhat awkward to deal with, and so we will simplify things by rewriting $Q_1(v)$. We can look at the entropy as a weighted sum of the logarithmic terms. Since, in $Q_1(v)$, these logarithmic terms are all equal (i.e. since $p_1(v,c)$ is the same for each colour c), we are free to change the weights, as long as their sum is approximately the same. In particular, since $C_i(v) = 1 + \mathrm{O}\left(\varDelta^{-\frac{1}{10}}\right) \times C_1(v)$, we can reweight the sum according to the vector $p_i(v)$ as follows:

$$\begin{aligned}
Q_1(v) &= -\sum_c p_1(v,c)\ln(p_1(v,c)) \\
&= -\sum_c p_i(v,c)\ln(p_1(v,c)) \pm O\left(\varDelta^{-\frac{1}{10}} Q_1(v)\right),
\end{aligned}$$

and so

$$Q_1(v) - Q_i(v) = \pm O\left(\Delta^{-\frac{1}{10}} Q_1(v)\right) + \sum_c p_i(v,c)\ln(p_i(v,c)) - \ln(p_1(v,c))$$
$$= \pm O\left(\Delta^{-\frac{1}{10}} \ln \Delta\right) + \sum_c p_i(v,c)\ln(p_i(v,c)/p_1(v,c)).$$

Each colour $c \in B_i(v)$ contributes $\hat{p}\ln(\hat{p}/\frac{1}{C}) > \hat{p} \times \frac{1}{12}\ln\Delta$ to the latter sum. Furthermore, if $p_i(v,c)$ is not zero then $p_i(v,c) > p_1(v,c)$ and so there are no negative terms in that sum. Therefore, we have

$$|B_i| \times \hat{p} \times \frac{1}{12}\ln\Delta \pm \mathrm{O}\left(\Delta^{-\frac{1}{10}}\ln\Delta\right) \le Q_1(v) - Q_i(v) \le \frac{1}{40}\ln\Delta,$$

and so

$$|B_i|\hat{p} \le \frac{3}{10} + o(1).$$

Since $i^* = O(\ln\Delta\ln\ln\Delta)$, $H_{i^*} < 2/C$ and so, as described in Sect. 13.2.3, our final stage will successfully complete the colouring. □

It only remains to prove Lemmas 13.6 and 13.7 which we will do in the next section.

13.3 Expectation and Concentration

Proof of Lemma 13.6. We know that (a) follows from (13.3), and we deal with the colours in $B_i(v)$ in a way that ensures that (13.3) continues to hold. Now, (b) is simply (13.5). We need to reprove this fact, taking into account the way that we deal with colours in $B_i(v)$. This is straightforward; we leave the tedious but routine details to the interested reader.

(c) Consider any vertex u. We let A_1 be the event that u is assigned at least one colour, A_2 be the event that u is assigned at least 2 colours, and A_3 be the event that some colour is assigned to both u and one of its neighbours. Clearly the probability that u retains a colour is at least $\mathbf{Pr}(A_1) - \mathbf{Pr}(A_2) - \mathbf{Pr}(A_3)$. Now,

$$\begin{aligned}\mathbf{Pr}(A_2) &\le \sum_{c_1 \ne c_2}\left(\frac{1}{\ln\Delta}\right)^2 p_i(v,c_1)p_i(v,c_2)\\ &< \left(\frac{C_i(v)}{\ln\Delta}\right)^2\\ &< \left(\frac{1}{\ln\Delta}\right)^2 + \mathrm{O}\left(\Delta^{-1/10}\right).\end{aligned}$$

As was shown in the proof of Theorem 13.1, $\sum_{c \notin B_i(v)} p_i(c) \ge \frac{7}{10} - o(1)$, so by the simplest case of Inclusion-Exclusion,

$$\mathbf{Pr}(A_1) \geq \sum_{c \notin B_i(v)} \frac{p_i(v,c)}{\ln \Delta} - \sum_{c_1 \neq c_2} \left(\frac{1}{\ln \Delta}\right)^2 p_i(v,c_1)p_i(v,c_2)$$
$$\geq \frac{7}{10 \ln \Delta} - \mathrm{o}\left(\frac{1}{\ln \Delta}\right).$$

Finally,

$$\mathbf{Pr}(A_3) \leq \sum_c \sum_{w \in N(u)} \left(\frac{1}{\ln \Delta}\right)^2 p_i(u,c)p_i(w,c)$$
$$= \left(\frac{1}{\ln \Delta}\right)^2 \sum_{w \in N(u)} H_i(u,w)$$
$$\leq \left(\frac{1}{\ln \Delta}\right)^2 \frac{\Delta}{C} + \mathrm{O}\left(\frac{\Delta}{C\Delta^{1/10}}\right)$$
$$< \frac{1}{150 \ln \Delta}.$$

Therefore, the probability that a particular uncoloured neighbour u of v retains a colour is at least $\frac{2}{3 \ln \Delta}$ and so (c) follows by Linearity of Expectation.

(d) Recall that for each $c \notin B_i(v)$, there is $K(c) \leq \mathrm{Keep}_i(v,c)$ such that

$$p_{i+1}(v,c) = \begin{cases} 0, & \text{with probability } 1 - K(c) \\ \frac{p_i(v,c)}{K(c)}, & \text{with probability } K(c)\,. \end{cases}$$

Therefore,

$$\mathbf{E}(p_{i+1}(v,c) \ln p_{i+1}(v,c)) = K(c)\left(\frac{p_i(v,c)}{K(c)} \ln(p_i(v,c)/K(c))\right)$$
$$= p_i(v,c)(\ln p_i(v,c) - \ln K(c)),$$

and so, since for each $c \in B_i(v)$, $p_{i+1}(v,c) = p_i(v,c) = \hat{p}$, we have

$$\mathbf{E}(Q_{i+1}(v)) \geq Q_i(v) - \sum_{c \notin B_i(v)} -p_i(v,c) \ln(K(c))$$
$$\geq Q_i(v) - \sum_c -p_i(v,c) \ln(K(c)).$$

Using the fact that $K(c) \leq \mathrm{Keep}_i(v,c) = \prod_{u \in N(v)}(1 - p_i(u,c)/\ln \Delta)$ along with the fact that $\ln(1-x) \geq -x - x^2$ for x sufficiently small, we have

$$\mathbf{E}(Q_{i+1}(v)) \geq Q_i(v) - \sum_c \sum_{u \in N(v)} \frac{p_i(v,c)p_i(u,c)}{\ln \Delta} \times \left(1 + \mathrm{O}\left(\ln^{-1} \Delta\right)\right)$$

$$= Q_i(v) - \frac{1}{\ln \Delta} \sum_{u \in N(v)} H_i(u,v) \times \left(1 + \mathrm{O}\left(\ln^{-1} \Delta\right)\right)$$
$$\geq Q_i(v) - \frac{2D_i}{C \ln \Delta},$$

as required. □

And finally, we outline the fairly simple concentration details:

Proof outline of Lemma 13.7. For each colour $c \in L_v$, we let $\mathcal{T}_c$ be the set of random choices which determine the value of $p_{i+1}(v,c)$. Note that for any distinct colours c_1, c_2, we have that $\mathcal{T}_{c_1}, \mathcal{T}_{c_2}$ are disjoint. So for each c, we can combine the choices in $\mathcal{T}_c$ into a single trial and these trials will be independent. By doing this, parts (a) (b) and (d) follow easily from the Simple Concentration Bound.

To prove part (c), we first use the Simple Concentration Bound to show that the number of neighbours of v which receive at least one colour is sufficiently concentrated. Then we prove that the number of these neighbours which are uncoloured is sufficiently concentrated. To do this, we let X_v be the number of neighbours of v which are uncoloured, and we let X_v' be the number of neighbours u of v such that (i) u is assigned a colour which is assigned to at most $\Delta^{1/10}$ neighbours of v and (ii) u is uncoloured. A straightforward application of Talagrand's Inequality implies that X_v' is sufficiently concentrated, and an easy first moment argument shows that $\mathbf{Pr}(X_v \neq X_v') < \Delta^{-6}$. (The first moment argument uses the Chernoff Bound and the fact that $\Delta \times \hat{p}$ is significantly less than $\Delta^{1/10}$.) This implies that X_v is sufficiently concentrated. We leave the details to the reader. □

Exercises

Exercise 13.1 Show that for every positive integer k, there exists a triangle free graph G with $\chi(G)$ =col$(G) = k$.

14. The List Colouring Conjecture

In this chapter we present a second application of an iterative variant of the Naive Colouring Procedure: Kahn's proof that the List Colouring Conjecture is asymptotically correct, i.e. that for any graph G of maximum degree Δ, $\chi'_\ell(G) = \Delta + o(\Delta)$ [89]. The proof that we present here is based on the refinement of Kahn's argument found in [123], where the $o(\Delta)$ term is tightened to obtain:

Theorem 14.1 *For any graph G with maximum degree Δ, $\chi'_\ell(G) \leq \Delta + 4\sqrt{\Delta}\log^4 \Delta$.*

Remark The exponent "4" is not optimal, and the reader will notice that it can be easily reduced. However, this approach will not remove the poly$(\log \Delta)$ term entirely, and so prove $\chi'_\ell(G) \leq \Delta + O(\sqrt{\Delta})$. Thus, the best known bound on the list chromatic index is still worse than even Hind's 1990 bound on the total chromatic number [78] – more evidence that the List Colouring Conjecture is significantly harder than the closely related Total Colouring Conjecture!

Kahn showed that his result also holds for k-uniform linear hypergraphs, for any constant k. (Recall that a hypergraph is k-uniform if every edge has size k, and it is linear if no pair of vertices lies in more than one edge.) In [123] the $o(\Delta)$ term is tightened to show that

Theorem 14.2 *For any k-uniform linear hypergraph H, $\chi'_\ell(H) \leq \Delta + \Delta^{1-\frac{1}{k}}\mathrm{poly}(\log \Delta)$.*

In both [89] and [123] the reader can find further generalizations to the case of those non-linear hypergraphs in which no pair of vertices lies in many edges. In [75], Häggkvist and Janssen use a different technique (including one probabilistic lemma) to prove that for graphs $\chi'_\ell(G) \leq \Delta + O(\Delta^{\frac{2}{3}} \log \Delta)$. Their argument does not extend to hypergraphs.

14.1 A Proof Sketch

14.1.1 Preliminaries

The proof of Theorem 14.1 is very similar to that of Theorem 12.1, and we will concentrate on the points at which it diverges from that proof. We carry out several iterations of the Naive Colouring Procedure, this time colouring the edges of our graph rather than the vertices. We prove that with positive probability we colour enough edges that the graph induced by the uncoloured edges can be easily coloured in one final stage.

As usual, we take Δ to be at least as large as some constant Δ_0, which we do not name, rather stating that it is large enough that several implicit inequalities hold. Also, we assume that G is Δ-regular, which is permissible by the construction in Sect. 1.5. Initially, each edge e of G has a list $\mathcal{L}_e$ containing $\Delta + 4\sqrt{\Delta}\log^4 \Delta$ colours which are acceptable for e. We will often use $L_e \subseteq \mathcal{L}_e$ to denote a set of colours that are available for e, at a particular point of our procedure.

We use $L(G)$ to denote the line graph of G, and so we are list colouring the vertices of $L(G)$. The maximum degree of $L(G)$ is easily seen to be $2\Delta - 2$, and so it is easy to list colour $L(G)$ greedily if each list has at least $2\Delta - 1$ colours. However, we wish to use lists of half this size. This leads us to the first main difference between our situation here, and that in Chap. 12:

Difference 1: *The number of colours initially available at each vertex is of the same asymptotic order as the degree.* Because of this, we will not need to use activation probabilities here. Unfortunately, there is a price to be paid for not using activation probabilities – we will not be able to use the Wasteful Colouring Procedure. In other words, we will remove a colour from the list L_e if that colour is *retained* by an incident edge, not if it is merely assigned to an incident edge. As a result, the expected value computations and the concentration proofs are significantly more complicated.

Remark Although we do not need to use activation probabilities here, we might have chosen to do so. This would have allowed us to use the Wasteful Colouring Procedure, and simplified some of the analysis, thus depriving the reader of learning how to analyze a more delicate procedure. In Exercise 14.2, the reader can look at the alternate approach.

14.1.2 The Local Structure

As we saw in Chap. 10, the Naive Colouring Procedure works well if a graph is sufficiently sparse. It is not hard to verify that each neighbourhood in $L(G)$ contains at most Δ^2 edges, and so Exercise 10.2 implies that the list chromatic number of $L(G)$ is at most $\Delta(2 - \frac{2}{e^6})$. Unfortunately, the sparseness of $L(G)$ alone will not be enough to reduce this bound all the way to $\Delta + o(\Delta)$, as desired, since there are graphs which are just as sparse but have chromatic number approximately $\sqrt{2}\Delta$ (see Exercise 14.1). So we will have to look more closely at the structure of the neighbourhoods in $L(G)$.

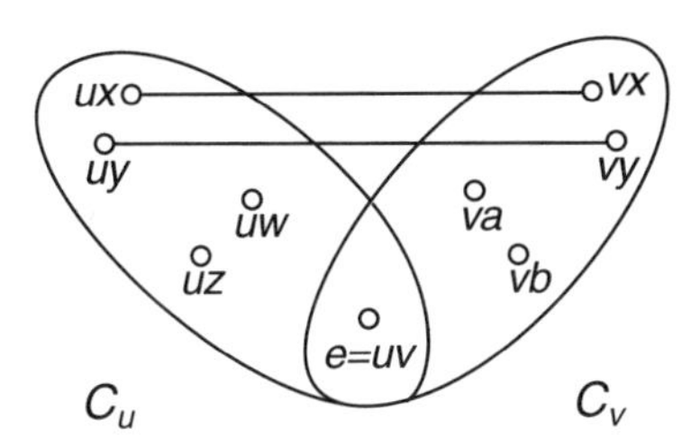

Fig. 14.1. The Local Structure

Note that for each vertex v of G the edges incident to v induce a clique of $L(G)$ which we denote by C_v. Furthermore, for each edge $e = uv$ of G, $C_u \cap C_v = e$ and the edges of $L(G)$ between $C_u - e$ and $C_v - e$ form a matching. This turns out to be the local structure which causes the Naive Colouring Procedure to perform well on line graphs. Intuitively, the evolution of the list of acceptable colours on an edge $e = uv$ depends heavily on the evolution of the lists on edges in $C_u - e$ and $C_v - e$ but is essentially independent of the evolution of all other lists. Furthermore, the evolution of the lists in $C_u - e$ is essentially independent of the evolution of the lists in $C_v - e$, because there are so few edges between these cliques.

Our analysis focuses on the set $\mathcal{C} = \{C_v | v \in V(G)\}$. At any point during our procedure, we use $L_v \subset \cup_{e \ni v} \mathcal{L}_e$ to denote a set of colours which do not yet appear on any edges in C_v. For any edge $e = uv$ the set of colours which are acceptable for e is $L_v \cap L_u$. For each $c \in L_v$, we let $T_{v,c}$ be the set of uncoloured edges of C_v for which c is still acceptable, i.e. the set of uncoloured edges $e = uv$ such that $c \in L_u$.

Since each pair of cliques C_u, C_v shares at most one vertex and has a relatively small number of edges between them, each list L_v will evolve nearly independently of the other lists. A similar property will hold for the sets $T_{v,c}$. These properties will be crucial to our analysis.

14.1.3 Rates of Change

As in Chap. 12, we will keep track of how these parameters change as we perform several iterations of our Naive Colouring Procedure. The second major difference here concerns the rates at which they change:

Difference 2: *The parameters decrease at approximately the same rate.* In Chap. 12, the sets corresponding to $T_{v,c}$ were initially much larger than the lists L_v, but the sizes of these sets dropped much more quickly than the sizes of the lists, and so eventually each L_v was significantly larger than each $T_{v,c}$. In this chapter, the sets drop at approximately the same rate, and this requires a more delicate analysis.

To gain an intuition as to why this is the case, we will consider an iteration where every list has size exactly L, and every $T_{v,c}$ has size almost L. The analysis for the general situation is essentially the same.

The first step is to compute the expected number of colours which remain in each L_e and the expected number of edges which remain in each $T_{v,c}$.

The probability that an edge e is assigned a particular colour $c \in L_e$ is exactly $\frac{1}{L}$. The probability that e then retains c, i.e. that c is assigned to no other vertex in $T_{v,c}$ or $T_{u,c}$, is $(1-\frac{1}{L})^{|T_{v,c}|+|T_{u,c}|} \approx \frac{1}{e^2}$. Since no two edges in $T_{v,c}$ can both retain c, the probability that c is removed from L_v, i.e. that exactly one edge in $T_{v,c}$ retains c, is approximately $|T_{v,c}| \times \frac{1}{L} \times \frac{1}{e^2} \approx \frac{1}{e^2}$.

Because there there are so few edges between the cliques in $\mathcal{C}$, it turns out that the colours retained by edges in two different elements of $\mathcal{C}$ are largely independent. In particular, for an edge $e = uv$ and colour $c \in L_e$ we will show that the probability that c is not retained by any edge in $C_u \cup C_v$ is approximately $(1-\frac{1}{e^2})^2$, and so:

14.3 *The expected number of colours which remain in L_e is approximately $|L_e|(1-\frac{1}{e^2})^2$.*

Recall that $T_{v,c}$ is the set of uncoloured edges $e = uv$ such that $c \in L_u$. If c remains in L_v then such an edge e remains in $T_{v,c}$ iff (i) e does not retain its colour $c' \neq c$, and (ii) c is not removed from L_u. Since $c' \neq c$, these two events are essentially independent. Furthermore, as shown above, each occurs with probability roughly $1-\frac{1}{e^2}$. Thus, we will be able to show that:

14.4 *The expected number of edges which remain in $T_{v,c}$, conditional on c remaining in L_v, is approximately $|T_{v,c}|(1-\frac{1}{e^2})^2$.*

Now, (14.3) and (14.4), along with a proof that the corresponding variables are highly concentrated, imply that in our iterative procedure $|T_{v,c}|$ and $|L_e|$ decrease at about the same rate, as claimed.

Fortunately, each $T_{v,c}$ is initially slightly smaller than each L_e. Since these sizes will decrease at approximately the same rate, we will be able to ensure that at the beginning of each iteration we still have that each $T_{v,c}$ is slightly smaller than each L_e. However, we will never be able to reach a situation where every $T_{v,c}$ is *significantly* smaller than each L_e and this makes it difficult to apply Theorem 4.3 at the end of the procedure.

To overcome this problem, we will have to perform a preprocessing step.

14.1.4 The Preprocessing Step

For each edge $e = uv$, we will choose a set of colours $\text{Reserve}_e \subseteq \mathcal{L}_e$ which we will save to use on e during the final stage of our procedure. To ensure that we can use these colours during our final stage, we will have to forbid every edge in $C_u \cup C_v$ from receiving any colours from Reserve_e during the main part of the procedure. The simplest way to do this is to remove Reserve_e from L_u and L_v.

Thus, for each edge $e' \in C_v$, we will remove $\text{Reserve}_{e'}$ from L_v. If these sets $\text{Reserve}_{e'}$ are chosen independently, then this could result in a very large number of colours being removed from L_v. To avoid this problem, we wish to ensure that the sets $\text{Reserve}_{e'}$ with $e' \in C_v$ are somewhat similar. The best way to do this is as follows:

For each vertex v, we will choose a set of colours Reserve_v. For each edge $e = uv$, we set $\text{Reserve}_e = \text{Reserve}_u \cap \text{Reserve}_v$. We then delete Reserve_v from L_v, thus ensuring that no edge in C_v can receive any colour from Reserve_v during our main procedure. We save the details as to how these sets are chosen until the next section (in fact, the next page).

We will choose these sets so that we remove at most $2\sqrt{\Delta}\log^4 \Delta$ colours from any $\mathcal{L}_e$. Since the size of $\mathcal{L}_e$ is $\Delta + 4\sqrt{\Delta}\log^4 \Delta$, we can afford to do this. Furthermore, each Reserve_e will have size at least $\frac{1}{2}\log^8 \Delta$.

For each vertex v and colour c, at any point during our procedure we will denote by $R_{v,c}$ the set of uncoloured edges $e = vw$ such that $c \in \text{Reserve}_w$. Initially, we will have $|R_{v,c}| = O(\sqrt{\Delta}\log^4 \Delta)$. To see how the size of $R_{v,c}$ evolves after several iterations of our procedure, observe that since each edge in $R_{v,c}$ remains uncoloured with probability approximately $(1 - \frac{1}{e^2})$, we have the following:

14.5 *The expected number of edges which remain in $R_{v,c}$ is approximately $|R_{v,c}|(1 - \frac{1}{e^2})$.*

As we will see, this variable is highly concentrated, and so $|R_{v,c}|$ decreases by a factor of about $(1-\frac{1}{e^2})$ per iteration. Thus it decreases at the square root of the rate at which $|L_e|$ and $|T_{v,c}|$ decrease. So by the time $|T_{v,c}|$ decreases from Δ to $\log^7 \Delta$, the size of each $R_{v,c}$ will have decreased to roughly

$$\sqrt{\Delta}\log^4 \Delta \times \left(\frac{\log^7 \Delta}{\Delta}\right)^{1/2} = \log^{7.5} \Delta.$$

Thus, when we finish our iterative procedure, for each colour c, each vertex and hence each edge will be incident with $O(\log^{7.5} \Delta)$ edges f with $c \in \text{Reserve}_f$. Since $|\text{Reserve}_e| \geq \frac{1}{2}\log^8 \Delta$, Theorem 4.3 guarantees that we will be able to complete our colouring by assigning to each uncoloured edge e, a colour from Reserve_e.

In the next 3 sections, we present the areas of our proof which differ from the proof of Theorem 12.1 – the preprocessing step and final stage, the expected value computations which yield (14.3), (14.4) and (14.5), and the concentration details. After that, it is straightforward for a reader familiar with Chap. 12 to complete the proof. In order to avoid repetitiveness, we will omit most of the remaining details and leave them as an exercise. If the reader is keen enough to work them out, she will be beset by a strong feeling of *deja vu.*

14.2 Choosing $Reserve_e$

Recall that we are given a Δ-regular graph G where each edge has a list $\mathcal{L}_e$ of $\Delta + 4\sqrt{\Delta}\log^4 \Delta$ available colours. For each vertex v, we choose a set of colours Reserve_v which cannot be used on any edge incident with v until the final stage. For each edge $e = uv$, we set $L_e = \mathcal{L}_e - (\text{Reserve}_u \cup \text{Reserve}_v)$ and we set $\text{Reserve}_e = \mathcal{L}_e \cap (\text{Reserve}_u \cap \text{Reserve}_v)$. Until the final stage, each e can only receive a colour from L_e. Thus, when we reach the final stage, all colours from Reserve_e will be available to e as none of them will have been used on an incident edge.

Lemma 14.6 *Given any Δ-regular graph G, along with a list of available colours $\mathcal{L}_e$ of size $\Delta + 4\sqrt{\Delta}\log^4 \Delta$ for each edge e, we can choose a set of colours* Reserve_v *for each vertex v, such that for each edge $e = uv$ and colour $c \in \text{Reserve}_v$:*

(a) $|\mathcal{L}_e \cap (\text{Reserve}_u \cup \text{Reserve}_v)| \leq 3\sqrt{\Delta}\log^4 \Delta$,
(b) $|\mathcal{L}_e \cap (\text{Reserve}_u \cap \text{Reserve}_v)| \geq \frac{1}{2}\log^8 \Delta$, *and*
(c) $|\{w \in N_v : c \in \text{Reserve}_w\}| \leq 2\sqrt{\Delta}\log^4 \Delta$.

Proof For each vertex v and colour c s.t. $c \in \mathcal{L}_f$ for some edge f incident with v, we place c into Reserve_v with probability $p = \log^4 \Delta/\sqrt{\Delta}$. For each edge e, vertex v and colour c we define A_e, B_e and $C_{v,c}$ to be the event that e violates condition (a), e violates condition (b), and v, c violate condition (c), respectively.

Note that

$$\begin{aligned}
\mathbf{E}(|\mathcal{L}(e) \cap (\text{Reserve}_u \cup \text{Reserve}_v)|) &\leq |\mathcal{L}(e)| \times 2p \approx 2\sqrt{\Delta}\log^4 \Delta \\
\mathbf{E}(|\mathcal{L}(e) \cap (\text{Reserve}_u \cap \text{Reserve}_v)|) &= |\mathcal{L}(e)| \times p^2 \approx \log^8 \Delta \\
\mathbf{E}(|\{w \in N_v : c \in \text{Reserve}_w\}|) &\leq \quad \Delta \times p \quad \approx \sqrt{\Delta}\log^4 \Delta.
\end{aligned}$$

Thus, it is a straightforward application of the Chernoff Bound to show that the probability of any one event is (much) less than $e^{-\log^2 \Delta}$. Furthermore, a straightforward application of the Mutual Independence Principle verifies that each event E is mutually independent of all those events which do not have a subscript at distance less than three from some subscript of E. Using the fact that for any vertex v, $C_{v,c}$ is defined for fewer than $2\Delta^2$ colours c, we see that each event is mutually independent of all but fewer than Δ^5 other events. Therefore, our lemma follows from the Local Lemma. We leave the details for the reader. □

14.3 The Expected Value Details

In this section, we carry out the expected value computations, proving (14.3), (14.4) and (14.5). Recall that we are performing a single iteration of our procedure (not necessarily the first). Our iterative analysis guarantees that at the beginning of this iteration, every list L_e has size at least L, and every set $T_{v,c}$ has size at most T where T is almost but not quite as large as L. Specifically, we will show that we can guarantee

14.7 $L - \sqrt{L}\log^2 \Delta \geq T \geq L - 4\sqrt{L}\log^4 \Delta$.

Our procedure terminates if L becomes too small, so we also have:

14.8 $L \geq \log^7 \Delta$.

We must be specific about how each iteration proceeds. As in Chap. 12, we remove colours from some of the lists L_e so that they all have size exactly L. Also, we will make use of equalizing coin flips which equalize (i) the probability of an edge retaining a colour and (ii) the probability of a colour c remaining in a list L_v.

Specifically, for each edge $e = (u, v)$ and colour $c \in L_e$, the probability that no edge incident to e is assigned c is

$$\begin{aligned} P(e,c) &= \left(1 - \frac{1}{L}\right)^{|T_{v,c}|+|T_{u,c}|-2} \\ &> e^{-(|T_{v,c}|+|T_{u,c}|-2)/L} - O\left(\frac{1}{L}\right) \\ &> e^{-2} \times \left(e^{\log^2 \sqrt{\Delta}/L} - O\left(\frac{1}{L}\right)\right) \\ &> e^{-2} \times \left(1 + \frac{\log^2 \sqrt{\Delta}}{L} - O\left(\frac{1}{L}\right)\right) \\ &> e^{-2}. \end{aligned}$$

In our equalizing flip, if e is assigned c and if no incident edge is assigned c then we remove c from e with probability $\mathrm{Eq}(e,c) = 1 - \frac{1}{e^2 P(e,c)} > 0$. This ensures that the probability that e retains c, conditional on e receiving c is precisely $P(e,c) \times (1 - \mathrm{Eq}(e,c)) = e^{-2}$.

Similarly, for every colour $c \in L_v$, the probability that c is not retained by any edge of $T_{v,c}$ is $Q(v,c) = 1 - \frac{1}{Le^2}|T_{v,c}| > 1 - e^{-2}$, since each edge of $T_{v,c}$ retains c with probability $\frac{1}{Le^2}$ and it is impossible for two of these edges to both retain c. If none of these edges retain c then we remove c from L_v with probability $\mathrm{Vq}(v,c) = 1 - \frac{1-e^{-2}}{Q(v,c)}$. This ensures that the probability that c remains in L_v is precisely $Q(v,c) \times (1 - \mathrm{Vq}(v,c)) = 1 - e^{-2}$.

In summary, an iteration proceeds as follows:

1. For each uncoloured edge $e = uv$, we remove $|L_e| - L$ colours from L_e (where these colours are chosen arbitrarily). Of course when we remove colour c from L_e, we also remove e from $T_{u,c}$ and $T_{v,c}$.
2. For each uncoloured edge e, assign to e a colour chosen uniformly at random from L_e.
3. Uncolour every edge e which is assigned the same colour as an incident edge.
4. If e is assigned a colour c and does not lose that colour in step 3 then uncolour e with probability $\text{Eq}(e, c)$.
5. For each vertex v and colour $c \in L_v$, if c is retained by some edge in $T_{v,c}$ then we remove c from L_v. If c is not retained by an edge in $T_{v,c}$ then we remove c from L_v with probability $\text{Vq}(v, c)$.

For convenience, we will use L_v to denote the contents of the set L_v after step 1, and we will use L_v' to denote the contents of L_v at the end of the iteration. We define $L_e, L_e', R_{v,c}, R_{v,c}'$ in the same manner.

For each vertex v and colour $c \in L_v$, we define $T_{v,c}$ to be the set of edges in $T_{v,c}$ after step 1. We define $T_{v,c}'$ to be the set of edges $uv \in T_{v,c}$ such that (i) uv is still uncoloured at the end of the iteration and (ii) $c \in L_u'$. (This is similar to our definition of $t'(v, c)$ in Chap. 12.) Note that if $c \in L_v'$ then $T_{u,v}'$ contains all uncoloured edges $e \in C_v$ such that $c \in L_e'$.

We now restate (14.3), (14.4) and (14.5) more precisely.

14.9 *For each vertex v and colour c,* $\mathbf{E}(|R_{v,c}'|) = (1 - \frac{1}{e^2})|R_{v,c}|$.

14.10 *For each edge e,* $\mathbf{E}(|L_e'|) \geq (1 - \frac{1}{e^2})^2 |L_e|$.

14.11 *For each vertex v and colour $c \in L_v$* $\mathbf{E}(|T_{v,c}'|) \leq (1 - \frac{1}{e^2})^2 |T_{v,c}| + 1$.

Remark We could remove the "+1" in this last equation by complicating the proof somewhat. Since this term is dwarfed by the $o(L)$ terms arising elsewhere in our proof, we do not bother.

Proof of (14.9). This is an immediate consequence of the fact that each edge retains the colour it is assigned with probability $\frac{1}{e^2}$. □

Proof of (14.10). Consider an edge $e = uv$. For each colour $c \in L_e$, we wish to show that the probability that c is in L_e', i.e. that $c \in L_u' \cap L_v'$, is at least $(1 - \frac{1}{e^2})^2$. Now, the probability that c is in L_u' is $(1 - \frac{1}{e^2})$, as is the probability that c is in L_v'. Thus, our desired bound would follow if these two events were independent. Unfortunately, the events are not independent because dependency is created by pairs of incident edges, one in C_u and one in C_v, and by edges which are incident to edges in both C_u and C_v.

However, we will see that such dependency is very minor and does not affect the probability that c is in L_e by a significant amount.

Let K_u, K_v be the events that no edge in $C_u - e$, respectively $C_v - e$ retains c. It turns out that the simplest way to compute $\mathbf{Pr}(K_u \cap K_v)$ is through the indirect route of computing $\mathbf{Pr}(\overline{K_u} \cup \overline{K_v})$ which is equal to $1 - \mathbf{Pr}(K_u \cap K_v)$. Now, by the most basic case of the Inclusion-Exclusion Principle, $\mathbf{Pr}(\overline{K_u} \cup \overline{K_v})$ is equal to $\mathbf{Pr}(\overline{K_u}) + \mathbf{Pr}(\overline{K_v}) - \mathbf{Pr}(\overline{K_u} \cap \overline{K_v})$.

We know that $\mathbf{Pr}(\overline{K_u}) + \mathbf{Pr}(\overline{K_v}) = \frac{|T_{u,c}|+|T_{v,c}|-2}{Le^2}$, so we just need to bound $\mathbf{Pr}(\overline{K_u} \cap \overline{K_v})$.

$\mathbf{Pr}(\overline{K_u} \cap \overline{K_v})$ is the probability that there is some pair of non-incident edges $e_1 = uw \in T_{u,c} - e$, and $e_2 = vx \in T_{v,c} - e$ such that e_1 and e_2 both receive and retain c. Note that, as e_1, e_2 are non-incident, $w \neq x$. Now, for each such pair we let R_{e_1,e_2} be the event that e_1 and e_2 both retain c.

$$\begin{aligned}\mathbf{Pr}(R_{e_1,e_2}) = \; & \frac{1}{L^2}\left(1 - \frac{1}{L}\right)^{|T_{u,c} \cup T_{v,c} \cup T_{w,c} \cup T_{x,c}|-2} \\ & \times (1 - \mathrm{Eq}(e_1,c))(1 - \mathrm{Eq}(e_2,c))\end{aligned}$$

Since

$$|T_{u,c} \cup T_{v,c} \cup T_{w,c} \cup T_{x,c}| < |T_{u,c}| + |T_{w,c}| - 1 + |T_{v,c}| + |T_{x,c}| - 1,$$

we obtain:

$$\begin{aligned}\mathbf{Pr}(R_{e_1,e_2}) &> \frac{1}{L^2} P(e_1,c)P(e_2,c)(1 - \mathrm{Eq}(e_1,c))(1 - \mathrm{Eq}(e_2,c)) \\ &= \frac{1}{L^2 \mathrm{e}^4}\end{aligned}$$

It is easy to see that there are at most T incident pairs $e_1 \in T_{u,c} - e, e_2 \in T_{v,c} - e$, as each edge in $T_{u,c} - e$ is incident with at most one edge in $T_{v,c} - e$. Therefore the number of nonincident pairs $e_1 \in T_{u,c} - e, e_2 \in T_{v,c} - e$, is at least $(|T_{v,c}| - 1)(|T_{u,c}| - 1) - T$.

For any two distinct pairs (e_1, e_2) and (e_1', e_2'), it is impossible for R_{e_1,e_2} and $R_{e_1',e_2'}$ to both hold. Therefore the probability that R_{e_1,e_2} holds for at least one non-incident pair, is equal to the sum over all non-incident pairs e_1, e_2 of $\mathbf{Pr}(R_{e_1,e_2})$ which by the above remarks yields:

$$\mathbf{Pr}(\overline{K_u} \cap \overline{K_v}) \geq \frac{(|T_{v,c}| - 1)(|T_{u,c}| - 1) - T}{L^2 \mathrm{e}^4}$$

Combining this with our bound on $\mathbf{Pr}(\overline{K_u}) + \mathbf{Pr}(\overline{K_v})$, we see that:

$$\begin{aligned}\mathbf{Pr}(K_u \cup K_v) &\geq 1 - \left(\frac{|T_{u,c}| + |T_{v,c}| - 2}{L\mathrm{e}^2}\right) + \left(\frac{(|T_{v,c}| - 1)(|T_{u,c}| - 1) - T}{L^2 \mathrm{e}^4}\right) \\ &\geq \left(1 - \frac{|T_{u,c}| - 1}{L\mathrm{e}^2}\right)\left(1 - \frac{|T_{v,c}| - 1}{L\mathrm{e}^2}\right) - \frac{1}{L\mathrm{e}^4} \\ &\geq \left(1 - \frac{|T_{u,c}|}{L\mathrm{e}^2}\right)\left(1 - \frac{|T_{v,c}|}{L\mathrm{e}^2}\right).\end{aligned}$$

Finally,

$$\begin{aligned}\mathbf{Pr}(c \in L_e^{'}) &= \mathbf{Pr}(K_u \cup K_v)(1-\mathrm{Vq}(u,c))(1-\mathrm{Vq}(v,c))\\ &\geq \left(1-\frac{|T_{u,c}|}{L\mathrm{e}^2}\right)(1-\mathrm{Vq}(u,c)) \times \left(1-\frac{|T_{v,c}|}{L\mathrm{e}^2}\right)(1-\mathrm{Vq}(v,c))\\ &= Q(u,c)(1-\mathrm{Vq}(u,c)) \times Q(v,c)(1-\mathrm{Vq}(v,c))\\ &= \left(1-\frac{1}{\mathrm{e}^2}\right)^2.\end{aligned}$$

(14.10) now follows from Linearity of Expectation. □

Proof of (14.11). We will show that for each $e = uv$ in $T_{v,c}$, we have: $\mathbf{Pr}(e \in T_{v,c}^{'}) \leq (1-\frac{1}{\mathrm{e}^2})^2 + \frac{1}{L}$. The result then follows from Linearity of Expectation, since $|T_{v,c}| \leq L$.

We define A to be the event that e does not retain its colour and B to be the event that no edge in $T_{u,c}$ retains c. We wish to bound $\mathbf{Pr}(A \cap B)$. Once again, we proceed in an indirect manner, and focus instead on $\overline{A} \cap \overline{B}$, showing that $\mathbf{Pr}(\overline{A} \cap \overline{B}) \leq \frac{|T_{u,c}|}{L\mathrm{e}^4} + \frac{1}{L}$, thus implying

$$\begin{aligned}\mathbf{Pr}(A \cap B) &= \mathbf{Pr}(A) - \mathbf{Pr}(\overline{B}) + \mathbf{Pr}(\overline{A} \cap \overline{B})\\ &\leq \left(1-\frac{1}{\mathrm{e}^2}\right) - \frac{|T_{u,c}|}{L\mathrm{e}^2} + \frac{|T_{u,c}|}{L\mathrm{e}^4} + \frac{1}{L}\\ &= \left(1-\frac{1}{\mathrm{e}^2}\right)\left(1-\frac{|T_{u,c}|}{L\mathrm{e}^2}\right) + \frac{1}{L}\end{aligned}$$

Therefore

$$\mathbf{Pr}(e \in T_{v,c}^{'}) = \mathbf{Pr}(A \cap B)(1-\mathrm{Vq}(u,c)) \leq \left(1-\frac{1}{\mathrm{e}^2}\right)^2 + \frac{1}{L},$$

as required.

For each colour $d \in L_e$ and edge $f = uw$ in $T_{u,c} - e$ we define $Z(d,f)$ to be the event that e retains d and f retains c. For each $d \neq c$, we have

$$\begin{aligned}&\mathbf{Pr}\ (Z(d,f)) = \frac{1}{L^2}\left(1-\frac{2}{L}\right)^{|(T_{v,d}\cap T_{w,c})+(T_{u,d}\cap T_{u,c})-e-f|}\\ &\times \left(1-\frac{1}{L}\right)^{|(T_{v,d}\cup T_{w,c})+(T_{u,d}\cup T_{u,c})-(T_{v,d}\cap T_{w,c})-(T_{u,d}\cap T_{u,c})-e-f|}\\ &\times (1-\mathrm{Eq}(e,d))(1-\mathrm{Eq}(f,c))\\ &\leq \frac{1}{L^2}\left(1-\frac{1}{L}\right)^{|(T_{v,d}\cup T_{w,c})+(T_{u,d}\cup T_{u,c})-e-f|+|(T_{v,d}\cap T_{w,c})+(T_{u,d}\cap T_{u,c})-e-f|}\\ &\times (1-\mathrm{Eq}(e,d))(1-\mathrm{Eq}(f,c)).\end{aligned}$$

Since

$$\begin{aligned}&|(T_{v,d}\cup T_{w,c})+(T_{u,d}\cup T_{u,c})\\&-e-f|+|(T_{v,d}\cap T_{w,c})+(T_{u,d}\cap T_{u,c})-e-f|\\&=|T_{v,d}-e|+|T_{u,d}-e-f|+|T_{u,c}-e-f|+|T_{w,c}-f|\\&\geq|T_{v,d}|+|T_{u,d}|+|T_{u,c}|+|T_{w,c}|-6,\end{aligned}$$

We obtain:

$$\begin{aligned}\mathbf{Pr}(Z(d,f))&\leq\frac{1}{L^2}\left(1-\frac{1}{L}\right)^{-2}P(e,d)P(f,c)(1-\mathrm{Eq}(e,d))(1-\mathrm{Eq}(f,c))\\&=\frac{1}{(L-1)^2\mathrm{e}^4},\end{aligned}$$

Since the events $Z(d,f)$ are disjoint, we have:

$$\begin{aligned}\mathbf{Pr}(\overline{A}\cap\overline{B})&=\mathbf{Pr}(e\text{ retains }c)+\sum_{d\in L_e-c,f\in T_{u,c}}\mathbf{Pr}(Z(d,f))\\&\leq\frac{1}{L\mathrm{e}^2}+\frac{L|T_{u,c}|}{(L-1)^2\mathrm{e}^4}\\&<\frac{|T_{u,c}|}{L\mathrm{e}^4}+\frac{1}{L}.\end{aligned}$$

□

14.4 The Concentration Details

In this section, we will show that the random variables from (14.9), (14.10), (14.11) are highly concentrated. In particular, we prove that there exists a constant $\beta>0$ such that for each a satisfying $\log\Delta<<a<\log^2\Delta$:

14.12 *For each* $e\in E(G)$, $\mathbf{Pr}\left(||L'_e|-\mathbf{E}(|L'_e|)|>a\sqrt{L}\right)\leq e^{-\beta a^2}$.

14.13 *For each vertex* $v\in V(G)$ *and* $c\in L_v$, $\mathbf{Pr}\left(||T'_{v,c}|-\mathbf{E}(|T'_{v,c}|)|>a\sqrt{L}\right)$ $\leq 6e^{-\beta a^2}$.

14.14 *For each* $v\in V(G)$ *and colour* c, *if* $R_{v,c}>\log^4\Delta$ *then* $\mathbf{Pr}\left(||R'_{v,c}|\right.$ $\left.-\mathbf{E}(|R'_{v,c}|)|>a\sqrt{|R_{v,c}|}\right)\leq e^{-\beta a^2}$.

Again, we will use Talagrand's Inequality, and so to prove that a variable X is concentrated, it will suffice to verify that for 2 particular constants k_1,k_2:

(A) Changing the outcome of a single random choice can affect X by at most k_1.
(B) For any s, if $X \geq s$ then there is a set of at most $k_2 s$ random choices whose outcomes certify that $X \geq s$.

By Talagrand's Inequality, if these conditions hold then $\mathbf{Pr}(|X - \mathbf{E}(X)| > a\mathbf{E}(X)^{1/2} \leq e^{-\beta a^2}$ for $\beta < \frac{1}{10k_1^2 k_2}$ and $\log \Delta << a \leq \sqrt{\mathbf{E}(X)}$. In all three proofs, the set of random choices made are the colour assignments and the equalizing coin flips, just as in Chap. 12. Of course, we can take β to be the minimum of the values for β yielded by the three proofs.

Proof of (14.14). This proof is very similar to the proofs of Lemmas 10.4 and 12.3, upon observing that changing the colour of an edge $e \in C_v$ from c_1 to c_2 can only cause at most one edge of $R_{v,c}$ to be uncoloured, since if more than one edge of $R_{v,c} - e$ were assigned c_2 then they would all be uncoloured regardless of e's change of colour. The reader who has progressed thus far should have no trouble completing the details. □

Proof of (14.12). Proving concentration here is more delicate than the corresponding result in the previous chapter because we are not using the Wasteful Colouring Procedure and so we must analyze the number of colours *retained* on neighbours of e rather than simply the number of colours assigned to these vertices. As with our expected value computation, an indirect approach is appropriate.

So for any $e = uv$, we let X be the number of colours from L_e which are assigned to and are retained by at least one edge in $C_u \cup C_v$. For $0 \leq k \leq j \leq 2$, we define $Y_{j,k}$ to be the number of colours which are assigned to an edge in *exactly* j of C_u, C_v and which are removed from an edge in at least k of C_u, C_v during step 3 or 4 of the main procedure. Similarly, we define $X_{j,k}$ to be the number of colours which are assigned to an edge in *at least* j of C_u, C_v and which are removed from an edge in at least k of C_u, C_v during step 3 or 4 of the main procedure. Note that $Y_{2,k} = X_{2,k}$ and for $j < 2$, $Y_{j,k} = X_{j,k} - X_{j+1,k}$. Making use of the very useful fact that for any w, if a colour is removed from at least one edge in C_w then it is removed from every edge in C_w to which it is assigned, we obtain that:

$$X = (Y_{2,0} - Y_{2,2}) + (Y_{1,0} - Y_{1,1}) = (X_{2,0} - X_{2,2}) + ((X_{1,0} - X_{2,0}) - (X_{1,1} - X_{2,1})).$$

It is straightforward to show that each $X_{j,k}$ is highly concentrated. First of all, changing the colour assigned to any one edge from c_1 to c_2 can only affect whether c_1 and/or c_2 are counted by $X_{j,k}$, and changing the decision to uncolour an edge in step 4 can only affect whether the colour of that edge is counted by $X_{j,k}$. Therefore, condition (A) from above is satisfied with $k_1 = 2$.

Secondly, if $X_{j,k} \geq s$, then there is a set of at most $s(j+k)$ outcomes which certify this fact, namely for each of the s colours, j edges on which that colour appears, along with k (or fewer) outcomes which cause k of those edges to be uncoloured. Therefore, Talagrand's Inequality implies that for each $X_{j,k}$ we have $\mathbf{Pr}\left(|X_{j,k} - \mathbf{E}(X_{j,k})| > t\sqrt{\mathbf{E}(X_{j,k})}\right) < e^{-\gamma t^2}$ for some constant $\gamma > 0$. By our expression for X above, if X differs from $\mathbf{E}(X)$ by at least $6t\sqrt{k}$ then some $X_{j,k}$ must differ from $\mathbf{E}(X_{j,k})$ by at least $t\sqrt{k}$. Since there are six variables $X_{j,k}$ the probability that this occurs is at most $6e^{-\gamma t^2}$. I.e. we have:

$$\mathbf{Pr}\left(|X - \mathbf{E}(X)| > a\sqrt{\mathbf{E}(X)}\right) < 6e^{-\gamma a^2}.$$

Let X' be the number of colours removed from L_u or L_v in step 5. That X' is highly concentrated follows easily from the Simple Concentration Bound. We leave it to the reader to fill in the details and, using the fact that $|L'_e| = |L_e| - (X + X')$, complete the proof of (14.12). □

Proof of (14.13). We again proceed indirectly. We let $A_{v,c}$ be the set of edges $e \in T_{v,c}$ which do not retain their colours. We let $B_{v,c}$ be the set of edges $e = uv$ in $A_{v,c}$ such that c is retained on some vertex of $T_{u,c}$. We let $C_{v,c}$ be the set of edges $e = uv$ in $A_{v,c} - B_{v,c}$ such that c is removed from L'_u because of an equalizing coin flip.

The proof that $|A_{v,c}|$ is highly concentrated is virtually identical to the proof of (14.14). The proof that the $|B_{v,c}|$ and $|C_{v,c}|$ are highly concentrated follows along the lines of the proof of (14.12). We leave the details as an exercise. Since, $T'_{u,v} = A_{v,c} - (B_{v,c} + C_{v,c})$, the desired result follows. □

14.5 The Wrapup

The remainder of the proof is along the same lines as the proof of Theorem 12.1. We apply the Local Lemma for each iteration of our procedure to show that with positive probability, the size of every set is within a small error of the expected size. An argument very similar to that in Lemma 12.5 shows that the accumulation of these small errors is insignificant and an extension of this argument establishes (14.7) inductively. By continuing our procedure until each L_e has size approximately $\ln^7 \Delta$, we have:

Lemma 14.15 *With positive probability, after several iterations of our procedure we have $|R_{v,c}| \leq 5\ln^{7.5} \Delta$ for every vertex v, edge e and colour c.*

Proof After i iterations, the size of each L_e is approximately $\Delta(1 - \frac{1}{e^2})^{2i}$ and the size of each $R_{v,c}$ is approximately $2\sqrt{\Delta}\log^4 \Delta(1-\frac{1}{e^2})^i$. If i is such that the first term is approximately $\ln^7 \Delta$, then the second term is approximately $2\ln^{7.5} \Delta$. We leave the remaining details as an exercise. □

By using the Reserve colours to complete our colouring, we can complete the proof of our main theorem.

Proof of Theorem 14.1. We will apply Theorem 4.3 to show that we can complete our colouring by assigning to each uncoloured edge e, a colour from the list Reserve_e. Recall that each of these lists has size at least $\ln^8 \Delta$. For any colour c, each uncoloured edge $e = uv$ has at most $|R_{u,c} \cup R_{v,c}| \leq 10 \ln^{7.5} \Delta$ incident uncoloured edges f with $c \subset \text{Reserve}_f$. Therefore Theorem 4.3 applies. □

14.6 Linear Hypergraphs

As mentioned earlier, these arguments extend to yield a bound on the list chromatic index of k-uniform linear hypergraphs, for any constant $k \geq 2$, i.e. hypergraphs in which every hyperedge has k vertices, and no two vertices lie in more than one hyperedge. We will highlight the 3 main parts of the proof of Theorem 14.1 which must be modified to yield Theorem 14.2. The rest of the extension should be straightforward to the reader who fully understands the material in this chapter.

First note that any such hypergraph H has the same local structure that was so helpful in our proof of Theorem 14.1, except of course that each hyperedge $e = v_1 v_2 \dots v_k$ now lies in k different Δ-cliques of $L(H)$: $C_{v_1}, \dots, C_{v_k}$.

Modification 1: The expected value computations. The probability that edge e retains colour $c \in L_e$ is now approximately $\frac{1}{L}\left(1 - \frac{1}{L}\right)^{kT} \approx \frac{1}{L}\mathrm{e}^{-k}$. Thus, the probability that exactly one edge in $T_{v,c}$ retains c is roughly e^{-k}, and so the probability that L_e retains c is approximately $(1-\mathrm{e}^{-k})^k$. Similarly, the probability that an edge e remains in $T_{v,c}$, given that c remains in L_v, is approximately $\left(1 - \mathrm{e}^{-k}\right)^k$. Thus our analogues of (14.3) and (14.4) are:

14.16 *The expected number of colours which remain in L_e is approximately $|L_e|\left(1 - \frac{1}{e^k}\right)^k$.*

14.17 *The expected number of edges which remain in $T_{v,c}$, conditional on c remaining in L_v, is approximately $|T_{v,c}|\left(1 - \frac{1}{\mathrm{e}^k}\right)^k$.*

To prove the precise versions of these statements, we again proceed indirectly, this time using a more general form of the Inclusion-Exclusion Principle. For example, in the proof of (14.16) for $e = v_1 v_2 \dots v_k$, we let K_i be the event that no edge of $C_{v_i} - e$ retains c, and we use the identity

$$\mathbf{Pr}(K_1 \cap \dots \cap K_k) = \sum_{j=1}^{k} \sum_{Z \subseteq \{1,\dots,k\}; |Z|=j} (-1)^{j+1} \mathbf{Pr}\left(\cap_{i \in Z} \overline{K_i}\right).$$

Modification 2: The concentration bounds. The arguments proving that these variables are concentrated become slightly more complex. In particular, for $1 \leq i \leq j \leq k$ we will be interested in $X_{j,i}$, the number of colours which were assigned to edges in at least j of the cliques containing e and removed from edges in at least i of these cliques. However, nothing really new is needed in this part of the argument.

Modification 3: The preprocessing step. The preprocessing step is also slightly different. Again we choose a set Reserve_v of colours for each vertex v in H, but now Reserve_e is the intersection of k of these sets. It is for this reason that we need to insist that each Reserve_v has $\Delta^{1-\frac{1}{k}}\text{poly}(\log \Delta)$ colours and hence insist that each edge initially has a set of $\Delta + \Delta^{1-\frac{1}{k}}\text{poly}(\log \Delta)$ colours.

Apart from these adjustments, the proof of Theorem 14.2 follows the lines of the proof of Theorem 14.1, and the reader is invited to work the details out himself (see Exercise 14.3) or consult [123] where they can be found.

Exercises

Exercise 14.1 Prove that for D arbitrarily large, there exists a $2D$-regular graph G such that every neighbourhood in G contains at most D^2 edges, and $\chi(G) \approx \sqrt{2}D$.

Exercise 14.2 Show how to prove Theorem 14.1 using the Wasteful Colouring Procedure, at each step activating an edge to receive a colour with probability $\frac{1}{\log \Delta}$.

Exercise 14.3 Prove Theorem 14.2, using the hints provided in Sect. 14.6.

Part VI

A Structural Decomposition

The pseudo-random colouring procedure we have been analyzing works best if there are many non-edges in each neighbourhood. In this part of the book, we present a technique for dealing with dense vertices, i.e. vertices which have very few non-edges in their neighbourhood. Our approach was foreshadowed in the proof of the Beutelspacher-Hering Conjecture. In that proof, we showed that the dense vertices in any minimal counterexample to the conjecture could be partitioned into disjoint sets each of which was either a clique or a clique and a vertex. This aided us in our analysis of the pseudo-random colouring procedure. In particular, it enabled us to show that we could construct a partial colouring which could be extended to a $\Delta - 1$ colouring provided we chose the order in which we coloured the uncoloured vertices carefully.

In the next chapter, we present a structural decomposition theorem which shows how to find, in a graph G, a family of disjoint dense sets which together contain all the dense vertices of G, and behave somewhat like Δ-cliques. We then show how this theorem can be used to prove a number of important results.

The approach taken here differs from that taken in the proof of the Beutelspacher-Hering Conjecture in the following two important respects.

(A) It is much more general. It applies to all graphs, not just minimal counterexamples to the Beutelspacher-Hering Conjecture. Furthermore, it applies to a range of definitions of dense; i.e. we can choose how many edges we will permit in the neighbourhood of a vertex which is not dense.
(B) We use the decomposition theorem in creating the partial colouring, not just in the analysis.

As an illustration of (B), consider the simple case in which we have only one dense set D which is a clique. Then rather than assign a colour to each vertex of D independently, we might well choose a random set of $|D|$ different colours and assign them to the vertices of D, using a uniformly random bijection. This procedure prohibits any conflicts within D. However, it also assigns each vertex of D a uniformily random colour. This second fact aids us in our analysis of the vertices not in D. More generally, we can assign colours on a dense set D by finding an appropriate optimal or near-optimal colouring of D and then permuting the colour class names.

Chaps. 17 and 18 contain the most important result obtained using this tool, a proof that there is a constant c such that every graph has a $\Delta + c$ total colouring. Chap. 16 discusses the first result obtained using this tool, a proof that the chromatic number is bounded by a convex combination of $\Delta + 1$ and ω. The next chapter contains a proof of the decomposition theorem and describes two methods for choosing appropriate optimal or near optimal colourings within a dense set.

15. The Structural Decomposition

15.1 Preliminary Remarks

In this chapter we will obtain a decomposition theorem for general graphs similar to that obtained in Chap. 11 for minimal counterexamples to the Beutelspacher–Hering Conjecture. That is, we will partition $V(G)$ into sets $D_1, \ldots, D_l$ and S where the D_i are "dense clique-like" sets and all the vertices in S are sparse.

A dense set D is like a Δ-clique in that:

(i) it has about Δ vertices,
(ii) most of the pairs of vertices within it are adjacent,
(iii) there are only a few edges between D and $G - D$,
(iv) a vertex is in D if and only if most of its neighbours are.

As in Chap. 11, in defining whether a vertex is dense or sparse we will consider $|E(N(v))|$. We will fix some threshold t and say that a vertex v of G is sparse if the subgraph induced by its neighbourhood contains fewer than t edges, and dense otherwise. The closer the threshold is to $\binom{\Delta}{2}$, the more clique-like the dense sets will be. Thus, we introduce a parameter into the definition which measures how dense these dense sets are.

The formal definition is given in the next section. We then present various techniques for partitioning these dense sets into stable sets which we will need when modifying the Naive Colouring Procedure as discussed in the introduction to this part of the book. Finally, we discuss briefly how this decomposition sheds light on the computational complexity of graph colouring.

15.2 The Decomposition

We say that a vertex v of G is *d-sparse* if the subgraph induced by its neighbourhood contains fewer than $\binom{\Delta}{2} - d\Delta$ edges. Otherwise, v is *d-dense* (we often simply use dense or sparse if d is clear from the context).

Remark If G is Δ-regular, then v is d-sparse precisely if there are at least $d\Delta$ edges in the subgraph of $\overline{G}$ induced by $N(v)$. For arbitrary graphs, v can have a clique as a neighbourhood and still be d-sparse, provided this clique has fewer than $\Delta - d$ vertices.

Note that if v is d-dense then there are at most $2d\Delta$ edges between $N(v) \cup v$ and $V - N(v) - v$.

Definition 15.1 *Let d be a positive integer and let G be a graph of maximum degree Δ. We say $D_1, \ldots, D_l, S$ form a* d-dense decomposition *of G if:*

(a) $D_1, \ldots, D_l, S$ are disjoint and partition V;
(b) every D_i has between $\Delta + 1 - 8d$ and $\Delta + 4d$ vertices;
(c) there are at most $8d\Delta$ edges between D_i and $V - D_i$;
(d) a vertex is adjacent to at least $\frac{3\Delta}{4}$ vertices of D_i if and only if it is in D_i;
(e) every vertex in S is d-sparse.

We shall call $D_1, \ldots, D_l$, *d-dense sets*, or simply dense sets. Note that although every vertex in S is sparse, not every sparse vertex need be in S.

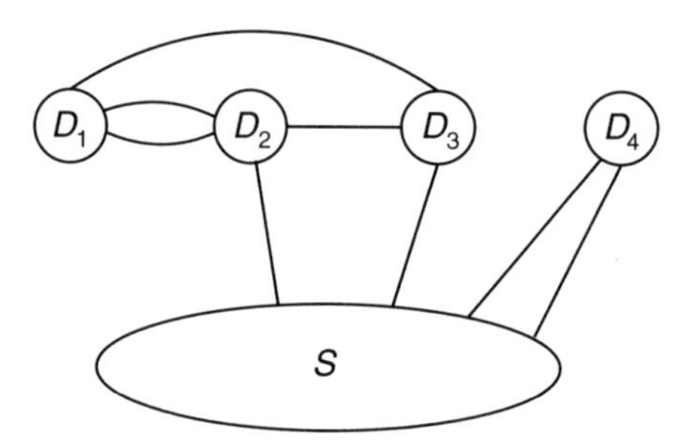

Fig. 15.1. A Dense Decomposition

A neighbour u of a vertex v in D_i is *internal* if u is also in D_i and *external* otherwise. We define the *external* and *internal neighbourhoods* of v similarly. We use Out_v to denote the external neighbourhood of v. The external neighbourhood of a subset X of D_i is the union of the external neighbourhoods of its elements.

The key result in this chapter is:

Lemma 15.2 *For every Δ and $d \leq \frac{\Delta}{100}$, every graph G of maximum degree Δ has a d-dense decomposition.*

Proof
For each d-dense vertex v of G, we define a set D_v as follows.

Step 0. Set $D_v = N(v) + v$.
Step 1. While there is at least one vertex y in D_v with $|N(y) \cap D_v| < \frac{3\Delta}{4}$ delete some such vertex from D_v.
Step 2. While there is at least one vertex y outside of D_v with $|N(y) \cap D_v| \geq \frac{3\Delta}{4}$ add some such vertex to D_v.

Note that this procedure does not alternate between additions and deletions. This ensures that the set D_v is uniquely defined, i.e. the order in which we delete vertices in Step 1 and add vertices in Step 2 is irrelevant. It is also immediate that:

15.3 *Every vertex in D_v has at least $\frac{3\Delta}{4}$ neighbours in D_v.*

15.4 *Every vertex outside of D_v has fewer than $\frac{3\Delta}{4}$ neighbours in D_v.*

Since $\binom{\Delta-2d}{2} < \binom{\Delta}{2} - d\Delta$, we also have:

15.5 *Every d-dense vertex has at least $\Delta - 2d$ neighbours.*

We also obtain:

15.6 $|N(v) - D_v| \leq 6d$

Proof Suppose that more than $6d$ vertices are deleted in Step 1, and let X be the set of the first $6d$ vertices deleted. Each vertex in X has at most $\frac{3\Delta}{4}$ neighbours in $(v \cup N(v)) - X$. So, since $|N(v)| \geq \Delta - 2d$, each vertex in X has at least $\frac{\Delta}{4} - 8d > \frac{\Delta}{6}$ non-neighbours in $(v \cup N(v)) - X$. But this contradicts the fact that there are at most $d\Delta$ pairs of non-adjacent vertices in $v \cup N(v)$, thereby proving the desired result. □

15.7 $|E(D_v, V - D_v)| \leq 8d\Delta$.

Proof Because v is d-dense, there are at most $2d\Delta$ edges out of $v \cup N(v)$. Deleting a vertex in Step 1 increases the number of edges between D_v and $G - D_v$ by at most $\frac{3\Delta}{4}$. In Step 2 the number of edges out of D_v always decreases. So, by (15.6), $|E(D_v, V - D_v)| \leq 2d\Delta + 6d(\frac{3\Delta}{4}) < 8d\Delta$. □

15.8 $|D_v - N(v)| \leq 4d$.

Proof The ith vertex of $V - N(v) - v$ added to D_v in Step 2 has at least $\frac{3\Delta}{4} - i$ neighbours in $v \cup N(v)$. So, since $|E(v \cup N(v), V - N(v) - v)| \leq 2d\Delta$ and $\sum_{i=1}^{4d} \left(\frac{3\Delta}{4} - i\right) > 2d\Delta$, fewer than $4d$ vertices of $V - N(v) - v$ are added to D_v in Step 2. □

Thus, we have:

15.9 $\Delta + 1 - 8d \leq |D_v| \leq \Delta + 4d$.

Combining (15.6), and (15.5), we see that $|N(v) \cap D_v| \geq \Delta - 8d \geq \frac{3\Delta}{4}$. So, applying (15.4), we obtain:

15.10 $v \in D_v$.

Because D_v is so close to $N(v)$, we can prove:

15.11 *If x and y are d-dense vertices and D_x intersects D_y then y is in D_x and x is in D_y.*

Proof Let x and y be two d-dense vertices such that D_x intersects D_y and let a be an element of $D_x \cap D_y$. Now, by (15.3) and (15.8), a is adjacent to at least $\frac{2\Delta}{3}$ neighbours of x. Similarly, a is adjacent to at least $\frac{2\Delta}{3}$ neighbours of y. Thus, $|N(x) \cap N(y)| \geq \frac{\Delta}{3}$.

Now, since x is a d-dense vertex, there are at most $2d\Delta \leq \Delta^2/50$ edges between $N(x) \cap N(y)$ and $N(y) - N(x)$. Since y is a d-dense vertex, there are at least $|N(x) \cap N(y)| \times |N(y) - N(x)| - \Delta^2/100$ edges between the two sets. So, $|N(x) \cap N(y)| \times |N(y) - N(x)| \leq \Delta^2/25$ and hence $|N(y) - N(x)| \leq \frac{\Delta}{8}$.

Thus, by (15.5), $|N(x) \cap N(y)| \geq \Delta - 2d - \frac{\Delta}{8} \geq \frac{5}{6}\Delta$. It follows from (15.4) and (15.6) that x is in D_y and y is in D_x.

□

By (15.11), for some integer l, we can greedily construct a sequence of d-dense vertices $x_1, \ldots, x_l$ and corresponding disjoint sets $D_1, \ldots, D_l$ with $D_i = D_{x_i}$ such that every d-dense vertex is in $\cup_{i=1}^{l} D_i$. We let S be the remaining (sparse) vertices of G. We have obtained the desired d-dense decomposition of G. □

Remark There are many variants of this technique for constructing our structural decomposition. For example, one could choose the dense set D_i to be D_v for the d-dense vertex v in $V - \cup_{j=1}^{i-1} D_j$ with the most edges in its neighbourhood. Also, for small d the constants 4 and 8 can be improved.

15.3 Partitioning the Dense Sets

Now consider a graph G of maximum degree Δ and a d-dense decomposition $D_1, \ldots, D_l, S$ of G for some $d \leq \frac{\Delta}{100}$. Since the number of edges from D_i to $G - D_i$ is small, we can almost colour the vertices of D_i independently of the rest of the graph. Thus, our first step in the modified naive colouring procedure will be to find a colouring of each D_i. Of course, we eventually need to deal with the interaction between D_i and $G - D_i$, so this colouring may need to be modified. In order to avoid confounding these original colourings with our final colouring of G, we prefer to think of the first step as partitioning each dense set into stable sets which we call *partition classes*. In this section, we present two different methods for choosing the partitions. Before we do so, a few general remarks about what kind of partitions we prefer are in order.

Our objective is to use as few colours as possible in colouring G. If we focus on just one dense set D_i in the partition and ignore the rest of the graph, then we want to minimize the number of partition classes so our best strategy would be to use an optimal colouring of D_i. However, we cannot ignore the

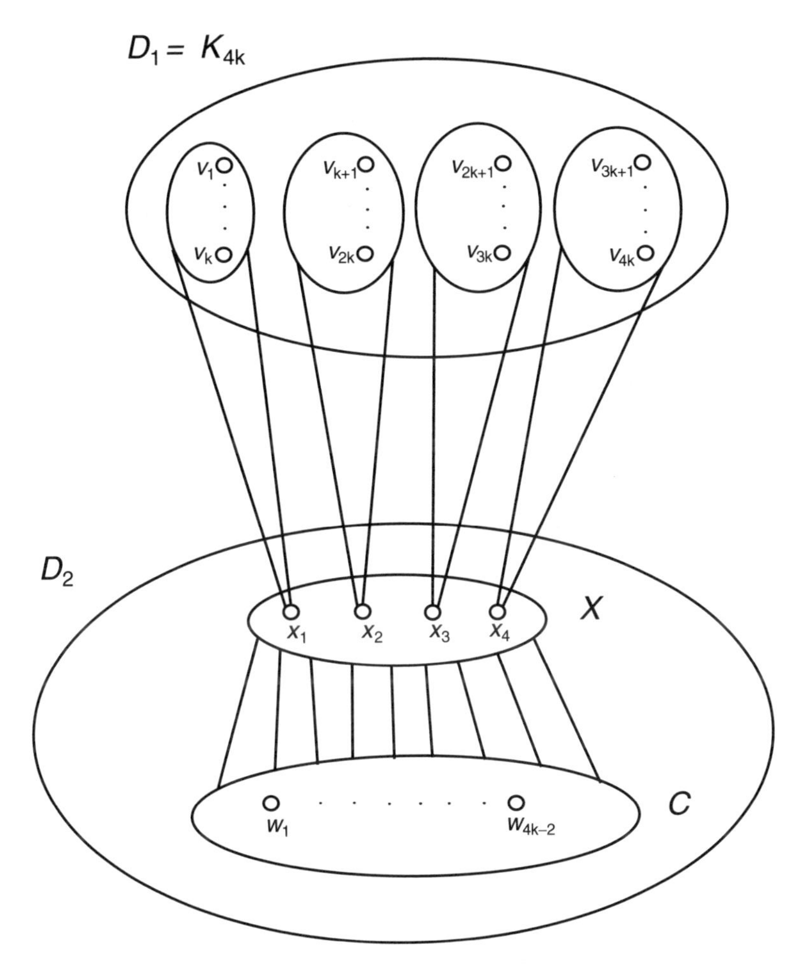

Fig. 15.2. G_c

rest of the graph. We also need to worry about the external neighbourhoods of the partition classes.

For example, suppose that for some positive integer k we are trying to $c = 4k$ colour the graph G_c of Fig. 15.2 which has a dense decomposition into two dense sets D_1, D_2 (i.e. $V(G) = V(D_1) \cup V(D_2)$) such that:

(a) D_1 is a c-clique,
(b) D_2 consists of a $(c-2)$-clique C, a stable set X with four vertices, and all possible edges between C and X,
(c) Each vertex of D_1 sees exactly one vertex of X and each vertex of X sees exactly k vertices of D_1.

We can c-colour this graph by first colouring D_1, then colouring the three vertices of X not adjacent to the vertex of D_1 of colour 1 with the colour 1, and finally colouring the remaining $c-1$ vertices of D_2 with the remaining $c-1$ colours. However, we run into a problem if we try to extend an optimal colouring of D_2 to a c-colouring of G. For in an optimal colouring of D_2, all of X receives the same colour and we will not be able to use this colour on D_1. Hence we will not be able to extend the partial colouring to a c-colouring of G.

We can avoid this type of problem by keeping the external neighbourhood of each partition class small. Clearly, if this were our only objective then it would be best to use a distinct partition class for each vertex of D_i. This of course, is the worst possible choice with respect to our goal of keeping the number of partition classes small. So, in partitioning D_i, we need to balance these two conflicting objectives.

In doing so, to bound the size of the external neighbourhoods of the partition classes, we begin by considering the size of the partition classes themselves. Note that if a partition class has at most two elements then its external neighbourhood contains at most $\frac{\Delta}{2}$ vertices. Thus, minimizing the number of partition classes of size exceeding two goes a long way towards our objective of keeping the external neighbourhoods of the partition classes small.

There are two natural ways to approach satisfying our two conflicting desires. We can either prioritize minimizing the number of partition classes of size exceeding two (by forbidding them entirely) or prioritize minimizing the total number of partition classes. The first approach, then, is to insist that there are no partition classes of size exceeding two and choose such a partition with the minimum number of partition classes. The second approach is to choose an optimal partition of the dense set into stable sets (i.e. a $\chi(D_i)$ colouring) in which we minimize the number of partition classes of size exceeding two.

We close this section by presenting two lemmas which illustrate these two techniques. We will use the first lemma in the next chapter. We discuss an application of a lemma similar to the second in the final section of this chapter.

When applying these lemmas to help obtain a c-colouring of a graph, we choose d so that $\Delta - d$ is much less than c. Thus, we can settle for partitioning D_i into $\Delta - d$ partition classes, as we thereby ensure that this initial colouring uses many fewer than c colours, and hence gives us room to manoeuvre.

In what follows, $D_1, \ldots, D_l, S$ is a d-dense decomposition of a graph G which is Δ-regular for some $\Delta \geq 100d$.

Lemma 15.12 *If $d \leq \frac{\Delta}{5000}$ then each D_i can be partitioned into classes of size one or two such that:*

(i) the number of partition classes is at most $max\{\frac{2}{3}(\Delta+1) + \frac{1}{3}\omega(D_i), \Delta - d\}$,

(ii) the external neighbourhood of each partition class contains at most $\frac{\Delta}{3}$ vertices, and

(iii) the number of non-singleton partition classes is at most 5d.

Proof We show first that there is a partition satisfying (i) and (iii). To this end, consider the size m_i of a maximum matching in $\overline{D_i}$. If $m_i > 5d$ let M be a matching of size $5d$ in $\overline{D_i}$. Otherwise, let M be a maximum matching in $\overline{D_i}$. Let C be $D_i - V(M)$. There is a partition of D_i into $|C|+|M|$ stable sets where each vertex of C forms a partition class of size 1 and the endpoints of each edge of M form a partition class of size 2. Clearly, this partition satisfies (iii). To prove that it satisfies (i), we need only show that $|C| + |M| \leq \max(\frac{2}{3}(\Delta + 1) + \frac{1}{3}\omega(D_i), \Delta - d)$.

We note that we can assume $|M| < 5d$ as otherwise

$$|C| + |M| = |D_i| - |M| \leq (\Delta + 4d) - 5d = \Delta - d$$

and we are done. So, M is a maximum matching of $\overline{D_i}$ and thus C must be a clique. Furthermore, we have: $|C| = |D_i| - 2|M| \geq (\Delta - 8d) - 10d \geq \Delta - 18d \geq 4982d$. Thus, in particular:

15.13 $|C| \geq 4982$.

Also,

15.14 *for each vertex v of M the number of neighbours of v in C is at least*

$$\frac{3\Delta}{4} - 2|M| \geq \frac{3\Delta}{4} - 10d \geq \frac{7\Delta}{10} \geq \frac{|C|}{2}.$$

Now, by the maximality of M, if xy is an edge of M then either x sees all of C, y sees all of C, or they both see all of $C - w$ for some vertex w. By (15.13), $2(|C| - 1) \geq \frac{3|C|}{2}$. So, by (15.14), we see that in either case, $|N(x) \cap C| + |N(y) \cap C| \geq \frac{3|C|}{2}$. Thus there are at least $\frac{3|M||C|}{2}$ edges between C and $V(M)$, and hence some vertex in C sees at least $\frac{3|M|}{2}$ vertices of M. So, $|C| + \frac{3|M|}{2} \leq \Delta + 1$. I.e., $\frac{2}{3}|C| + |M| \leq \frac{2}{3}(\Delta + 1)$. Also, $|C| \leq \omega(D_i)$. Combining these results yields $|C| + |M| \leq \frac{2}{3}(\Delta + 1) + \frac{1}{3}\omega(D_i)$. Thus we can indeed find a partition of D_i into classes of size at most two which satisfies (i) and (iii).

As we remarked earlier, the fact that each partition class contains at most two vertices implies that the external neighbourhood of each class in such a partition has at most $\frac{\Delta}{2}$ elements. We see now that we can actually choose our partition so that these external neighbourhoods have at most $\frac{\Delta}{3}$ elements, i.e. so that (ii) holds.

To this end, we consider a partition satisfying (i) and (iii) which has the minimum number of partition classes whose external neighbourhoods have more than $\frac{\Delta}{3}$ elements. If there is any such partition class U, then we know that it has size two, contains a vertex x which has at least $\frac{\Delta}{6}$ external neighbours, and another vertex y which has at least $\frac{\Delta}{12}$ external neighbours. We will obtain a contradiction by (a) matching x with some vertex z of D_i

which was a singleton stable set and has fewer than $\frac{\Delta}{12}$ external neighbours and (b) making y a singleton partition class. To this end, we note first that there are at least $\frac{\Delta}{6} - 8d$ non-neighbours of x in D_i, by 15.1(b). We note further that all but at most $10d$ of the vertices of D_i form singleton colour classes. Furthermore, since $d \leq \frac{\Delta}{5000}$ by assumption, it follows from 15.1(c) that there are fewer than $\frac{\Delta}{12}$ vertices in D_i with more than $\frac{\Delta}{12}$ external neighbours. Thus, we can indeed find the desired z, indeed we have more than $\frac{\Delta}{12} - 18d$ choices.

□

Our second approach works particularly well when d is small:

Lemma 15.15 *If $d \leq \frac{\sqrt{\Delta}}{50}$ then each D_i has a partition into max $(\chi(D_i), \Delta - d)$ classes such that the external neighbourhood of each partition class contains at most $\frac{\Delta}{2}$ vertices.*

Proof We say a dense set is *matchable* if there is a matching of size $6d$ in its complement. Now, for each matchable D_i, we let $k_i = |D_i| - 6d$ which, by 15.1(b), is less than $\Delta - d$ and we take a k_i-colouring of D_i consisting of $6d$ colour classes with 2 elements and $|D_i| - 12d$ singleton colour classes. As each vertex in D_i has at most $\frac{\Delta}{4}$ external neighbours, this yields the desired partition of D_i. It remains to prove the result for dense sets which are not matchable. We observe first that:

15.16 *A non-matchable D_i has chromatic number at least $\Delta - 26d$. Furthermore, it has an optimal colouring in which the singleton colour classes form a clique with at least $\Delta - 32d$ elements, all of whose vertices are adjacent to every vertex of each colour class with more than two elements.*

Proof Since D_i has at least $\Delta - 8d$ vertices, in any $\Delta - 26d$ colouring of D_i, there must be at least $18d$ vertices in non-singleton colour classes. But each non-singleton colour class U contains a matching in the complement of D_i with $\lfloor \frac{|U|}{2} \rfloor \geq \frac{|U|}{3}$ edges. The first statement follows. Since D_i is not matchable, at most $6d$ of the colour classes in an optimal colouring are non-singleton. Furthermore, in an optimal colouring of any graph, the singleton colour classes form a clique. By choosing an optimal colouring with the maximal number of two element colour classes, we ensure that there is no edge of $\overline{D_i}$ joining a singleton colour class and a colour class with $k \geq 3$ elements, as otherwise we could replace these two colour classes with one of size two and another of size $k - 1$. Thus, the second statement also holds. □

15.17 *For a colouring of a non-matchable dense set as in (15.16), the external neighbourhood of each colour class has at most $\frac{\Delta}{2}$ elements.*

Proof If the colour class has at most two elements then clearly its external neighbourhood has at most $\frac{\Delta}{2}$ elements. If the colour class has more than two vertices then, by (15.16), each of its vertices has at least $\Delta - 32d$ internal neighbours and hence at most $32d$ external neighbours. Furthermore, the colour class can contain at most $|D_i| - (\Delta - 32d) \leq 36d$ vertices. Thus, its external neighbourhood has at most $(32d)(36d)$ elements which yields the desired bound. □

Combining (15.16) and (15.17), we see that the desired result also holds for non-matchable dense sets, thereby completing the proof of the lemma.

□

15.4 Graphs with χ Near Δ

15.4.1 Generalizing Brooks' Theorem

As we discussed in Chap. 1, it is computationally difficult to compute or even approximate χ. However, Brooks' Theorem shows that it is easy to determine whether $\chi = \Delta + 1$ as follows. If $\Delta = 2$, we need only check if the graph is bipartite; if Δ is at least 3, we need only check if any component of G induces a $(\Delta+1)$-clique. It is natural to ask how close χ must be to $\Delta+1$ to ensure that computing χ is easy. The following result, due to Emden–Weinert, Hougardy, and Kreuter [38], shows that χ must be within $\sqrt{\Delta}$ of $\Delta + 1$.

Theorem 15.18 *For any fixed Δ, determining whether a graph G of maximum degree Δ has a $(\Delta + 1 - k)$-colouring is NP-complete for any k such that $k^2 + k > \Delta$, provided $\Delta + 1 - k$ is at least 3.*

In [124] Molloy and Reed prove the following complementary result (see [121] for an earlier result along the same lines):

Theorem 15.19 *For any fixed sufficiently large Δ, determining if a graph G of maximum degree Δ has a $(\Delta+1-k)$-colouring is in P for any k such that $k^2 + k \leq \Delta$.*

The reason $k^2 + k = \Delta$ is the threshold at which $(\Delta + 1 - k)$-colouring becomes computationally difficult is that it is at this point that the chromatic number begins to be determined by global rather than local factors. The proof of the following theorem will be discussed in the next subsection.

Theorem 15.20 *For every Δ, k, and N with $k \leq \Delta - 1$ and $k^2 + k \geq \Delta$, there exists a graph G with maximum degree Δ such that $\chi(G) > \Delta + 1 - k$, $|V(G)| \geq N$, and for every proper subgraph H of G, $\chi(H) \leq \Delta + 1 - k$.*

This is essentially a tight result, as the following establishes.

Theorem 15.21 *For all sufficiently large Δ and k such that $k^2 + k < \Delta$, there exists an $N = N(\Delta, k)$ such that if G has maximum degree Δ and $\chi(G) > \Delta + 1 - k$ then there exists a subgraph H of G with at most N vertices such that $\chi(H) > \Delta + 1 - k$.*

Theorem 15.21 was the key ingredient in the proof of Theorem 15.19. In fact, Theorem 15.21 implies Theorem 15.19 directly for the case $k^2 + k < \Delta$ because as Δ is fixed we can use exhaustive search to check all $\binom{n}{N(\Delta,k)} < n^{N(\Delta,k)}$ subgraphs of size $N(\Delta, k)$ to see if they can be k-coloured. In fact, as discussed below, the algorithm is much faster: it runs in linear time.

In [124], Molloy and Reed conjecture that Theorem 15.21 actually holds for all values of Δ and hence so does Theorem 15.19.

The difficulty in proving Theorem 15.21 is the dense vertices. If G has no $1000\sqrt{\Delta}$-dense vertices then we can use the Naive Colouring Procedure to obtain a $(\Delta+1-k)$-colouring (cf. Theorem 10.5). So, in proving the theorem, we set $d = 1000\sqrt{\Delta}$ and consider a d-dense decomposition $D_1, \ldots, D_l, S$ of G. We construct a partition of D_i into $\chi(D_i)$ partition classes each with a small external neighbourhood by modifying the proof of Lemma 15.15 (see Exercises 15.3–15.5).

For $k^2 + k < \Delta - k$, it turns out that if each D_i is $(\Delta+1-k)$-colourable, then by taking advantage of these partitions, we can use a variant of the Naive Colouring Procedure to obtain a $(\Delta + 1 - k)$-colouring of G. This proves Theorem 15.21 with $N(\Delta, k) = \Delta + 4000\sqrt{\Delta}$ as this is an upper bound on the size of each D_i. (The proof is quite complicated and is along the lines of that discussed in the next chapter.) The proof of the remaining case $\Delta - k \leq k^2 + k < \Delta$ is similar in spirit, but requires a somewhat larger value for N.

Returning to the algorithmic consequences of the theorem, we note that for $k^2 + k < \Delta - k$, we can find a d-dense decomposition in linear time. We can then determine the chromatic number of each dense set, which takes constant time for each of the $O(n)$ dense sets (since Δ is a constant). Thus, we can determine if G has a $(\Delta + 1 - k)$-colouring in linear time. For the remaining values of k where $k^2 + k \leq \Delta$, a simple linear time preprocessing step transforms them into graphs for which the scheme will work.

15.4.2 Blowing Up a Vertex

We complete this chapter by proving Theorem 15.20.

Central to the proof is the notion of a *reducer*. A (c,r)*-reducer* in G consists of a clique C on $c - 1$ vertices, a stable set X on r vertices, and all possible edges between C and X. We further insist that every vertex of C is non-adjacent to all of $V - (C \cup X)$.

Clearly, in any c-colouring of a (c, r)-reducer (C, X) of G, all the vertices of X receive the same colour. Conversely, since there are no edges from C to $V-(C\cup X)$, any c-colouring of $G-X$ in which all of X receives the same colour

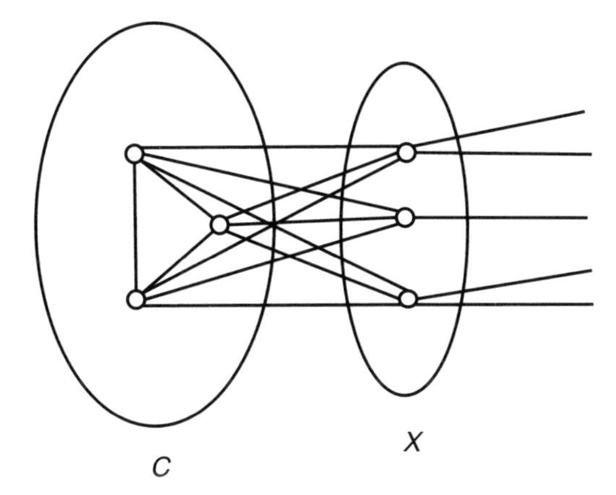

Fig. 15.3. A good $(4, 3)$-reducer

can be extended to a c-colouring of G. Thus letting G' be the graph obtained from $G-(C \cup X)$ by adding a vertex adjacent to $\{w | \exists u \in X \ s.t. \ uw \in E(G)\}$, we see that G is c-colourable if and only if G' is. We call G' the *reduction* of G by (C, X).

Note that each vertex in C has $c + r - 2$ neighbours in total and each vertex in X has $c - 1$ neighbours in $C \cup X$. We call a reducer *good* if each vertex of X has at most $r - 1$ neighbours in $G - (C \cup X)$ and hence at most $c+r-2$ neighbours in total. If $c = \Delta+1-k$ and $r = k+1$ then every vertex in a good (c, r)-reducer has degree at most Δ.

It is easy to see that for any vertex v of degree at most $r(r - 1)$ in any graph G', for any $c \geq 2$ there is a graph G such that G' is the reduction of G by a good (c, r)-reducer (C, X) such that $G' - v = G - (C \cup X)$.

We shall apply such anti-reductions, for k satisfying $k^2 + k \geq \Delta$ and $k \leq \Delta - 1$, with $c = \Delta + 1 - k$ and $r = k + 1$. If we do so, starting with a $(c + 1)$-clique and arbitrarily choosing a vertex to play the role of v, we construct larger and larger graphs of maximum degree at most Δ, which have no c-colouring. As the reader may verify (see Exercise 15.1), for each graph H we construct, every subgraph of H has a c-colouring. This sequence of graphs demonstrates the truth of Theorem 15.20.

In the same vein, we can use reducers to reduce the problem of c-colouring to that of c-colouring graphs of maximum degree Δ (see Exercise 15.2). This proves Theorem 15.18.

Exercises

Exercise 15.1 For the construction of G' from G discussed in Sect. 15.4.2, show that if for every edge e of G, $G - e$ is c-colourable then for every edge f of G', $G' - f$ is c-colourable.

Exercise 15.2 Prove Theorem 15.18 as follows:

Fix a Δ, k with $k^2 + k > \Delta$ and $c = \Delta + 1 - k \geq 3$. Use a construction similar to that given in Sect. 15.4.2 to obtain from any graph G, a new graph G^* with maximum degree Δ such that (i) G is c-colourable if and only if G^* is, and (ii) $|C^*|$ is polynomial in $|G|$.

Exercise 15.3 Show that if Δ is sufficiently large and $D_1, \ldots, D_l, S$ is a $\lceil 20\sqrt{\Delta} \rceil$-dense-decomposition of G and $\chi(D_i) \leq \Delta - \sqrt{\Delta}$, then there is a partition of D_i into $\Delta - \lceil \sqrt{\Delta} \rceil + 2$ stable sets each of which has an external neighbourhood containing at most $\frac{3\Delta}{4}$ vertices.

Exercise 15.4 Show that if Δ is sufficiently large, $k^2 + k \leq \Delta$, and $D_1, \ldots, D_l, S$ is a $\lceil 20\sqrt{\Delta} \rceil$-dense-decomposition of G then there is a partition of each D_i into $\max\{\chi(D_i), \Delta+1-k\}$ stable sets each of which has an external neighbourhood containing at most Δ vertices.

Exercise 15.5 Show that for Δ sufficiently large, if $k^2+k \leq \Delta$, $D_1, \ldots, D_l, S$ is a $\lceil 20\sqrt{\Delta} \rceil$-dense-decomposition of G, and G has no good $(\Delta+1-k), (k+1)$-reducer then there is a partition of each D_i into $\max\{\chi(D_i), \Delta+1-k\}$ stable sets each of which has an external neighbourhood containing at most $\Delta - k$ vertices.

Exercise 15.6 As described in Sect. 15.4, Molloy and Reed have shown that for Δ sufficiently large, if $k^2+k < \Delta-k$, and if $D_1, \ldots, D_l, S$ is any $(1000\sqrt{\Delta})$-dense-decomposition of a graph G with maximum degree Δ, then G is $(\Delta + 1 - k)$-colourable iff every D_i is $(\Delta + 1 - k)$-colourable.

Use this fact to provide an alternate proof of Theorem 11.1.

See [50, 51] for some further results along these lines.

16. ω, Δ and χ

In this chapter, we continue our analysis of the relationship between ω, Δ, and χ. To do so, we need to modify our naive colouring procedure so as to take advantage of the decomposition result described in the last chapter. The analysis of this variant of the Naive Colouring Procedure requires a strengthening of Talagrand's Inequality obtained by McDiarmid.

The analysis of the modified Naive Colouring Procedure, including the description of McDiarmid's Inequality is contained in Sects. 16.1 to 16.4. These results will also be needed in Chap. 18.

Two basic bounds on the chromatic number of a graph are $\omega \leq \chi$ and $\chi \leq \Delta+1$. The lower bound on χ equals the upper bound when both bounds are tight, that is, if G is a clique. On the other hand, Brooks' Theorem says that if $\Delta \geq 3$ then the upper bound is tight only if the lower bound is tight, i.e. if $\omega = \Delta + 1$. In the same vein, we saw, in Chap. 11, that for sufficiently large Δ, if $\omega < \Delta$ then $\chi < \Delta$. It is natural to ask how quickly, if at all, χ must decrease as ω moves further away from Δ. For example, asking if the decrease in χ must be at least a constant fraction of the decrease in ω is equivalent to asking if we can bound χ by a convex combination of ω and $\Delta + 1$. I.e:

Is there an $a > 0$ such that:

16.1 *For* $\Delta \geq 3, \forall G, \quad \chi(G) \leq (1-a)(\Delta+1) + a\omega.$

As we saw in Exercise 3.6, for every n there are graphs with $\chi = \frac{n}{2}$ and $\omega = o(n)$. Thus, if (16.1) is to hold then a must be at most $\frac{1}{2}$.

In [132], Reed conjectured that (modulo a round-up) this necessary condition is in fact sufficient to ensure that (16.1) holds. Specifically, he posed:

Conjecture 16.2 $\forall G, \quad \chi \leq \left\lceil \frac{1}{2}(\Delta+1) + \frac{1}{2}\omega \right\rceil.$

Remark Trivially, if (16.1) holds for a it also holds for any $a' \leq a$ since $\omega(G) \leq \Delta(G) + 1$.

An example due to Kostochka [100] shows that this conjecture is false if we fail to round up, even for large Δ. His example is the line graph of the multigraph G_k obtained from a five cycle by taking k copies of each edge. Clearly, $\omega(L(G_k)) = 2k$, $\Delta(L(G_k)) = 3k-1$, and $\chi(L(G_k)) = \lceil \frac{5k}{2} \rceil$. Fig. 16.1 depicts G_5; Fig. 11.1 depicts $L(G_3)$.

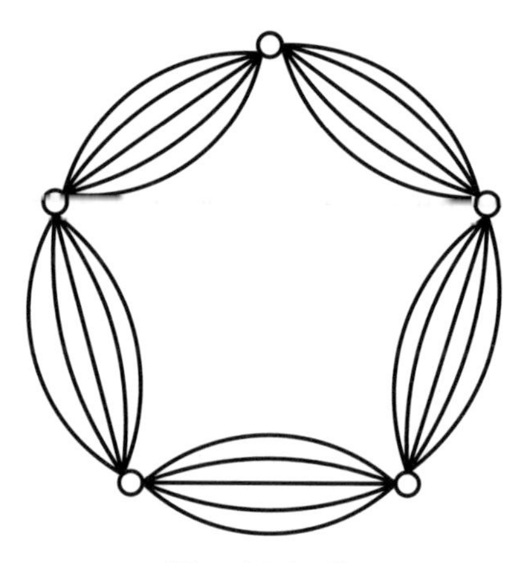

Fig. 16.1. G_5

Partial evidence for Reed's conjecture is given in [132], where he answers our first question in the affirmative:

Theorem 16.3 $\exists a > 0$ *such that* $\chi \leq (1-a)(\Delta+1) + a\omega$.

In fact he proves a stronger result which provides further evidence for his conjecture:

Theorem 16.4 $\exists \zeta > 0$ *such that if* $\omega \geq (1-\zeta)(\Delta+1)$ *then* $\chi \leq \frac{1}{2}(\Delta+1) + \frac{1}{2}\omega$.

Theorem 16.4 implies Theorem 16.3 with $a = \frac{1}{2}\zeta$, as follows: If $\omega \geq (1-\zeta)(\Delta+1)$ then $\chi \leq \frac{1}{2}(\Delta+1) + \frac{1}{2}\omega \leq (1-a)(\Delta+1) + a\omega$ (where the second inequality holds because $\omega \leq \Delta + 1$). If $\omega < (1-\zeta)(\Delta+1)$ then we form G' by adding to G a new component consisting of a $(1-\zeta)(\Delta+1)$-clique. Then $\chi(G) \leq \chi(G') \leq \frac{1}{2}(\Delta+1) + \frac{1}{2}\omega(G') \leq (1-\frac{1}{2}\zeta)(\Delta+1) \leq (1-\frac{1}{2}\zeta)(\Delta+1) + \frac{1}{2}\zeta\omega(G)$.

Even further evidence for the conjecture is presented in Chap. 21 where we will show that every graph has a fractional $(\frac{\Delta+1+\omega}{2})$-colouring. In this chapter, we prove the following weakening of Theorem 16.3 whose proof contains most of the key ideas needed in the proof of Theorems 16.3 and 16.4 but avoids many of the technical complications.

Theorem 16.5 *There exists a* $\delta > 0$ *and* Δ_0, *such that for all* $\Delta \geq \Delta_0$ *if* G *has maximum degree* Δ *and clique number* $\omega \leq \Delta - \log^{10}\Delta$ *then* $\chi \leq (1-\delta)(\Delta+1) + \delta\omega$.

Remark Theorem 16.3 implies Theorem 16.5 with $\Delta_0 = 1$ and $\delta = a$.

We work through the details of the proof in the next four sections. In the final section of the chapter, we discuss the additional complications required to prove Theorems 16.3 and 16.4.

We set $\epsilon = \frac{(\Delta+1-\omega)}{\Delta+1}$. Thus, we want to colour G with $(1-\epsilon\delta)(\Delta+1)$ colours. This is easy to do for graphs without $\epsilon\Delta$-dense vertices using the Naive Colouring Procedure. Indeed, Theorem 10.5 implies that Theorem 16.5 holds with $\delta = e^{-6}$ for such graphs.

The decomposition theorem introduced in the last chapter was designed specifically to handle the dense vertices. After applying this procedure to a graph G with $\omega \leq (1-\epsilon)(\Delta+1)$, we can colour each dense set separately using $c \leq (1-\epsilon\delta)(\Delta+1)$ colours, by applying Lemma 15.12. The challenge is to avoid conflicts between these colourings of the dense sets, and to assign colours to the vertices in S as well. To do so, we must modify the Naive Colouring Procedure significantly. In the next four sections, we present and analyze the modified procedure. In Sect. 16.5, we use it prove Theorem 16.5. We remark that we will present a second application of this modified procedure in Chap. 18.

16.1 The Modified Colouring Procedure

Suppose that we have a graph G with a d-dense decomposition $D_1, \ldots, D_\ell, S$ for some $d \leq \frac{\Delta}{6000}$. Suppose further that each dense set D_i has a partition $CP_i = \{U_1^i, \ldots, U_{|CP_i|}^i\}$ into independent sets satisfying the conditions of Lemma 15.12. We wish to colour G using c colours for some c which exceeds $\max\{|CP_i| : 1 \leq i \leq l\}$. In our initial discussion, we assume c actually exceeds this maximum by at least $\alpha\Delta$ for some small but positive α.

We want to obtain a partial colouring in which every vertex has a sufficiently large number of repeated colours in its neighbourhood by applying a modified Naive Colouring Procedure. As before, the vertices in S are assigned a uniformly random colour. Now, however, for each dense set D_i we will use the colouring given by CP_i. In order to make this colouring look reasonably random to vertices outside D_i, we use a random subset of $|CP_i|$ of our c colours and a random bijection between these colour class names and the partition classes in CP_i.

For each vertex v in S, there will still be many colours repeated on $N(v)$ because v is sparse. If a vertex v in a dense set has many external neighbours, then it turns out that v is also reasonably sparse and hence there will again be many repeated colours on $N(v)$. Those vertices in dense sets which have few external neighbours are slightly harder to deal with. However, if a vertex v in D_i has fewer than $\frac{\alpha\Delta}{2} - 1$ external neighbours, then either (a) it has degree less than $c-1$ and so is easy to colour or (b) it is adjacent to at least $\frac{\alpha\Delta}{2}$ pairs of vertices which form partition classes of size 2 in D_i. In the second case, since we insist that all such pairs of vertices are assigned the same colour by the Naive Colouring Procedure, it is likely that there will be many repeated colours in $N(v)$.

In the general case, when some CP_i has size very close to c and so we do not have a spare $\alpha\Delta$ colours to play with, things get a bit trickier. However our analysis still relies on combining the same two approaches: (i) using sparseness to deal with vertices in S and the vertices in dense sets with large external neighbourhoods, and (ii) taking advantage of the fact that the colour assignment on D_i agrees with CP_i to deal with the remaining vertices. Thus, we use the following variant of our Naive Colouring Procedure:

1. For each vertex $v \in S$, assign to v a uniformly random colour from $\{1, \ldots, c\}$.
2. For each dense set D_i, choose a uniformly random permutation π_i of $\{1, \ldots, c\}$ and for $1 \leq j \leq |CP_i|$, assign the vertices of U_j^i the colour $\pi_i(j)$.
3. For each vertex $v \in S$, if any neighbour of v is assigned the same colour as v then uncolour v.
4. For each D_i and partition class $U_j^i \in CP_i$, if any vertex in the external neighbourhood of U_j^i is assigned the colour $\pi_i(j)$ assigned to U_j^i, then uncolour all the vertices in U_j^i.

In the next three sections, we will see how we can analyze this variant of the Naive Colouring Procedure to obtain results similar to those obtained in the first application that we saw in Chap. 10.

In this chapter and those following, in order to ease our discussion, each vertex of S will also be referred to as a *partition class*.

16.2 An Extension of Talagrand's Inequality

Recall that Talagrand's Inequality only applies to a sequence of independent random trials. We have often considered situations in which, for each vertex v of our graph, the assignment of a colour to v is a trial. Since these colour assignments were independent, we could apply Talagrand's Inequality. However, in the present variant of the Naive Colouring Procedure, these colour assignments are *not* independent, as there is clearly dependency between the colours assigned to two vertices in the same dense set. Fortunately, McDiarmid [115] has developed an extension of Talagrand's Inequality which is applicable in this situation.

His extension deals with a series of random trials and random permutations. In this context, a *choice* is defined to be either (a) the outcome of a trial or (b) the position that a particular element gets mapped to in a permutation.

McDiarmid's Inequality [115] *Let X be a non-negative random variable, not identically 0, which is determined by n independent trials $T_1, \ldots, T_n$ and m independent permutations $\Pi_1, \ldots, \Pi_m$ and satisfying the following for some $c, r > 0$:*

1. *changing the outcome of any one trial can affect X by at most c;*
2. *interchanging two elements in any one permutation can affect X by at most c;*
3. *for any s, if $X \geq s$ then there is a set of at most rs choices whose outcomes* certify *that $X \geq s$,*

then for any $0 \leq t \leq \mathbf{E}(X)$,

$$\mathbf{Pr}(|X - \mathbf{E}(X)| > t + 60c\sqrt{r\mathbf{E}(X)}) \leq 4\mathrm{e}^{-\frac{t^2}{8c^2 r\mathbf{E}(X)}}.$$

Remark

1. Again, c and r are typically small constants and we usually take t to be asymptotically much larger than $\sqrt{\mathbf{E}(X)}$ so the $60c\sqrt{r\mathbf{E}(X)}$ term is negligible.
2. Talagrand [148] proved this result for a single random permutation, i.e. the case $n = 0, m = 1$.

We will apply this inequality for various random variables X determined by the colours assigned by our procedure. To do so, we consider the colour assigned to a vertex in S to be one of the random trials, and we consider the assignment of colours to a dense set D_i to be one of the random permutations.

Note that interchanging 2 elements of a permutation will have one of the following effects: (i) the colours assigned to two partition classes are interchanged, or (ii) the colour assigned to a partition class is changed to a colour previously unused on its dense set. Furthermore, the colour assigned to U_j^i is certified by the choice of $\pi_i(j)$. Therefore, in order to apply McDiarmid's Inequality it is enough to verify that:

1. making a change of type (i) or (ii) above, or changing the colour of a vertex in S, can affect X by at most c, and
2. if $X \geq s$ then there is a set of at most rs partition classes whose colour assignments certify that $X \geq s$.

We will see an example of such a usage in Sect. 16.4.

16.3 Strongly Non-Adjacent Vertices

In Chap. 10, we showed that a single application of the Naive Colouring Procedure to a sparse graph will, with positive probability, result in a partial colouring where every vertex has several repeated colours in its neighbourhood. An important part of that proof rested on the fact that for each vertex v, there were many pairs of neighbours $u, w \in N(v)$ which were non-adjacent and thus eligible to retain the same colour. Now that we are dealing with dense sets, the situation is more complicated, even for sparse vertices.

In particular, it is quite possible that two non-adjacent vertices u, w in the neighbourhood of a sparse vertex v are *not* eligible to retain the same colour. For example, perhaps u, w lie in the same dense set D_i and do not form a partition class. Or perhaps u lies in a dense set D_i and there is some neighbour x of w such that $\{u, x\}$ forms a partition class. Since u and x will receive the same colour during our procedure, it is not possible for u and w to retain the same colour.

We say that two partition classes are *non-adjacent* if they do not lie in the same dense set and there is no edge between them (recall that the vertices of S are partition classes). We say two vertices are *strongly non-adjacent* if they are in non-adjacent partition classes. In order to show that a vertex will probably have many repeated colours in its neighbourhood, we will typically first show that it has many strongly non-adjacent pairs in its neighbourhood.

Actually, this still may not quite be enough to ensure that there are many repeated colours in the neighbourhood of a vertex. For, there are some strongly non-adjacent pairs of vertices which are very unlikely to retain the same colour. The extreme example is a pair $\{x, y\}$ of strongly non-adjacent vertices such that $N(x) \cup N(y)$ contains all the vertices of some dense set D_i with $|CP_i| = c$. In this case, every colour is assigned to either a neighbour of x in D_i or a neighbour of y in D_i. So, no colour can be retained by both x and y. Similarly, if every partition class in such a D_i is adjacent either to the partition class containing x or the partition class containing y then x and y cannot retain the same colour.

We say that a pair x and y of strongly non-adjacent vertices is a *monocolourable pair* if for each dense set D_i which contains neither x nor y there are at most $\frac{7\Delta}{8}$ partition classes of CP_i which are adjacent either to the partition class containing x or to the partition class containing y.

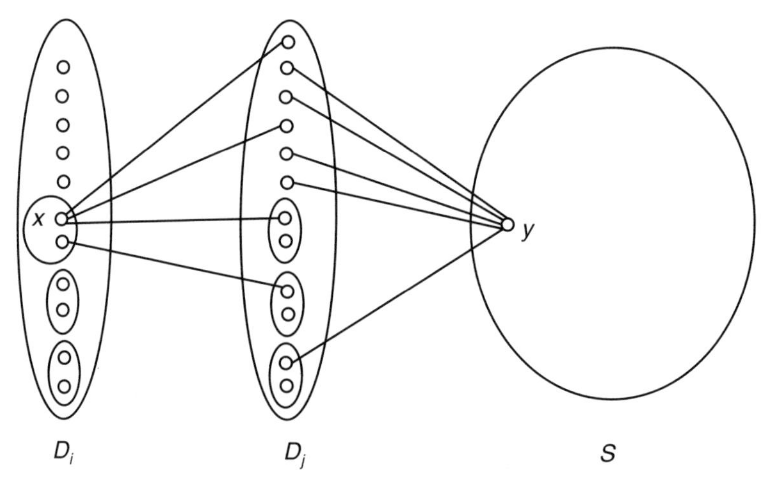

Fig. 16.2. (x, y) is not a monocolourable pair

We turn now to proving that certain sparse vertices have many monocolourable pairs of neighbours. In doing so, it will be convenient to assume the graph is Δ-regular. This is no real restriction, as our standard trick allows us to reduce most of the problems we are interested in to the regular case.

So, consider integers d and Δ with $d \leq \frac{\Delta}{6000}$ and a d-dense decomposition, $D_1, \ldots, D_l, S$ of a Δ-regular graph G. By definition, every vertex $v \in S$ has at least $d\Delta$ non-adjacent pairs of vertices in its neighbourhood. As we remarked earlier, those vertices in dense sets which have large external neighbourhoods are also reasonably sparse. In particular, if we define S' to be the set of vertices in dense sets which have at least $\frac{d}{3}$ external neighbours, then it is straightforward to show that each vertex in S' has at least $\frac{d\Delta}{15}$ pairs of non-adjacent vertices in its neighbourhood.

We will choose a partition CP_i of each D_i satisfying Properties (i)–(iii) of Lemma 15.12, and show that this ensures that the vertices in $S \cup S'$ will have many pairs of strongly non-adjacent vertices in their neighbourhoods:

Lemma 16.6 *Every vertex in $S \cup S'$ has at least $\frac{d\Delta}{16}$ pairs of strongly non-adjacent neighbours.*

Even better, for the partition we have chosen, we can show that such vertices actually have many monocolourable pairs of neighbours.

Lemma 16.7 *Every vertex in S' has at least $\frac{d\Delta}{16}$ monocolourable pairs of neighbours.*

Lemma 16.8 *Every vertex in S has at least $\frac{d\Delta}{16}$ monocolourable pairs of neighbours.*

We prove the first two of these three lemmas in Sect. 16.7 at the end of the chapter. The proof of the third result is given through the series of exercises: Exercise 16.1–Exercise 16.5.

16.4 Many Repeated Colours

Here, we prove that one of the most useful properties of the Naive Colouring Procedure also holds for our variant. Namely, we will show that with positive probability, we will produce a colouring in which every sufficiently sparse vertex, i.e. every vertex in $S \cup S'$, has many repeated colours in its neighbourhood. Recall that we are considering a d-dense decomposition for some $d \leq \frac{\Delta}{6000}$, and we are using the canonical partitions CP_i guaranteed by Lemma 15.12.

Lemma 16.9 *If Δ is sufficiently large, $d \geq (\log \Delta)^{10}$, and G is Δ-regular then after a single application of our procedure using $c \geq (1 - \frac{1}{100})\Delta$ colours, with positive probability we obtain a partial colouring such that:*

for each vertex $v \in S \cup S'$, the difference between the number of neighbours of v which retain a colour and the number of colours retained on $N(v)$ is at least $\frac{d}{10^7} - d^{\frac{2}{3}}$.

The proof of this lemma follows along the same general lines as the proof of Theorem 10.5. For each vertex v, we let X_v be the number of colours assigned to at most ten vertices of $N(v)$ and retained by at least two of them. We show:

Lemma 16.10 *For each $v \in S \cup S'$, $\mathbf{E}(X_v) \geq \frac{d}{10^7}$.*

Lemma 16.11 *For each v, $\mathbf{Pr}(|X_v - \mathbf{E}(X_v)| > \log \Delta \sqrt{\mathbf{E}(X_v)}) < \Delta^{-6}$.*

Given these two lemmas, Lemma 16.9 follows from a straightforward application of the Local Lemma, as we now show.

Proof of Lemma 16.9. For each vertex $v \in S \cup S'$, we define A_v to be the event that $X_v < \frac{d}{10^7} - d^{\frac{2}{3}}$. It suffices to prove that with positive probability none of these events hold. By Lemmas 16.10 and 16.11, since $d \geq (\log \Delta)^{10}$, the probability of each A_v is at most Δ^{-6}.

As in the proof of Theorem 10.2, each event A_v is determined by the colours assigned to vertices near v. In particular, letting B_v be the set of partition classes adjacent to the partition class containing v and C_v be the set of partition classes adjacent to some element of B_v, we see that A_v depends only on the colours assigned to $E_v = B_v \cup C_v$. The difference in this situation is that the colour assignments are no longer independent, and in particular, there is dependency between the colours assigned to classes of the same dense set. Thus, for two events A_u, A_v to be dependent, we do not need E_u and E_v to intersect – it suffices for there to exist a dense set containing an element of both E_u and E_v. However, these are the only ways in which two events can be dependent, and it is straightforward to confirm that each A_v is mutually independent of all but at most $O(\Delta^5)$ other events. Our lemma now follows from the Local Lemma. □

It only remains to prove Lemmas 16.10 and 16.11.

Proof of Lemma 16.10. This follows along the same lines as the proof of Lemma 10.3, but the details are a little more delicate. To begin, we recall the crucial difference between the two proofs.

The basic step in both proofs is to compute the probability that two neighbours of v retain the same colour. As we have already seen, this probability differs under the two procedures. In particular, under the original procedure this probability is at least $\frac{1}{c}(1 - \frac{1}{c})^{2\Delta-2}$ for any pair of non-adjacent neighbours. whilst under the new procedure it is only non-zero for strongly non-adjacent vertices. A more subtle, and at this juncture more important, difference is that the probability that two strongly non-adjacent vertices retain the same colour depends heavily on the intersection of their neighbourhoods with the various dense sets. This was what motivated our definition of a monocolourable pair.

For example, if u and w are a pair of non-adjacent vertices in S, then we can colour u and w independently of the rest of the graph. So, we see that letting Y be the probability that a specific colour j is not assigned to $N(u) \cup N(w)$ (note that this probability is independent of the choice of j by symmetry), we have that the probability u and w retain the same colour is $\frac{Y}{c}$. More generally, for any pair of strongly non-adjacent vertices u and w let Z_{uw} be the set of partition classes adjacent either to the partition class containing u or the partition class containing w and not in the same dense set as u or w. Then, letting Y_{uw} be the probability that a specific colour is not assigned to Z_{uw}, we have that the probability u and w retain the same colour is $\frac{Y_{uw}}{c}$ (we spell out the details below). The first step in the proof will be to bound the Y_{uw}.

To this end, fix a set Z of vertices and a colour j. We will bound the probability that j is not assigned to any vertex of Z. For each dense set D_i, we denote by z_i the number of elements of CP_i that Z intersects. We use z_0 to denote the number of vertices in $S \cap Z$. We define

$$\theta(Z) = \left(1 - \frac{1}{c}\right)^{z_0} \times \prod_{i \geq 1} \left(1 - \frac{z_i}{c}\right)$$

16.12 *The probability that j is not assigned to any class of Z is $\theta(Z)$*

Proof We only need to verify that the probability that no vertex in $Z \cap D_i$ is assigned j is $1 - \frac{z_i}{c}$. But this is clear. □

Suppose that $|Z| = r$, and that for each $i \geq 1$ we have $z_i \leq t$. Define $a = \lfloor \frac{r}{t} \rfloor$ and $b = r - at$. Subject to these constraints, it is easy to see from (16.12) that $\theta(Z)$ is maximized at $z_0 = 0, z_1 = \ldots = z_a = t, z_{a+1} = b$ and so:

16.13 $\theta(Z) \geq (1 - \frac{b}{c}) \times (1 - \frac{t}{c})^a$.

We now consider two strongly non-adjacent vertices u and w. Let Y_{uw} and Z_{uw} be defined as above. Note that u and w both retain colour j if and only if:

(i) u and w are both assigned j, and
(ii) no element of Z_{uw} is assigned j.

Furthermore, these two events are independent, the first occurs with probability $\frac{1}{c^2}$ and the second occurs with probability Y_{uw}. Since there are c colours to choose for j, we do obtain, as we thought:

16.14 *the probability that u and w retain the same colour is $\frac{Y_{uw}}{c}$.*

In the same vein, we have:

16.15 *If u and w are a pair of strongly non adjacent vertices and X is a set of vertices such that no element of X is in a dense set containing either u or w then the probability that u and w retain the same colour and this colour is not assigned to any of X is $\frac{\theta(Z_{uv} \cup X)}{c}$.*

Combining (16.14) with (16.12) and (16.13) we will obtain:

16.16 *For any monocolourable pair, $\{u, w\}$ the probability that u and w retain the same colour is at least $\frac{1}{200c}$.*

To prove (16.16), we set $Z = Z_{uw}$ and note that $|Z| \leq 2\Delta$ and since uv is a good pair: $\forall i$, $|Z_{uv} \cap D_i| \leq \frac{7\Delta}{8}$. So applying (16.13) with $r = 2\Delta$, $t = \frac{7\Delta}{8}$, $a = 2$ and $b = \frac{\Delta}{4}$ yields $Y_{uw} \geq \frac{1}{200}$ (since $c \geq \left(1 - \frac{1}{100}\right)\Delta$).

Using (16.15) instead of (16.14) we will obtain the following result:

16.17 *For any monocolourable pair, $\{u, w\}$ in the neighbourhood of a vertex v, the probability that u and w retain the same colour and this colour is assigned to at most 10 neighbours of v, is at least $\frac{1}{2000c}$.*

To prove (16.17), we consider the set Z' which is the union of $N(u) \cup N(v) \cup N(w)$. We let A be the union of the three dense sets which have the largest intersection with Z'. We also consider the set X consisting of those vertices of $N(v) - A$ which are not in a dense set containing either u or w. We will show that the probability that u and w retain the same colour and that this colour is not assigned to any of X is at least $\frac{1}{2000c}$. Such a colour is assigned to at most 2 vertices in each of the 3 dense sets comprising A, and at most one other vertex in each of the partition classes containing u, w. Thus it is assigned to at most 10 neighbours of v and so (16.17) follows.

To prove our probability bound, we need to apply (16.13) to $Z^* = Z_{uw} \cup X$. Clearly, $|Z^*| \leq |Z'| \leq 3\Delta$. We claim that $Z^* \cap D_i \leq \frac{7\Delta}{8}$ for every i. As $Z^* \subseteq Z'$ this is trivially true if $D_i \cap Z' \leq \frac{3\Delta}{4}$. Since Z' has at most 3Δ vertices, if $|Z' \cap D_i| > \frac{3\Delta}{4}$ then $D_i \subseteq A$ and so $D_i \cap X = \emptyset$. Therefore, $Z^* \cap D_i = Z_{uw} \cap D_i$. By the definition of "monocolourable pair", $|Z_{uw} \cap D_i| \leq \frac{7\Delta}{8}$, and so the claim holds. Thus, we can apply (16.13) with $r = 3\Delta$, and $t = \frac{7\Delta}{8}$ to obtain (16.17).

With (16.17) in hand, it is straightforward to complete the proof of the lemma. We consider a vertex v of $S \cup S'$ and let X'_v be the set of colours which are assigned to at most 10 neighbours of v and retained by at least two of them. We consider the number N_v of monocolourable pairs $\{u, w\}$ for v which are assigned a colour in X'_v. By (16.7), (16.8), and (16.17), $N_v \geq \frac{d\Delta}{16} \times \frac{1}{2000c}$. Since N_v is clearly at most $45X_v$, the lemma follows. □

Proof of Lemma 16.11. Our proof here is reminiscent of that used to prove the concentration results in Chap. 10. Instead of proving the concentration of X_v directly, we will focus on a number of related variables. So, for i between 2 and 10 we define:

(a) X_v^i to be the number of colours assigned to exactly i neighbours of v and retained by at least two of them,
(b) Y_v^i to be the number of colours assigned to at least i neighbours of v, and
(c) Z_v^i to be the number of colours assigned to at least i neighbours of v and either assigned to at least $i+1$ neighbours of v or removed from at least $i-1$ of them.

We note that $X_v = \sum_{i=2}^{10} X_v^i$ and $X_v^i = Y_v^i - Z_v^i$. Thus, it suffices to prove the following concentration bounds, which hold for any $t > \sqrt{\mu \log \mu}$, where μ is the expected value of the variable considered.

$$\textit{Claim 1:} \quad \mathbf{Pr}\left(\left|Y_v^i - \mathbf{E}(Y_v^i)\right| > t\right) < 4e^{-\frac{t^2}{350\mathbf{E}(X_v)}}.$$

$$\textit{Claim 2:} \quad \mathbf{Pr}\left(\left|Z_v^i - \mathbf{E}(Z_v^i)\right| > t\right) < 4e^{-\frac{t^2}{700\mathbf{E}(X_v)}}.$$

We prove both of these claims using McDiarmid's Inequality.

Proof of Claim 1. The value of Y_v^i is determined by the colour assignments made to the neighbours of v. Furthermore, carrying out either of changes (i), (ii) from Sect. 16.2 can affect Y_v^i by at most 2, as such a change can only affect whether the at most two colours involved are counted by Y_v^i. Furthermore, to certify that Y_v^i is at least s we need only specify $10s$ colour assignments; for each of s colours, we specify $i \leq 10$ assignments of the colour to neighbours of v. So, we can indeed apply McDiarmid's Inequality to prove our claim. □

Proof of Claim 2. As with Y_v^i, carrying out either of changes (i), (ii) from Sect. 16.2 can affect Z_v^i by at most 2, as such a change can only affect whether the at most two colours involved are counted by Z_v^i. Furthermore, to certify that Z_v^i is at least s we need only specify at most $19s$ colour assignments. For each of s colours, we first specify $i \leq 10$ assignments of the colour to neighbours of v. We then specify either an $(i+1)$st assignment on the neighbourhood or $i-1$ assignments which cause $i-1$ neighbours of v to lose this colour. So, we can again apply McDiarmid's Inequality to prove our claim. □

16.5 The Proof of Theorem 16.5

Recall that we defined ϵ so that $\omega = (1-\epsilon)(\Delta+1)$. We prove Theorem 16.5 for $\epsilon \leq \frac{1}{1000}$, $\delta = 10^{-9}$ and Δ_0 large enough to satisfy certain implicit inequalities. The theorem then follows for arbitrary ϵ with $\delta = 10^{-12}$, since for $\epsilon > \frac{1}{1000}$, if $\omega \leq (1-\epsilon)(\Delta+1)$ then $\omega \leq \left(1 - \frac{1}{1000}\right)(\Delta+1)$ and so $\chi \leq \left(1 - \frac{1}{1000} \times 10^{-9}\right)(\Delta+1) \leq \left(1 - \epsilon \times 10^{-12}\right)(\Delta+1)$.

Using our standard reduction (from Sect. 1.5), we can reduce the problem to the case of regular graphs. So, let G be a Δ-regular graph for some $\Delta \geq \Delta_0$ with $\omega(G) \leq (1-\epsilon)\Delta$. We set $c = \lfloor (1-\epsilon\delta)(\Delta+1) \rfloor$, i.e. c is the number of colours we will use in our colouring. We set $d = \lceil 10^8 \epsilon\delta(\Delta+1) \rceil$ and consider a d-dense decomposition $D_1, \ldots, D_l, S$ of G. Since, $d \leq \frac{\Delta}{6000}$, for each $1 \leq i \leq l$ we can find a partition CP_i of D_i satisfying the conditions of Lemma 15.12. We note that for each i, $\omega(D_i) \leq (1-\epsilon)(\Delta+1) < \Delta - 8d$. So Lemma 15.12 guarantees that $|CP_i| \leq \Delta - d$.

Now, by Lemma 16.9, we can find a partial c-colouring of G so that:

(1) For every vertex v in $S \cup S'$, the number of coloured neighbours of $N(v)$ exceeds the number of colours which appear on $N(v)$ by at least $\frac{d}{10^8} \geq \epsilon\delta(\Delta+1)$.
(2) For every two vertex partition class U, either both vertices of U receive the same colour or neither is coloured.

We will complete our partial colouring of G greedily paying particular attention to the order in which we colour the remaining vertices. Condition (1) implies that the vertices in $S \cup S'$ will pose no problems so we leave them to last. We focus instead on the *core vertices*, defined to be those vertices in dense sets with fewer than $\frac{d}{3}$ external neighbours (or equivalently $V - S - S'$). Condition (2) implies:

16.18 *At most $|CP_i|$ colours appear on D_i.*

Since $|CP_i| + \frac{d}{3} \leq c - 1$, we can certainly colour the core vertices provided we ensure that (16.18) continues to hold.

To this end, whilst colouring the core vertices, we shall insist that at all times:

16.19 *For each partition class U there is at most one colour used on the vertices of U.*

So, rather than considering the core vertices one at time, we still consider any pair forming a partition class as a unit. When we colour such a pair $\{x, y\}$ in D_i, we give x and y the same colour. This will be a colour which appears neither on D_i nor on any external neighbour of x or y. Now, by (16.19) there are at most $|CP_i| \leq \Delta - d$ colours appearing on D_i. Furthermore, by definition, each of x and y has at most $\frac{d}{3}$ external neighbours. So, since $\frac{d}{3} > \epsilon\delta(\Delta+1)$, there will certainly be a colour with which we can colour $\{x, y\}$. Thus, we can indeed extend our partial colouring to the core vertices, hence the theorem is proved. □

16.6 Proving the Harder Theorems

The only differences between Theorem 16.3 and Theorem 16.5 is the former must hold (a) when ω is arbitrarily close to Δ not just when $\Delta-\omega \geq (\log\delta)^{10}$, and (b) for all $\Delta \geq 3$ not just those larger than some Δ_0.

The second difference is unimportant, as we can deal with it using a simple trick which we have applied many times already. Specifically, if the theorem holds for $\delta = r$ for Δ exceeding some fixed Δ_0, then it holds for $\Delta \geq 3$ with $\delta = \min\left(r, \frac{1}{\Delta_0+1}\right)$, since for $\Delta \leq \Delta_0$ the result reduces to Brooks' Theorem.

We could not deal with graphs which contain arbitrarily large cliques, using the techniques presented in this chapter, because most vertices in a dense set containing a clique of size exceeding $\Delta - (\log\Delta)^{10}$ have very small external neighbourhoods. Too small, for us to be able to control the number of repeated colours in such a vertex's neighbourhood by applying the Local Lemma. However, in this case, the dense sets we are interested in are so close to cliques that there are very few edges out of any one of them, and hence very little interaction between them. This allows us to deal with them separately with a different kind of randomized fix-up procedure. We note that our analysis in Sect. 18.4 is along the same lines. In this way, we can extend our proof of Theorem 16.5 to a proof of Theorem 16.3.

We turn now to the proof of Theorem 16.4. We start by recalling the differences between Theorems 16.4 and 16.3:

1. in Theorem 16.4, we require $\delta = \frac{1}{2}$;
2. Theorem 16.4 only has to hold for $\omega \geq (1-\zeta)(\Delta+1)$.

To extend our proof of Theorem 16.3 to one of Theorem 16.4, we only need to deal with the first difference, since the second only helps us.

The proof is similar to that which we have just presented. Once again, we need to pseudo-randomly find a partial colouring in which all the core vertices are coloured whilst ensuring there are many repeated colours in the neighbourhoods of the noncore vertices. There is little change in our treatment of the noncore vertices.

To deal with the first difference, we need to take a much closer look at the core vertices. We make the following adjustments in the proof.

(A) If a core vertex receives the same colour as a noncore neighbour we uncolour only the noncore vertex. If two adjacent core vertices receive the same colour, we only uncolour the one with the larger external neighbourhood.

(B) Because singleton partition classes of the CP_i are easy to deal with when we are trying to complete a colouring so that (16.19) holds, we uncolour a constant fraction of these at random, only recolouring them after all the core vertices of the non-singleton stable sets have been coloured.

(C) We colour the uncoloured core vertices with smaller external neighbourhoods before those with larger external neighbourhoods

(D) For each core vertex v, we compute a bound on the number of colours available for v when we come to colour it, by estimating the number of colours retained in its external neighbourhood but not its internal neighbourhood. Modification (C) is useful here as it tells us that many of the external neighbours of v left uncoloured in the initial pseudo-random partial colouring are still uncoloured so will not contribute to this sum.
(E) We sometimes use stable sets of size 3 in the CP_i, this provides us with a slightly better colouring and a bit of slack to play with.

Remark We note that the first two adjustments increase the effect the structural decomposition has on the random colouring procedure. It is this interaction between the two proof techniques which has proven to be especially powerful.

For details of the proof, see [132].

16.7 Two Proofs

In this section we prove Lemmas 16.7 and 16.8. We need a definition.

Definition For a vertex v in a dense set D_i, we use Out'_v to denote the vertices of $G - D_i$ which are not strongly non-adjacent to v.

Proof of Lemma 16.6.

Case 1: $v \in S'$ and v has more than $44d$ external neighbours.

We know that v has at least $\frac{3\Delta}{4}$ internal neighbours and $44d$ external neighbours. Every external edge from D_i corresponds to at most 4 pairs of vertices, consisting of one internal and one external neighbour, which are not strongly non-adjacent. Thus, by 15.1(c), the number of such pairs which are strongly non-adjacent is at least $\frac{3\Delta}{4}(44d) - 32d\Delta = (\frac{3}{4}44d - 32d)\Delta > \frac{d}{16}\Delta$.

Case 2: $v \in S'$ and v has fewer than $44d$ external neighbours.

We know that v has at least $\Delta - 44d$ internal neighbours. Since there are at most $5d$ non-singleton elements of CP_i, at least $\Delta - 54d > \frac{19\Delta}{20}$ of these neighbours are singleton elements of CP_i. Obviously every $w \in N(v) \cap S$ is non-adjacent and hence strongly non-adjacent to at least $\frac{19\Delta}{20} - \frac{3\Delta}{4} = \frac{\Delta}{5}$ of these vertices. For any external neighbour w of v in some dense set D_j, the external neighbourhood of the partition class containing w has at most $\frac{\Delta}{3}$ elements so w is strongly non-adjacent to at least $\frac{19\Delta}{20} - \frac{\Delta}{3} > \frac{\Delta}{5}$ of these vertices. The fact that v has at least $\frac{d}{3}$ external neighbours now yields the desired result.

Case 3: $v \in S$.

Let $t_i = |N(v) \cap D_i|$, and $T = \sum_{i=1}^{\ell} t_i = |N(v) - S|$. Since $v \in S$, there are at least $d\Delta$ pairs of non-adjacent vertices in $N(v)$. If at least $\frac{d}{2}\Delta$ of them lie entirely within S, then we are done as each such pair is also strongly non-adjacent. Otherwise, we must have $\Delta \times |N(v) - S| > \frac{d}{2}\Delta$, and so $T > \frac{d}{2}$.

If for any i, $t_i \geq 160d$, then, as v has at least $\frac{\Delta}{4}$ neighbours outside D_i, it follows as in Case 1 that the number of strongly non-adjacent pairs of vertices in $N(v)$ is at least $\frac{160d\Delta}{4} - 32d\Delta > \frac{d\Delta}{16}$.

If $t_i < 160d$ for all i, then each $u \in D_i \cap N(v)$ is strongly non-adjacent to at least $\Delta - t_i - 2(\frac{\Delta}{3}) > \frac{\Delta}{4}$ vertices of $N(v) - D_i$, and so $N(v)$ has more than $\frac{1}{2}(T \times \frac{\Delta}{4}) = \frac{d}{16}\Delta$ pairs of strongly non-adjacent vertices. □

Proof of Lemma 16.7. To begin, we note that for any vertex w not in some D_j, the partition class containing w is adjacent to at most $\frac{3\Delta}{4}$ partition classes in D_j. The following observation is an immediate consequence of this fact:

16.20 *If a vertex u in D_i forms a singleton element of the canonical partition and has at most $\frac{\Delta}{8}$ external neighbours and w is a vertex strongly non-adjacent to u then $\{u, w\}$ forms a monocolourable pair.*

Now, let v be a vertex in $S' \cap D_i$. We note that because of our bound on the edges out of D_i all but at most $64d$ vertices of D_i have fewer than $\frac{\Delta}{8}$ external neighbours.

Case 1: v has more than $44d$ external neighbours.

We know that there are at least $\frac{3\Delta}{4} - 10d$ internal neighbours of v which form singleton partition classes. Further, there is a set S of at least $\frac{3\Delta}{4} - 74d$ of these neighbours which have fewer then $\frac{\Delta}{8}$ external neighbours. Every external edge from D_i causes at most 4 pairs consisting of one internal and one external neighbour to not be strongly non-adjacent. Thus, by 15.1(c), the number of such strongly non-adjacent pairs consisting of a vertex of S and an external neighbour of v is at least $(\frac{3\Delta}{4} - 74d)(44d) - 32d\Delta > \frac{d\Delta}{16}$.

Case 2: v has fewer than $44d$ external neighbours.

As in the proof of Case 2 of Lemma 16.6 we obtain that every external neighbour of v is strongly non-adjacent to at least $\frac{\Delta}{5}$ internal neighbours of v which form singleton elements of CP_i. Since there are at most $8d\Delta$ edges out of D_i, all but $64d$ of these $\frac{\Delta}{5}$ internal neighbours have at most $\frac{\Delta}{8}$ external neighbours. The result then follows from (16.20) as v has at least $\frac{d}{3}$ external neighbours. □

Exercises

Exercise 16.1 Using (16.20), show that for any vertex $v \in S$ such that for some i we have $|N(v) \cap D_i| \geq 200d$, there are at least $d\Delta$ monocolourable pairs in $N(v)$.

Exercise 16.2 Show that for any vertex v in D_i there are at most $\frac{2\Delta}{3}$ vertices outside D_i which are not strongly non-adjacent to v.

Exercise 16.3 Show that for any vertex v there are at most $8(8d\Delta)(\frac{7\Delta}{8} - \frac{3\Delta}{4})^{-1} = 512d$ partition classes which contain vertices which are strongly non-adjacent to v but do not form a monocolourable pair with v.

Exercise 16.4

(a) Use Exercise 16.2 and Exercise 16.3 to show that for any vertex v in S such that $\forall i$, $|N(v) \cap D_i| \leq 200d$ there are at least $\sum_{i=1}^{l} |N(v) \cap D_i| \frac{\Delta}{12}$ monocolourable pairs in $N(v)$ which contain a vertex from a dense set.

(b) Deduce that if $v \in S$ and $|N(v) - S| \geq d$ then there are at least $\frac{d\Delta}{16}$ monocolourable pairs in $N(v)$.

Exercise 16.5 We consider a vertex v of S s.t. $|N(v) - S| \leq d$.

(a) Show that if u is a vertex in $N(v) \cap S$ which is strongly non-adjacent to at least $\frac{\Delta}{3}$ other vertices of $N(v) \cap S$ then v forms a monocolourable pair with at least $\frac{\Delta}{6}$ vertices of $N(v) \cap S$.

(b) Show that for any non-adjacent pair of vertices in $N(v) \cap S$ which is not monocolourable, at least one of the vertices is non-adjacent to at least $\frac{\Delta}{3}$ vertices of $N(v) \cap S$.

(c) Use (a) and (b) to deduce that if $N(v) \cap S$ contains at least r pairs of non-adjacent vertices then it contains $\frac{r}{12}$ monocolourable pairs. Combine this fact with Exercise 16.4(a) to deduce that $N(v)$ contains at least $\frac{d\Delta}{16}$ monocolourable pairs.

17. Near Optimal Total Colouring I: Sparse Graphs

17.1 Introduction

In the next two chapters, we present the main result of [119]:

Theorem 17.1 *There exists an absolute constant C such that every graph with maximum degree Δ has total chromatic number at most $\Delta + C$.*

As usual, we will prove the result for Δ exceeding some absolute constant Δ_0. Since every graph with maximum degree $\Delta < \Delta_0$ has total chromatic number at most $2\Delta < \Delta + \Delta_0$, this clearly implies that the total chromatic number of any graph G is at most $\Delta(G)$ plus an absolute constant. We do not specify Δ_0; rather we just assume that Δ is large enough to satisfy various inequalities scattered throughout the proof. Also as usual, we will only prove the result for regular graphs, as the construction in Sect. 1.5 allows us to reduce our theorem to this case.

We will make no attempt to optimize the constant C. In [119] we provide a proof for $C = 10^{26}$, so long as $\Delta \geq \Delta_0$. By modifying the proof slightly and being careful one can obtain $C = 500$, and by being even more careful one could probably obtain $C = 100$. However, this technique will almost certainly not yield a value of C which is close to the target of 2 from the Total Colouring Conjecture.

Our approach is similar to that taken in Chap. 7. We will start with an arbitrary $\Delta + 1$ edge colouring, Ψ, of G (which is guaranteed to exist by Vizing's Theorem). We will then carefully choose a compatible vertex colouring using the same $\Delta + 1$ colours. As described at the beginning of Chap. 9, we cannot hope to find a vertex colouring which does not conflict at all with Ψ, so we must settle for one which does not conflict very much. As in Chap. 7, any edge e such that one of the endpoints of e receives the same colour as e will be called a *reject edge*. We will refer to the graph R induced by the reject edges as the *reject graph*. We will find a vertex colouring such that the maximum degree of R is at most $C-2$. By Vizing's Theorem, we can recolour the edges of R using $\Delta(R)+1 \leq C-1$ new colours, thus obtaining a $\Delta + C$ total colouring of G.

For any vertex v, since Ψ is a proper edge colouring, v can have the same colour as at most one reject edge incident to v. All other reject edges incident to v must have the same colour as their other endpoints. It turns out to be convenient to use a slight abuse of notation and define the *reject degree* of v in any full or partial vertex colouring of G to be the number of edges $e = uv$ such that u has the colour $\Psi(e)$. Because $\deg_R(v)$ is at most the reject degree of v plus one, it will suffice to find a vertex colouring for which no vertex has reject degree greater than $C - 3$.

We will begin by discussing intuitively why we might be able to find such a vertex colouring using the Naive Colouring Procedure. Suppose we carry out a single iteration of the Naive Colouring Procedure, and consider, for any vertex v, the reject degree of v in the resulting proper colouring. For any edge $e = uv$, the probability that u receives $\Psi(e)$ is $\frac{1}{\Delta+1}$. Therefore, we see that the reject degree of v is dominated by $\text{BIN}(\Delta, \frac{1}{\Delta+1})$. Thus, by Exercise 2.12, for any constant K, the probability that the reject degree of v exceeds K is no more than roughly $\frac{1}{K!}$, and so for large values of C we expect that a very small (albeit significant) proportion of the vertices will have reject degree greater than $C - 3$.

If a vertex has too high a reject degree, then we will rectify the problem by uncolouring all neighbours of v whose colours are contributing to this reject degree. The probability that a particular vertex is uncoloured by this rule is a very small constant, much smaller than the probability that it is uncoloured because a neighbour receives the same colour. Thus, this increase in the probability of a vertex being uncoloured should not significantly affect the performance of the Naive Colouring Procedure. In particular, we should still be able to use the procedure to iteratively obtain our desired vertex colouring.

As we have seen, the Naive Colouring Procedure often works well when G is sparse. In this chapter, we will present a relatively short proof of Theorem 17.1 for sufficiently sparse graphs G, along the lines discussed above. That is, we prove:

Theorem 17.2 *For any $0 \le \epsilon < 1$ there exists $\Delta_\epsilon, C_\epsilon$ such that for every ϵ-sparse graph G with maximum degree $\Delta \ge \Delta_\epsilon$, we have $\chi_T(G) \le \Delta + C_\epsilon$.*

Proving Theorem 17.1 for graphs which have dense vertices is more complicated. We will need to apply the variant of the Naive Colouring Procedure discussed in the last two chapters. In the next chapter we will show how to modify our proof to yield a complete proof of Theorem 17.1.

As described above, to prove Theorem 17.2 we first fix an arbitrary $(\Delta+1)$-edge colouring Ψ of G. We then find a vertex colouring on the same colour set such that the reject degree of every vertex is bounded. We will find this vertex colouring by applying several iterations of our Naive Colouring Procedure, modified slightly to ensure that the reject degrees remain bounded.

To begin, we analyze the first iteration. We will show that with positive probability, this iteration will produce a partial colouring with bounded reject degree such that every vertex has many repeated colours in its neighbourhood. We will then see that these repeated colours allow us to continue for many more iterations, keeping the reject degree bounded. As usual, we will continue until the vast majority of the vertices have been coloured, and then we complete the colouring using a different sort of final step. In the next chapter we will use a similar, albeit much more complicated, approach.

Our situation is not nearly as sensitive as that in Chaps. 12, 13 and 14, and so our analysis will not need to be as delicate. In fact our analysis will be closer in spirit to the relatively simple proof of Theorem 10.2.

Note that we can still assume that G is regular since the construction in Sect. 1.5 creates no new triangles and so preserves sparseness.

17.2 The Procedure

For each i, we will ensure that during iteration i, no vertex has its reject degree increase by more than T, where T is a constant to be defined below. After some fixed (positive) number I of iterations, so much of the graph will be coloured that we will be able to be much more strict about the number of reject edges appearing. We will ensure that no vertex can have its reject degree increase by more than one during a single iteration, and furthermore, that if its reject degree increases by one, then it will not increase at all in any subsequent iteration. Thus, at the end of our procedure, no vertex will have reject degree greater than $IT + 1$.

For each vertex v, we will maintain a list L_v of colours which do not yet appear on the neighbourhood of v. Initially, of course, $L_v = \{1, \ldots, \Delta + 1\}$. After iteration I, we will also maintain a list of colours F_v which are forbidden from being assigned to v in order to keep the reject degree of some neighbours of v from increasing further. Until iteration I, $F_v = \emptyset$.

Thus, we will perform several iterations of the following procedure:

1. Assign to each uncoloured vertex v a uniformly random colour from L_v.
2. Uncolour any vertex which receives the same colour as a neighbour.
3. In iteration $i \leq I$:
   For any vertex v which has more than T neighbours u for which u is assigned the colour $\Psi(u, v)$ in this iteration, we uncolour all such neighbours.

   In iteration $i > I$:
   a) Uncolour any vertex v which receives a colour from F_v.
   b) For any vertex v which has more than 1 neighbour u for which u receives the colour $\Psi(u, v)$, we uncolour all such neighbours.
   c) For any vertex v which has at least one neighbour u for which u receives the colour $\Psi(u, v)$, we place $\Psi(w, v)$ into F_w for every $w \in N_v$.
4. For any vertex v which retained its colour γ, we remove γ from L_u for each neighbour u of v.

We will continue for a number of iterations to be specified later, and then we complete our colouring with a final phase which we describe in Sect. 17.4. Note that we are being a little overcautious here. For example, in Step 3 we are counting neighbours which were assigned a bad colour regardless of whether they were uncoloured in Step 2. This will simplify the proof at the cost of an increase in our constant C_ϵ.

17.3 The Analysis of the Procedure

To begin, we focus on what happens to the neighbourhood of a particular vertex v during the first iteration. We define A_v to be the number of colours γ such that exactly 2 neighbours of v receive γ and are not uncoloured in Step 2. The same analysis as in the proof of Theorem 10.2 yields that for $\zeta = \frac{\epsilon}{2e^6}$, we have:

17.3 $\mathbf{Pr}(A_v < \zeta\Delta) < e^{-a\Delta}$, *for a particular constant* $a > 0$.

We need to show that with high probability very few of these neighbours lose their colour during Step 3, We define B_v to be the number of neighbours of v which are uncoloured in Step 3.

Lemma 17.4 $\mathbf{Pr}\left(B_v \geq \frac{\zeta}{2}\Delta\right) \leq e^{-b\Delta}$ *for a particular constant* $b > 0$.

Proof Consider any $u \in N_v$. We will bound the probability that u is uncoloured in Step 3. For any particular neighbour w of u, the probability that u and T other neighbours of w are each assigned the same colour as the edge joining them to u is at most $\binom{\Delta-1}{T}\left(\frac{1}{\Delta+1}\right)^{T+1}$. Therefore, the probability that this happens for some neighbour of u is at most

$$\Delta\binom{\Delta-1}{T}\left(\frac{1}{\Delta+1}\right)^{T+1} < \frac{1}{T!} < \frac{\zeta}{4}$$

for T sufficiently large in terms of ζ. Therefore, $\mathbf{E}(B_v) < \frac{\zeta}{4}\Delta$.

To complete the proof, we must show that B_v is sufficiently concentrated. This follows easily from Talagrand's Inequality after observing that changing the colour assigned to any one vertex can affect B_v by at most $T + 1$, and that we can always find a certificate for B_v consisting of the vertices which produce all the reject edges. We leave the details to the reader. □

A straightforward application of the Local Lemma proves that with positive probability, the first iteration produces a partial colouring with bounded reject degree for which every vertex has $\frac{\zeta}{2}$ repeated colours in its neighbourhood.

Next we prove that given such a partial colouring, the remaining iterations of the procedure will, with positive probability, complete it to a colouring with bounded reject degree. To do so, it is no longer necessary that the graph be sparse – this was only required to analyze the performance of the first iteration.

So we now turn our attention to the next iterations of our procedure, which we will analyze in a significantly different way than we analyzed the first iteration. The crucial fact which makes our analysis work is:

17.5 *At the beginning of every iteration after the first, for every uncoloured vertex v, the size of L_v exceeds the number of uncoloured neighbours of v by at least $\rho\Delta$ where $\rho = \zeta/2$.*

This follows easily from the fact that each such v has at least $\rho\Delta$ repeated colours in its neighbourhood.

(17.5) guarantees a significant gap between the size of L_v and the number of uncoloured neighbours of v. This gap will be very useful. For one thing, it implies that $|L_v|$ is always at least $\rho\Delta$. Thus, during any iteration after the first, the probability that a newly coloured vertex is not uncoloured during Step 2 is at least $(1 - 1/(\rho\Delta))^\Delta \approx \mathrm{e}^{-\frac{1}{\rho}}$.

So we need only bound the probability that a newly coloured vertex v is uncoloured in Step 3. For each iteration $2 \leq i \leq I$, this probability is at most $\binom{\Delta}{T}(\frac{1}{L_v})^T$ which, because of the aforementioned gap guaranteed by (17.5), is at most $\binom{\Delta}{T}(\frac{1}{\rho\Delta})^T$. So if we choose a constant T which is sufficiently large in terms of ρ then the probability that v is uncoloured during Step 3 is at most $\frac{1}{2}\mathrm{e}^{-\frac{1}{\rho}}$. Using this and the Local Lemma, it follows easily that with positive probability, for every vertex w, the proportion of uncoloured neighbours of w which retain their colours during any iteration after the first is at least $\frac{1}{4}\mathrm{e}^{-\frac{1}{\rho}}$. So, for any choice of I, we can prove inductively that for each $1 \leq i \leq I$ the degree of the subgraph induced by the uncoloured vertices at the end of iteration i is at most:

$$D_i = \left(1 - \frac{1}{4}\mathrm{e}^{-\frac{1}{\rho}}\right)^i \Delta.$$

(Note that $D_1 = \left(1 - \frac{1}{4}\mathrm{e}^{-1/\rho}\right)\Delta$ which is less than $(1-\rho)\Delta$ since $\rho \leq \frac{1}{2e^6}$, so this bound does indeed hold for $i = 1$).

We want to show that the same is also true for each iteration $i > I$. To do so, we again show that the probability a vertex v is uncoloured during Step 3 is at most $\frac{1}{2}\mathrm{e}^{-1/\rho}$. To begin, we bound the probability that v is uncoloured in Step 3(a). We must first bound $|F_v|$. Again, we use the fact that each L_v will always have size at least $\rho\Delta$. For any vertex u, the probability that some neighbour w of u receives the colour $\Psi(u,w)$ during iteration j is at most $\frac{1}{\rho\Delta} \times D_j$ since for each such w, $\frac{1}{\rho\Delta} \geq \frac{1}{|L_w|}$. Thus, the expected number of neighbours u of v for which $\Psi(u,v)$ enters F_v during iteration j is at most $\Delta \times \frac{1}{\rho\Delta} \times D_j = D_j/\rho$. We will be able to ensure that with positive probability, at most $2D_j/\rho$ colours enter each F_v during iteration j, and so we define for each $i > I$:

$$F_i = \sum_{j=I+1}^{i-1} 2D_j/\rho.$$

An easy computation shows that, if we take I sufficiently high, each F_i is less than $\frac{1}{4}e^{-\frac{1}{\rho}} \times \rho\Delta$. Therefore, the probability that v is uncoloured in Step 3(a) is at most $\frac{|F_v|}{|L_v|} < \frac{1}{4}e^{-\frac{1}{\rho}}$.

The probability that v is uncoloured in Step 3(b) is at most $\frac{D_i}{\rho^2\Delta}$. This is because there are at most ΔD_i choices for a path v, a, b where b is an uncoloured vertex, and there is at most a $(\frac{1}{\rho\Delta})^2$ probability that v, b receive the colours $\Psi(a,v), \Psi(a,b)$. By choosing I to be sufficiently large, we ensure that D_i is so small that this probability is also at most $\frac{1}{4}e^{-\frac{1}{\rho}}$. Therefore, the probability that v is uncoloured in either Step 3(a) or 3(b) is at most $\frac{1}{2}e^{-\frac{1}{\rho}}$. This is enough to show that with positive probability, the degree of the subgraph induced by the uncoloured vertices is at most D_i for iterations $i > I$.

We will continue for i^* iterations, where i^* is the minimum value of i such that $D_i \leq \sqrt{\Delta}$.

We now specify our choice for I and fill in the details of the proof that our procedure works as desired.

We set $r = (1 - \frac{1}{4}e^{-\frac{1}{\rho}})$. Since r is less than one, $\sum_{i=1}^{\infty} r^j$ is a constant. We denote this constant by Q. Then $D_i = r^i\Delta$, $i^* = \left\lceil \log_r \frac{1}{\sqrt{\Delta}} \right\rceil$ and, whatever our choice of I, for $i \geq I$ we have $F_i \leq \frac{2r^I\Delta}{\rho}Q$. We set $I_1 = \left\lceil \log_r \frac{1}{4}e^{-\frac{1}{\rho}}\rho^2 \right\rceil$. We insist $I \geq I_1$ which implies that for $i \geq I$ we have $D_i \leq \frac{1}{4}e^{-\frac{1}{\rho}}\rho^2\Delta$. We set $I_2 = \left\lceil \log_r \frac{\rho^2}{8Q}e^{-\frac{1}{\rho}} \right\rceil$. We insist $I \geq I_2$ which implies that for $i \geq I$ we have $F_i \leq \frac{1}{4}e^{-\frac{1}{\rho}} \times \rho\Delta$.

Lemma 17.6 *If we apply our algorithm with $I = \max\{I_1, I_2, 2\}$ then with positive probability, at the end of each iteration $1 \leq i \leq i^*$, every vertex has at most D_i uncoloured neighbours, and each F_v has size at most F_i.*

Our proof is by induction on i, and so we assume that at the beginning of iteration i, each vertex v has at most D_{i-1} uncoloured neighbours, and each F_v has size at most F_{i-1}. The facts that at least $\rho\Delta$ vertices are coloured in the first iteration (see (17.5)) and that $F_v = \emptyset$ until iteration $I > 1$, establish the base case $i = 1$, so we assume $i \geq 2$.

As noted earlier, the probability that a vertex u is uncoloured in Step 2 is at most $1 - e^{-\frac{1}{\rho}}$.

For $i \leq I$, using the same calculation as in the proof of (17.5) we see that for the probability that u is uncoloured in Step 3 is at most

$$\Delta\binom{D_i}{T}\left(\frac{1}{\rho\Delta}\right)^{T+1} < \frac{1}{\rho}\left(\frac{e}{\rho T}\right)^T < \frac{1}{2}e^{-\frac{1}{\rho}},$$

for T sufficiently large in terms of ρ.

As described above, for $i \geq I$ a similar calculation shows that the probability u is uncoloured in Step 3(b) is at most

$$\Delta \times D_i \left(\frac{1}{\rho\Delta}\right)^2 < \frac{1}{4}\mathrm{e}^{-\frac{1}{\rho}}.$$

Furthermore, the probability that u is uncoloured in Step 3(a) is at most

$$|F_u|/|L_u| \leq F_i/(\rho\Delta) \leq \frac{1}{4}\mathrm{e}^{-\frac{1}{\rho}}.$$

We let A, B be the number of uncoloured neighbours of v at the beginning of iteration i and at the end of iteration i, respectively. By our earlier calculations, $\mathbf{E}(B) \leq (1 - \frac{1}{2}\mathrm{e}^{-\frac{1}{\rho}})A$.

If $A \leq D_i$ then $B \leq D_i$. If $A > D_i$, then a straightforward application of Talagrand's Inequality, combining ideas from the applications in the proof of Theorem 10.2 and Lemma 17.4 shows that B is sufficiently concentrated that

$$\mathbf{Pr}\left(B \geq \left(1 - \frac{1}{4}\mathrm{e}^{-\frac{1}{\rho}}\right)A\right) < \mathrm{e}^{-aD_i},$$

for a particular constant $a > 0$.

We let C be the number of colours which enter F_v during iteration i. As discussed earlier, $\mathbf{E}(C) \leq D_i/\rho\Delta$. A simple application of Talagrand's Inequality shows that C is sufficiently concentrated that

$$\mathbf{Pr}(C \geq 2D_i/\rho\Delta) < \mathrm{e}^{-bD_i},$$

for a particular constant $b > 0$.

Lemma 17.6 now follows from a straightforward application of the Local Lemma. We leave the details for the reader. □

17.4 The Final Phase

At this point, we have a partial colouring in which each vertex v has at most $\sqrt{\Delta}$ uncoloured neighbours, has reject degree at most $IR+1$ and has a list L_v of at least $\rho\Delta$ available colours. It will be convenient for all the lists to be the same size, so for each vertex v, we arbitrarily delete colours from L_v so that $|L_v| = \rho\Delta$.

We will now show that we can complete the colouring so that no vertex has its reject degree increased by more than 1, thus proving Lemma 17.8 and Theorem 17.2.

It is tempting to simply say that the maximum degree of the uncoloured subgraph is much less than the size of the lists, and so it is easy to complete the colouring greedily. Unfortunately, it is not easy to greedily complete the colouring while at the same time preventing the reject degrees from growing too high. A similar problem occurs if we take our usual approach of applying Theorem 4.3. Thus we must adopt a new approach.

For each uncoloured vertex v, we will choose a subset of colours from L_v, which we call *candidates* for v. We will show that with positive probability, we can choose a candidate for each uncoloured vertex to complete the colouring. In particular, we say that a candidate c is a *good candidate* if:

(i) c is not a candidate for any neighbour of v; and
(ii) there is no candidate c' of any uncoloured vertex v' such that v, v' have a common neighbour u where $\Psi(u,v) = c$ and $\Psi(u,v') = c'$.

If we assign a good candidate to every vertex, then condition (i) ensures that we will have a proper colouring of G, and condition (ii) ensures that no reject degree will increase by more than 1. Thus, it suffices to prove:

Lemma 17.7 *It is possible to choose a set of candidates for each uncoloured vertex v so that every such vertex has at least one good candidate.*

Proof For each uncoloured v, we choose 20 uniform random candidates from L_v. We define A_v to be the event that none of these candidates are good.

We want to obtain a bound on $\mathbf{Pr}(A_v)$. To do so, we first expose the candidates for all vertices other than v. After doing so, we define:

$$\begin{aligned} \text{Bad}_1 &= \{c \in L_v : c \text{ is a candidate for some neighbour of } v\} \\ \text{Bad}_2 &= \{c \in L_v : \text{ choosing } c \text{ for } v \text{ would violate condition (ii)}\}. \end{aligned}$$

We set $\text{Bad} = \text{Bad}_1 \cup \text{Bad}_2$, and we define B to be the event that $|\text{Bad}| \leq 60\sqrt{\Delta}$.

Next, we choose the 20 candidates for v. A candidate is good iff it does not belong to Bad. Therefore,

$$\mathbf{Pr}(A_v|B) \leq \left(\frac{60\sqrt{\Delta}}{\rho\Delta}\right)^{20} < \frac{\Delta^{-9}}{2}.$$

So it will suffice to bound $\mathbf{Pr}(\overline{B})$.

Since v has at most $\sqrt{\Delta}$ uncoloured neighbours, $|\text{Bad}_1| \leq 20\sqrt{\Delta}$. For any colour $c \in L_v$, $c \in \text{Bad}_2$ iff the the unique vertex u joined to v by an edge of colour c has another uncoloured neighbour which also produces a reject edge to u. Since u has at most $\sqrt{\Delta}$ uncoloured neighbours, this happens with probability at most $\frac{20\sqrt{\Delta}}{(\rho\Delta)}$ and so $\mathbf{E}(|\text{Bad}_2|) \leq \frac{20\sqrt{\Delta}}{\rho\Delta} \times |L_v| = 20\sqrt{\Delta}$. A simple application of McDiarmid's Inequality shows that $|\text{Bad}_2|$ is highly concentrated: For each vertex w, we can choose the candidates for w by taking a random permutation of L_w, and choosing the first 20 colours. Exchanging two members of a permutation can, at worse, change one of the candidates, and this can affect $|\text{Bad}_2|$ by at most 1. Furthermore, it is easy to find a suitable certificate. This yields that $\mathbf{Pr}(\overline{B}) < \Delta^{-10}$, which implies that

$$\mathbf{Pr}(A_v) \leq \mathbf{Pr}(\overline{B}) + \mathbf{Pr}(A_v|B) < \Delta^{-9}.$$

A_v is mutually independent of all events A_w where w is a distance more than 4 from v, i.e. of all but at most Δ^4 events. Since $\Delta^4 \times \Delta^{-9} < \frac{1}{4}$, the desired result follows from the Local Lemma. □

This completes the proof of Theorem 17.2.

Recall that we only used the fact that G is sparse once in our analysis. Specifically, we needed this fact to show that the first iteration yields a partial colouring with many repeated colours in each neighbourhood. Thus, if we were given such a partial colouring, then we would not require G to be sparse. In other words, the same proof used in this chapter will yield the following lemma, which will be useful in the next chapter.

Lemma 17.8 *For every $\rho > 0$, there exists $C = C(\rho)$ and $\Delta(\rho)$ such that the following holds: Consider any graph G with maximum degree $\Delta \geq \Delta(\rho)$, any edge colouring of G, and any partial colouring of G where every uncoloured vertex has $\rho\Delta$ colours appearing at least twice in its neighbourhood. The partial colouring can be completed to a colouring such that the maximum reject degree does not increase by more than C.*

18. Near Optimal Total Colouring II: General Graphs

18.1 Introduction

In the previous chapter, we proved that for any constant $\epsilon > 0$, graphs in which every vertex is $\epsilon\Delta$-sparse have a $\Delta + C(\epsilon)$ total colouring. In this chapter, we will show how to modify that proof to handle graphs that include dense vertices, thereby proving Theorem 17.1. To do so, we make use of the decomposition from Chap. 15.

Recall that we are assuming G to be a Δ-regular graph for some sufficiently large Δ.

We fix a particular small constant, say 10^{-10}, and we set ϵ to be the largest rational less than 10^{-10} such that $\epsilon\Delta$ is an integer. Then we consider an $\epsilon\Delta$-dense decomposition of G into $D_1, \ldots, D_t, S$ along with the corresponding partitions $CP_1, \ldots, CP_t$ yielded by Lemma 15.12.

As in Chap. 16, we refer to each member of CP_i as a *partition class* and, for convenience, we also refer to each vertex in S as a partition class. Thus, most partition classes have exactly one vertex and the rest have exactly two vertices.

Once again, our approach is to fix an arbitrary $(\Delta+1)$-edge colouring and, using the same $\Delta+1$ colours, construct a vertex colouring with bounded reject degree. To do so, we use a three step process. As in the last chapter, we first perform one iteration of our standard colouring procedure, deleting the colour on any vertex which conflicts with a neighbour or contributes to the reject degree of a neighbour with high reject degree. This implies that the maximum reject degree in the partial colouring obtained is bounded. We show that with positive probability this partial colouring also satisfies certain properties which will allow us to complete it while maintaining a (larger) bound on the reject degrees. In the second step, we extend the partial colouring to most of the vertices in the dense sets. In the third step we complete the partial colouring by colouring all remaining uncoloured vertices, all of which are sparse.

The second step of our process has two phases. In the first, we are mainly interested in ensuring that the colouring behaves as we require within each dense set. In the second, we need to deal with the interaction between each dense set and the rest of the graph. Thus, we can think of our procedure as a four phase process, which we describe below.

For each vertex v in a dense set D_i, we define the *internal reject degree* of v to be the number of its neighbours $u \in D_i$ such that u has the same colour as the edge uv.

While every vertex in S is sparse by the definition of our decomposition, not all vertices in the dense sets are necessarily dense. For example, vertices which have sufficiently large outneighbourhoods will be sparse. In particular, we define

$$S' = \{v \notin S : |\text{Out}_v| \geq 12\epsilon\Delta\}.$$

We claim that each $v \in S'$ is sparse. To see this, note that $N(v)$ includes at least $\frac{3}{4}\Delta$ vertices in D_i and at least $12\epsilon\Delta$ vertices outside of D_i. By part (c) of Definition 15.1 of a dense decomposition, at most $8\epsilon\Delta^2$ of the $9\epsilon\Delta^2$ potential edges between these two sets are present. Therefore $N(v)$ contains at least $\epsilon\Delta^2$ non-edges.

For each dense set D_i, a *partition respecting colouring* of D_i is an assignment of colours to the partition classes of CP_i, where each class gets a different colour. By the definition of CP_i, this must yield a colouring of D_i. We define a *partition respecting partial colouring* in the same manner.

Phase I: We choose an initial partial colouring.

We do this using a technique similar to that employed to find the random partial colouring of Chap. 16. We give each sparse vertex a uniform colour between 1 and $\Delta + 1$. For each dense set D_i, we essentially choose a random partition respecting colouring with each such colouring equally likely. We uncolour vertices involved in conflicts and also those which contribute to the reject degree of vertices whose reject degree is high. As in Chap. 17, we will ensure that each vertex $v \in S$ has at least $\alpha\Delta$ repeated colours in its neighbourhood, for some constant $\alpha > 0$. In fact, we extend this condition to the vertices of S' as well, which we can do since they are also sparse. This will allow us to complete the colouring of $S \cup S'$ in the fourth phase by applying Lemma 17.8. We will also impose some further conditions which allow us to deal with the other uncoloured vertices in phases II and III.

Phase II: We assign a colour to every uncoloured vertex in each dense set.

We ensure that at the end of Phase II, we have a partition respecting colouring of each dense set D_i such that the maximum reject degree within each dense set is bounded by a constant.

The only problem with the partial colouring produced in this phase is that the interaction between the colouring of a dense set and the rest of the graph may not be perfect. There will be a small number of vertices in dense sets which were coloured during this phase and whose colour is problematic. For example, a vertex $v \in D_i$ might have the same colour as a neighbour outside of D_i or as the edge joining v to such a neighbour. Such vertices (as well as some others coloured in this phase) are said to be *temporarily coloured*, and must be recoloured later. If a vertex has a colour and it is not temporarily coloured, then it is said to be *truly coloured.* If a vertex $v \in D_i$ is temporarily coloured with the colour of an external edge uv, then since v is eventually recoloured, this edge-vertex conflict will eventually be resolved. So we will not need to be concerned with any external reject edges being formed during this phase.

We carry out Phase II as follows: For each D_i, we will choose a completion of the partial colouring obtained in Phase I uniformly from amongst all such completions which yield a partition respecting colouring with bounded internal reject degree. The fact that the number of edges from D_i to the rest of the graph is so small will allow us to ensure that with high probability the number of temporarily coloured vertices in D_i will be small. Applying the Local Lemma we will obtain that with positive probability, the number of temporarily coloured vertices in each D_i is indeed small. This fact, along with the conditions we ensured held in the first phase and a new one that we enforce in this phase, will allow us to recolour the temporarily coloured vertices not in S' in the third phase.

The analysis of Phase II is fairly simple. However, it is not immediately clear how to get a handle on the uniform distribution over the set of partition respecting completions with bounded reject degree. Doing so requires some interesting new techniques.

Phase III: We modify the colourings on each dense set to deal with the conflicts between it and the rest of the graph. In other words, we recolour the temporarily coloured vertices. At the end of this phase, all vertices outside of $S \cup S'$ will have been coloured definitively.

The further the chromatic number of D_i is from Δ, the more room to manoeuvre we will have when modifying the colouring of D_i. Thus, the real difficulty in this stage is in dealing with dense sets D_i for which $\chi(D_i) \approx \Delta$, which we refer to as *ornery sets* (a precise definition follows in Sect 18.2.1). This phase has two steps: in the first, we focus exclusively on the most problematic vertices (the *kernels*) of the ornery sets; in the second, we deal with all the other vertices of the dense sets. The second step is relatively easy – it uses a simple greedy colouring procedure similar to many that we have seen already. The first step is more difficult.

In this first step, we recolour a problematic vertex by randomly choosing a suitable vertex in the same dense set, and switching the colours of the two vertices. Much of the work involves proving that (a) there will be many suitable vertices to switch with, and (b) the number of vertices which need recolouring is small enough that these switches have a limited total effect on the overall colouring. This is very similar to, but more complicated than, our treatment of the final phase in the previous chapter. Readers may prefer to skip over the technical details.

Phase IV We complete our colouring by colouring all the remaining uncoloured vertices in $S \cup S'$ by applying Lemma 17.8.

We choose constants C_1, C_2, C_3, C_4 which are sufficiently large in terms of $1/\epsilon$, and we ensure that during each Phase i, no reject degree will increase by more than C_i. Therefore, at the end of our procedure, no vertex will have reject degree greater than $C_1+C_2+C_3+C_4$. Setting $C = C_1+C_2+C_3+C_4+3$ completes the proof.

Having outlined how each phase proceeds, we turn to the formal details.

18.2 Phase I: An Initial Colouring

In this section, we will precisely define the properties required of the output of Phase I and sketch a proof that we can find a partial colouring with the required properties.

18.2.1 Ornery Sets

In Chap. 16 we discussed the complications involved in extending the proof of Theorem 16.5 to a proof of Theorem 16.4. We mentioned that any dense set D_i with $|CP_i|$ very close to the number of colours being used is particularly troublesome. In this subsection, we discuss these sets more precisely.

We say that a dense set D_i is *ornery* if

$$|CP_i| \geq \Delta + 1 - \log^4 \Delta.$$

While all dense sets are close to being cliques, ornery sets are particularly close. For one thing, most of their vertices have very small outneighbourhoods. More specifically, defining the *kernel*, K_i, of D_i to be the set of vertices with at most $\log^6 \Delta$ neighbours in $G - D_i$, we have:

Lemma 18.1 *For each ornery D_i,*

(a) $|D_i| < \Delta + \log^5 \Delta$,
(b) $|D_i - K_i| < \log^5 \Delta$, *and*
(c) $|E(D_i, G - D_i)| < \Delta \log^7 \Delta$

We defer the proof of this lemma to Sect. 18.6.

Recolouring the ornery dense sets during Phase III will be particularly delicate, and so we must impose some conditions on the output of Phase I to facilitate that part of our procedure. For one thing, if a colour appears on too many edges of $E(D_i, G - D_i)$, we will not use that colour on D_i. Similarly, we must restrict the colours that can appear on vertices which are outside of D_i but which have a reasonably large number of neighbours in D_i. We make the following definitions for each ornery set D_i:

$$\begin{aligned}
\text{Big}_i &= \{v \notin D_i : |N(v) \cap D_i| > \Delta^{7/8}\} \\
\text{Overused}_i &= \text{the set of colours which appear on at least} \\
&\quad \Delta - \tfrac{\Delta}{\log^{10} \Delta} \text{ external edges of } D_i \\
\text{Oftenused}_i &= \text{the set of colours which appear on at least} \\
&\quad \tfrac{\Delta}{\log \Delta} \text{ external edges of } D_i.
\end{aligned}$$

In Sect. 18.6, we will prove the following bounds on the sizes of these sets:

Lemma 18.2 *For any ornery S_i,*

(a) $|CP_i| \leq \Delta + 1 - \text{Overused}_i$;
(b) $|\text{Overused}_i| \leq \log^4 \Delta$;
(c) $|\text{Oftenused}_i| \leq \log^8 \Delta$;
(d) $|\text{Big}_i| \leq \Delta^{1/8} \log^7 \Delta$.

In order to facilitate Phase III, we will insist on the following restrictions when we carry out Phases I and II:

Restriction 1: *No vertex in D_i can receive a colour from* Overused_i.

When we switch colours within D_i during Phase III, we will insist that no new external reject edges are created. Restriction 1 will be very helpful, since without it, colours from Overused_i would be difficult to switch, as they appear on external edges incident to so many vertices.

Note that Lemma 18.2(a) permits this restriction.

Restriction 2: *No two vertices in* Big_i *can receive the same colour, unless they form a partition class in another dense set.*

If we did not impose this restriction, then it is conceivable that by appearing on only 2 vertices of Big_i, each with more than $\frac{\Delta}{2}$ neighbours in D_i, a colour could be, in effect, forbidden from being used on any vertex of D_i. (Note that if 2 vertices of Big_i form a partition class in another dense set, then neither of them has more than $\frac{\Delta}{4}$ neighbours in D_i.) If this happened to several colours, then it could be very difficult to colour all of D_i.

Restriction 3: *No vertex in* Big_i *can be coloured with any colour from* Oftenused_i.

Thus, if a colour is, in effect, forbidden from being switched onto many vertices of D_i because they are incident to external edges with that colour, then it cannot also be forbidden from many vertices in D_i by appearing on just one vertex in Big_i. Without this restriction, it could be very difficult to use such a colour during the switching process in Phase III.

By Lemma 18.2, $\text{Big}_i, \text{Overused}_i$ and Oftenused_i are small. As we shall see, this implies that enforcing these restrictions has a negligible effect on our analysis. (See, for example, the discussion following Lemmas 18.5 and 18.6.)

18.2.2 The Output of Phase I

At the end of Phase I, we will have a partial colouring of G such that each partition class of a dense set is either completely uncoloured or coloured using a single colour. Restrictions 1, 2, 3 will hold, and the reject degree of each vertex will be at most C_1. The following six additional properties will also hold:

The first one will allow us to apply Lemma 17.8 in Phase IV.

(P1.1) For each $v \in S \cup S'$, there are at least $\frac{\epsilon}{4\times 10^7}\Delta$ colours which appear twice in $N(v)$.

In the second step of Phase III, we will colour vertices of D_i with moderate outdegree in a greedy manner. When we come to colour such a vertex v, we will want to do so without creating any external reject edges. So the set of colours which won't be available for v will include the set of colours which appear on $N(v)$ *and* the set of colours which do not appear on $N(v)$ but which appear on external edges from v. (There will also be a few other forbidden colours, but we don't need to go into such details now.) As usual, we will ensure that there will be at least one colour available for v by ensuring that v has several repeated colours in its neighbourhood. The following bound will suffice:

(P1.2) For each $v \in D_i - S'$ such that $|\mathrm{Out}_v| \geq \log^3 \Delta$ and v is not in the kernel of an ornery set, the number of colours that appear in both Out_v and $N(v) \cap D_i$ exceeds the number of colours on external edges of v which do *not* appear on $N(v)$ by at least $5\epsilon|\mathrm{Out}_v|$.

To facilitate the next two phases, we need to ensure that there are many uncoloured vertices in each dense set. This will give us some room to manoeuvre while completing the colouring. We define U_i to be the set of uncoloured vertices in D_i. We shall ensure that the following property holds:

(P1.3) For every D_i, $|U_i| \geq \zeta\Delta$ where $\zeta = \left(\frac{1}{5C_1}\right)^{C_1}$.

As we described in Sect. 18.2, in order to be able to eventually use a colour on some vertex in an ornery set D_i, we can't have that colour appearing in the neighbourhood of too many vertices of D_i.

(P1.4) For each ornery D_i and each colour c, the number of vertices in D_i having an external neighbour outside of Big_i with colour c is at most $\Delta^{31/32}$.

Finally, there are two similar technical conditions which allow us to complete the colouring of the ornery dense sets in Phase III.

(P1.5) For each ornery D_i and partition class w not in D_i there are at least $\frac{2\Delta}{\log \Delta}$ uncoloured non-neighbours of w in D_i.

(P1.6) For each ornery D_i and colour $c \notin \text{Overused}_i$ there are at least $\frac{\Delta}{(\log \Delta)^{11}}$ uncoloured vertices in D_i which are not incident to an external edge of colour c.

18.2.3 A Proof Sketch

The random process which we consider is an amalgamation of the processes used in the previous two chapters. As in Chap. 16, we assign a uniformly random colour to each member of S, and a uniformly random permutation of colours to the partition classes of each CP_i. Here, the colours that we use for CP_i are a uniformly random subset of $\{1, \dots, \Delta + 1\} - \text{Overused}_i$, since Restriction 1 forbids us from using any colours from Overused_i. (For convenience, we define $\text{Overused}_i = \emptyset$ for each non-ornery D_i.)

We uncolour a vertex in S, or a partition class in a dense set, if it receives the same colour as a neighbour. As in Chap. 17, we will also uncolour any vertex in S and any partition class in a dense set which contributes to a neighbour having too high a reject degree. We also uncolour vertices which violate Restrictions 2 and 3. More specifically, the procedure runs as follows:

1. For each vertex $v \in S$, assign to v a uniformly random colour from $\{1, \dots, \Delta + 1\}$.
2. For each dense set D_i, we choose a uniformly random permutation π_i of $\{1, \dots, \Delta + 1\} - \text{Overused}_i$ and for $1 \le j \le |CP_i|$, we assign the colour $\pi_i(j)$ to the vertices of the jth partition class of D_i.
3. For each vertex v, if any neighbour of v is assigned the same colour as v then we uncolour v. If v is in a partition class of size 2, then we also uncolour the other member of that class.
4. For each vertex u, if u has reject degree greater than C_1, then for every neighbour v of u such that v has the same colour as the edge uv, we uncolour v. If v is in a partition class of size 2, then we also uncolour the other member of that class.
5. For each vertex $v \in \text{Big}_i$, we uncolour v if it receives a colour from Oftenused_i or if it receives the same colour as some other $u \in \text{Big}_i$ not in the same partition class as v. If v is in a partition class of size 2, then we also uncolour the other member of that class.

The analysis which allows us to claim that properties (P1.1) through (P1.6) hold is very similar to that given in the last two chapters. The following technical lemmas will be useful. First note that:

18.3 *For any vertex v and colour c, the probability that v is assigned c is less than $\frac{2}{\Delta}$.*

Proof If $v \in S$ then the probability is $\frac{1}{\Delta+1}$. If $v \in D_i$ then the probability is at most $\frac{1}{\Delta+1-|\mathrm{Overused}_i|}$ which is less than $\frac{2}{\Delta}$ by Lemma 18.2. □

Our first technical lemma generalizes (18.3) to larger sets of vertices and colours.

Lemma 18.4 *Given any list of partition classes, $w_1, \ldots, w_t$ and colours $c_1, \ldots, c_t$, the probability that each w_i is assigned c_i is at most $\left(\frac{3}{\Delta}\right)^t$.*

We leave the easy proof to the reader (it can be found in [119]). The next two symmetric lemmas show that the uncolouring caused by high reject degrees or Restrictions 2 and 3 is relatively insignificant. They are very similar to Lemma 17.4.

Lemma 18.5 *For any set X of at most 2Δ partition classes, the probability that more than $\frac{\epsilon}{10^{10}}|X|$ of these parts are uncoloured in Steps 4 and 5 is at most $\mathrm{e}^{-\alpha|X|}$, for an absolute constant $\alpha > 0$.*

Lemma 18.6 *For any dense set D_i and any set Y of colours, the probability that more than $\frac{\epsilon}{10^{10}}|Y|$ of these colours are removed from vertices of D_i in Steps 4 and 5 is at most $\mathrm{e}^{-\alpha|Y|}$, for an absolute constant $\alpha > 0$.*

The proofs are nearly identical to that of Lemma 17.4. The main difference is that because some of our random choices are random permutations, we must use McDiarmid's Inequality. The other difference between these proofs and that of Lemma 17.4 is that here we must account for vertices uncoloured in Step 5. But this effect is negligible since each vertex v can be in Big_j for at most $\Delta^{1/8}$ ornery sets D_j. Therefore, by Lemma 18.2(c,d), the probability of v being uncoloured in Step 5 is at most $O(\Delta^{1/4}\log^7 \Delta/\Delta)$, and so Step 5 only increases the probability of v being uncoloured by $o(1)$.

With these tools in hand, we now turn our attention to bounding the failure probabilities of our properties.

Consider any vertex $v \in S \cup S'$. We will bound the probability that (P1.1) fails for v. It follows in exactly the same manner as the proof of Lemma 16.9 that with probability greater than $1-\Delta^{-10}$, there are at least $\frac{\epsilon}{2\times 10^7}\Delta$ colours which are each assigned to a monocolourable pair of vertices in $N(v)$, and which are not assigned to any neighbours of those two vertices. The only difference between our present situation and that in Lemma 16.9 is that here some of these colours might be removed in Steps 4 and 5. However, Lemma 18.5 implies that with probability at least $1-\Delta^{-10}$, fewer than $\frac{\epsilon}{4\times 10^7}\Delta$ vertices in $N(v)$ will be uncoloured in those steps. Therefore, the probability that (P1.1) fails for v is at most $2\Delta^{-10} < \Delta^{-9}$.

A similar argument shows that the probability that (P1.2) fails for some particular v is at most Δ^{-9}. We defer the somewhat lengthy details to the end of this section.

The remaining four properties are much easier to deal with. We leave the following as an exercise for the reader:

18.7 *For any vertex $v \in D_i$, the probability that, before uncolouring, v has internal reject degree greater than C_1 is at least ζ.*

(ζ is not the optimal constant here. Depending on the proof approach that the reader chooses, she might very well obtain a much higher probability.)

To simplify our proof, we will consider a modified uncolouring rule. We suppose that we only uncolour a vertex if it contributes to a neighbour having high *internal* reject degree. In fact, for each vertex v with internal reject degree greater than C_1, instead of uncolouring every internal neighbour contributing to this high reject degree, we only uncolour the first $C_1 + 1$ such neighbours, under some arbitrary predetermined ordering. Of course, the probability of (P1.3) failing under this modified rule is smaller than the probability under the real uncolouring rule. So this is a valid way to bound the latter probability.

Under this modified uncolouring rule, (18.7) immediately implies that $\mathbf{E}(|U_i|) \geq \zeta \times |D_i| \times (C_1 + 1)$. Furthermore, it is easy to see that exchanging the colours on two partition classes can affect $|U_i|$ by at most $4(C_1 + 1)$. A straightforward application of McDiarmid's Inequality implies that the probability of (P1.3) failing under the modified uncolouring rule is less than Δ^{-9}. Therefore, the probability of it failing under the real uncolouring rule is also less than Δ^{-9}.

A similar argument proves that the probability of (P1.5) or (P1.6) failing for a particular D_i and vertex v or colour c is at most Δ^{-9}.

We turn now to (P1.4). We consider any specific ornery set D_i and colour c, and we let Z denote the number of vertices $v \in D_i$ which have an external neighbour outside of Big_i which is assigned c. By Lemma 18.1(c), the total number of edges from D_i to $G - D_i$ is at most $\Delta \log^7 \Delta$, and each such edge has less than a $\frac{2}{\Delta}$ chance of its endpoint outside of D_i receiving the colour c. Therefore, $\mathbf{E}(Z) \leq \Delta \log^7 \Delta \times \frac{2}{\Delta} \leq 2 \log^7 \Delta$.

For each vertex $v \in S$ which has a neighbour in D_i, we consider the colour assignment to v as a random trial. For each of the dense sets D_j which are joined by an edge to D_i, we consider the entire assignment of colours to D_j as a single random trial. Since we are only counting vertices not in Big_i, and since any one such vertex has fewer than $\Delta^{7/8}$ neighbours in D_i, the outcome of the trial determining the colour assignment to a particular vertex in S can affect Z by at most $\Delta^{7/8}$. Since at most 2 vertices in any D_j can receive c, changing the outcome of the trial determining the entire assignment of some D_j can affect Z by at most $2\Delta^{7/8}$. A straightforward application of Talagrand's Inequality yields that $\mathbf{Pr}(Z > \Delta^{31/32}) \leq e^{-\Omega(\Delta^{3/16}/\log^7 \Delta)} < \Delta^{-9}$. Of course, the probability that D_i, c violate (P1.4) is even smaller, since some vertices assigned colour c might be uncoloured.

Now a straightforward application of the Local Lemma, of the sort that we have seen countless times already in this book, implies that with positive probability, we will successfully complete Phase I.

It only remains to deal with (P1.2). The rest of this section is occupied with doing so. Many readers may wish to skip these details, as they are similar to many proofs found earlier in this book, and move directly to the more interesting material in the next few sections.

Consider some $v \notin S'$ in some dense set D_i. We must bound the probability of (P1.2) failing for v. To do so, we define Y to be the number of colours which appear in both Out_v and $N(v) \cap D_i$ at the end of Step 3, and we define Z to be the number of colours on external edges from v which appear on $N(v) \cap D_i$ at the end of Step 3. If

(i) $Y + Z \geq |\mathrm{Out}_v|(1 + 10\epsilon)$;
(ii) fewer than $\epsilon|\mathrm{Out}_v|$ external neighbours of v are uncoloured during Steps 4 and 5;
(iii) fewer than $\epsilon|\mathrm{Out}_v|$ of the colours assigned to external neighbours of v are removed from vertices in D_i during Steps 4 and 5; and
(iv) fewer than $\epsilon|\mathrm{Out}_v|$ of the colours on external edges from v are removed from vertices in D_i during Steps 4 and 5,

then (P1.2) holds for v. Conditions (ii) and (iv) hold with sufficiently high probability by Lemmas 18.5, 18.6.

The proof that (iii) holds with sufficiently high probability is along the same lines as the proof of Lemma 18.6. We consider exposing our colour assignment in two steps. In the first step, we expose the colour assignments to the external neighbours of v. If C_1 is sufficiently large in terms of $1/\epsilon$ then with high probability, many fewer than $\frac{\epsilon}{2C_1}|\mathrm{Out}_v|$ members of D_i will have reject degree greater than $\frac{C_1}{2}$ after this step; we denote the set of such members by T. Let Ψ be the set of colours assigned to Out_v. In the second step, we expose the rest of the colouring. Virtually the same argument as that used for Lemma 18.6 implies that, with high probability, fewer than $\frac{\epsilon}{2}|\mathrm{Out}_v|$ colours in Ψ will appear on a vertex in D_i which causes a neighbour to have reject degree increase by more than $\frac{C_1}{2}$ during the second step. If a vertex in D_i has a colour from Ψ removed in Step 4, then either (a) it contributed to a reject degree increasing by more than $\frac{C_1}{2}$ in the second step, or (b) it was one of the fewer than $\frac{C_1}{2} \times \frac{\epsilon}{2C_1}|\mathrm{Out}_v|$ vertices which contributed to a vertex from T having its reject degree increase by less than $\frac{C_1}{2}$ in the second step. Therefore, with sufficiently high probability, fewer than $\frac{3\epsilon}{4}|\mathrm{Out}_v|$ vertices of D_i lose a colour from Ψ during Step 4. It is straightforward to show that, with high probability, fewer than $\frac{\epsilon}{4}|\mathrm{Out}_v|$ vertices of D_i lose a colour from Ψ during Step 5. This implies that (iii) holds with sufficiently high probability.

Thus, it will be enough to prove that with high probability condition (i) holds.

We start by showing that with high probability, Y is large. First we show that there are many monocolourable pairs consisting of one vertex in Out_v and another in $N(v) \cap D_i$. By Lemma 15.12(iii) and Definition 15.1(b), CP_i has at least $\Delta+1-13\epsilon\Delta$ partition classes of size 1, and at most $64\epsilon\Delta$ of these partition classes have outdegree greater than $\frac{\Delta}{8}$ since, by Definition 15.1(c), the total outdegree of D_i is at most $8\epsilon\Delta$. Consider any vertex $x \in \mathrm{Out}_v$. Recall that x has fewer than $\frac{3}{4}\Delta$ neighbours in D_i and if x belongs to a partition class of size two, then by Lemma 15.12 that class has fewer than $\frac{1}{3}\Delta$ neighbours in D_i. Therefore, there are at least $\Delta + 1 - 13\epsilon\Delta - 64\epsilon\Delta - \frac{3}{4}\Delta > \frac{\Delta}{5}$ vertices $y \in D_i$ such that (i) y is a singleton partition class, (ii) $|\mathrm{Out}_y| \le \frac{\Delta}{8}$, and x, y are a strongly non-adjacent pair. By (16.20), x, y are a monocolourable pair. Therefore, there are at least $\frac{\Delta}{5}|\mathrm{Out}_v|$ such monocolourable pairs. From this collection, we can choose at least $\frac{\Delta}{20}|\mathrm{Out}_v|$ monocolourable pairs such that no partition class intersects more than one of the pairs. By the same argument as in the proof of Lemma 16.9 (more specifically, the arguments for Lemma 16.11 and (16.17)), with probability greater than $1 - \mathrm{e}^{-\Omega(|\mathrm{Out}_v|)} > 1 - \Delta^{-10}$, at the end of Step 3, at least $\frac{1}{4000\Delta} \times \frac{\Delta}{20}|\mathrm{Out}_v|$ of these pairs are such that (a) both vertices have the same colour, and (b) that colour does not appear on any other partition classes intersecting $N(v)$. This implies that $Y \ge \frac{1}{80{,}000}|\mathrm{Out}_v| \ge 100\epsilon|\mathrm{Out}_v|$.

Next we prove that with high probability, Z is large. We denote by X the set of colours on the external edges from v. Let Z_1 be the number of partition classes in CP_i which are adjacent to v and which are assigned a colour from X in Step 2. Let Z_2 be the number of such classes which are uncoloured in Step 3. Thus, $Z = Z_1 - Z_2$. To simplify the discussion, we first assume that D_i is not ornery.

$|N(v) \cap D_i| = \Delta - |\mathrm{Out}_v| > (1-12\epsilon)\Delta$. Therefore, since at most $5\epsilon\Delta$ partition classes of CP_i have size 2, v is adjacent to at least $(1-17\epsilon)\Delta$ partition classes of CP_i. Thus, $\mathbf{E}(Z_1) \ge (1-17\epsilon)\Delta \times \frac{|X|}{\Delta+1} > |\mathrm{Out}_v| \times (1-18\epsilon)$. A very simple application of McDiarmid's Inequality (or Azuma's Inequality) shows that $\mathbf{Pr}(Z_1 < |\mathrm{Out}_v| \times (1-20\epsilon)) < \Delta^{-10}$.

By (18.3) and the fact that the total number of external edges from D_i is at most $8\epsilon\Delta^2$, the expected number of these edges whose endpoints are both assigned the same colour from X is at most $8\epsilon\Delta^2 \times |X| \times \left(\frac{2}{\Delta}\right)^2 = 32\epsilon|\mathrm{Out}_v|$. Thus, $\mathbf{E}(Z_2) \le 32\epsilon|\mathrm{Out}_v|$. McDiarmid's Inequality implies that $\mathbf{Pr}(Z_2 > 40\epsilon|\mathrm{Out}_v|) < \Delta^{-10}$. To see this, note that changing a colour assignment on S or interchanging two colour assignments in some D_j can affect Z_2 by at most 2. Furthermore, if $Z_2 \ge s$ then it is easy to find a set of s colour assignments which certify this, namely those assignments that cause the s colour classes to be uncoloured in Step 3.

In the case where D_i is ornery, we have to account for the fact that any colour in $X \cap \mathrm{Overused}_i$ cannot be assigned to D_i. However, since $|\mathrm{Overused}_i| < \log^4 \Delta$ and $|\mathrm{Out}_v| > \log^7 \Delta$ (since $v \notin K_i$), this represents a negligible number of colours – we omit the details.

Therefore, $\mathbf{Pr}(Z < (1-60\epsilon)|\text{Out}_v|) < 2\Delta^{-10}$, and so $\mathbf{Pr}(Y+Z < (1+10\epsilon)|\text{Out}_v|) < 3\Delta^{-10}$ which is small enough to show that the probability of (P1.2) failing is at most Δ^{-9}.

18.3 Phase II: Colouring the Dense Sets

We now complete the partition respecting partial colouring of each dense set D_i to a partition respecting colouring of D_i with bounded internal reject degree. To do so, we use Property (P1.3) which says that, setting $\zeta = (\frac{1}{5C_1})^{C_1}$, for each i, the set U_i of uncoloured partition classes in D_i has size at least $\zeta\Delta$. As we see below, this implies that our partial colouring of D_i can be completed to a partition respecting colouring of D_i in which each vertex has internal reject degree at most $\frac{12}{\zeta}$, and such that Restriction 1 continues to hold (i.e. no colour in Overused_i appears on D_i). We actually allow the reject degree to be even higher, choosing for each D_i a uniform element of the set Υ_i of partition respecting completions such that:

(i) the internal reject degrees are at most $C_2 = \frac{100}{\zeta}$, and
(ii) no colour in Overused_i appears on D_i.

Now, the union of the Υ_i need not be a colouring, as the colour of a vertex may conflict with the colour of an external neighbour. Also, some vertices may have high reject degree because of the colours of external neighbours. To deal with such problems, we say that any vertex which is assigned, during this phase, the same colour as an external neighbour or an edge to an external neighbour, is *temporarily coloured.* To ensure that Restrictions 2 and 3 hold on the truly coloured vertices, we also say that a vertex in Big_i assigned a colour which is either in Overused_i or assigned to another vertex of Big_i not in the same partition class, is temporarily coloured. Note then that if in Phase II, v is assigned a colour that an external neighbour u kept in Phase I, then v is temporarily coloured but u is not.

In phases III and IV, we will recolour the temporarily coloured vertices along with some of the other vertices coloured in this phase. In order to ensure we can do so, we need to bound the number of temporarily coloured vertices. We define Temp_i to be the set of uncoloured vertices in D_i, and for each $a \geq 1$, we define $\text{Temp}_i(a)$ to be the vertices in Temp_i with external degrees at most a. We shall ensure that the following property holds:

(P2.1) For all i and a with $\log^3 \Delta \leq a \leq 10^{-6}\zeta^2\Delta$, $|\text{Temp}_i(a)| \leq 300\zeta^{-1}a$.

We never recolour a vertex coloured in Phase I, so (P1.1) and (P1.2) will hold throughout the process. As we will see, it is straightforward to combine these properties with (P2.1) and Lemma 17.8 to show that we can recolour all of the temporarily coloured vertices, except those in ornery dense sets which have low outdegree.

To deal with such temporarily coloured vertices in ornery dense sets, we need to impose one further condition, reminiscent of (P1.4)

(P2.2) For each ornery D_i and each colour c, the number of vertices in D_i having an external neighbour outside of Big_i assigned colour c in this phase is at most $\Delta^{31/32}$.

We complete this section by showing that with positive probability, our choice of colour assignment is such that (P2.1) and (P2.2) hold. To do so, we show that for each D_i, the probability that either of these properties fails is very small. It is then straightforward to apply the Local Lemma to obtain the desired result.

18.3.1 Υ_i is Non-Empty

In this section, we prove the following:

Lemma 18.8 *Every partition respecting partial $(\Delta+1)$-colouring of a dense set D_i which leaves at least $\zeta\Delta$ partition classes uncoloured and for which the maximum (internal) reject degree is at most $\frac{12}{\zeta}$ can be completed to a partition respecting $(\Delta+1)$-colouring of D_i with maximum (internal) reject degree at most $\frac{12}{\zeta}$.*

Proof For any completion of our partial colouring to a colouring of D_i, we let $\mathrm{rej}(v)$ be the (internal) reject degree of v. We consider a partition respecting completion γ of our partial colouring to a $(\Delta+1)$-colouring of D_i which minimizes $\sum_{v\in D_i} \max\left(0, \mathrm{rej}(v) - \frac{12}{\zeta}\right)$. If this sum is 0, we are done. Otherwise, we choose a vertex u of maximum reject degree. Since $\mathrm{rej}(u) > \frac{12}{\zeta}$, there is at least one neighbour of u contributing to the reject degree of u which was uncoloured in the original partial colouring. We let v be such a neighbour, and we let w be the partition class containing v. We let c be the colour assigned to w. We are going to swap the colour of w with that of some other partition class w' which was originally uncoloured. An appropriate choice of w' will yield a new partition respecting completion which contradicts our choice of γ.

Let S_1 be the set of vertices of reject degree exceeding $\frac{11}{\zeta}$. Let S_2 be the set of partition classes joined to S_1 by an edge of colour c. We note that $|S_1| \le \frac{|D_i|\zeta}{11} \le \frac{\zeta\Delta}{10}$. Thus $|S_2| \le \frac{\zeta\Delta}{10}$.

Let X be the set of edges between w and the elements of S_1 and let Y be the set of colours appearing on X. Note that $|Y| \le |X| \le 2|S_1| \le \frac{\zeta\Delta}{5}$.

Thus, there are at least $\frac{7\zeta\Delta}{10}$ partition classes which were uncoloured under the original partial colouring which are neither in S_2 nor coloured using a colour in Y. It is easy to verify that swapping the colour on w with the colour on such a colour class w' does not increase the reject degree of any vertex in S_1, nor does it increase the reject degree of any vertex outside

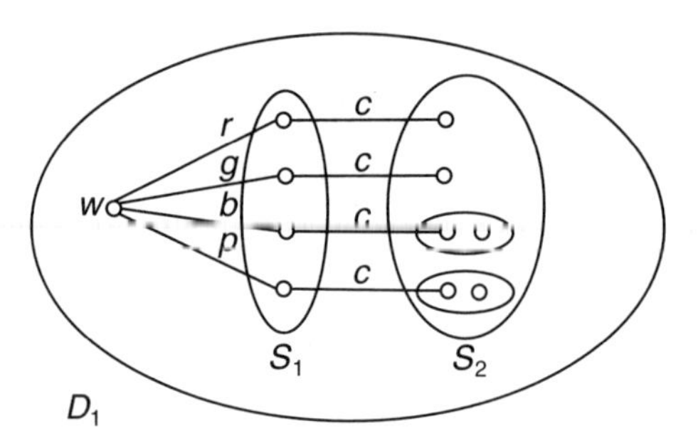

Fig. 18.1. $Y = \{r, g, b, p\}$

of S_1 by more than 2. Furthermore, it decreases the reject degree of u by one. Thus, this new colouring does indeed contradict our choice of γ. □

We note that the same result holds for ornery dense sets via the same proof, even if we impose Restriction 1 on our partial colouring and extension.

18.3.2 Our Distribution is Nearly Uniform

We will show that the probability that a partition class in U_i is assigned any specific colour c when we choose a uniform element of Υ_i is not much greater than the probability it is assigned this colour if we simply choose a uniform partition-respecting extension regardless of the resulting reject degrees.

Lemma 18.9 *For any uncoloured $w \in CP_i$ and $c \in \{1, \dots, \Delta+1\}$, the probability that c is assigned to w is at most $\frac{2}{|U_i|}$.*

Proof Given any assignment in Υ_i under which w gets c, for any other colour c', we say that c and c' can be *switched* if assigning w the colour c' and assigning the (at most one) partition classes assigned c' the colour c, results in another assignment of Υ_i. Following the lines of Lemma 18.8, we will show that c can be switched with at least $\frac{9|U_i|}{10}$ other colours.

To this end, let S_1 be the set of vertices of reject degree at least $\frac{100}{\zeta} - 2$. Let S_2 be the set of partition classes joined to S_1 by an edge of colour c. We note that $|S_1| \leq \frac{|D_i|\zeta}{99} \leq \frac{\zeta\Delta}{88} \leq \frac{|U_i|}{88}$. Thus $|S_2| \leq \frac{|U_i|}{88}$.

Let X be the set of edges between w and the elements of S_1 and let Y be the set of colours appearing on X. Note that $|Y| \leq |X| \leq 2|S_1| \leq \frac{|U_i|}{44}$.

Thus, there are at least $\frac{85|U_i|}{88} > \frac{9|U_i|}{10}$ partition classes of U_i which are neither in S_2 nor coloured using a colour in Y. It is easy to verify that swapping the colour on w with the colour on such a colour class w' does not increase the reject degree of any vertex in S_1. Furthermore, it increases the reject degree of any other vertex by at most 2. Thus, we can indeed switch c with at least $\frac{9|U_i|}{10}$ other colours.

It is easy to see that no colour assignment can be obtained via a single switch from two distinct assignments of Υ_i in which w receives c. Therefore, the number of assignments of Υ_i in which w does not get c is at least $\frac{9}{10}|U_i|$ times the number of assignments in which w receives c. So, the probability that w receives c is at most $\frac{10}{9|U_i|} < \frac{2}{|U_i|}$ as claimed. □

In fact, a nearly identical argument shows that the same bound holds, even when conditioning on a prescribed set of colour classes receiving a prescribed set of colours, as long as that set is not too large. The proof is essentially identical to that of Lemma 18.9 and we omit it.

Lemma 18.10 *Consider any $t \leq \frac{|U_i|}{200}$ and any set of t partition classes $w_1, \ldots, w_t \in CP_i$ not coloured in Phase I, and $c_1, \ldots, c_t \in \{1, \ldots, \Delta+1\}$. For any other originally uncoloured partition class w and colour c, the probability that w receives c, conditional on the event that each w_i receives c_i is at most $\frac{2}{|U_i|}$.*

This immediately yields the following:

Corollary 18.11 *For any $t \leq \frac{|U_i|}{200}$ and any originally uncoloured $w_1, \ldots, w_t \in CP_i$ and $c_1, \ldots, c_t \in \{1, \ldots, \Delta+1\}$, the probability that each w_i receives c_i is at most $(\frac{2}{\zeta\Delta})^t$.*

18.3.3 Completing the Proof

For each ornery dense set D_i and $\log^3 \Delta \leq a \leq 10^{-6}\zeta^2\Delta$, we let $A_{i,a}$ be the event that D_i violates (P2.1) for this value of a, and B_i be the event that D_i violates (P2.2). We show below that the probability of any of these events is at most Δ^{-10}. It is straightforward to verify that each event is mutually independent of all but at most Δ^9 other events. So, a straightforward application of the Local Lemma will imply that, with positive probability, Phase II will be successful.

We actually prove that our bound on the probability of $A_{i,a}$ failing, holds given any choice of assignments on the other dense sets. Summing over all such assignments then yields the desired result.

We consider separately the set S_1 of vertices put into $\mathrm{Temp}_i(a)$ because of Restrictions 2 and 3, the set S_2 of vertices put into $\mathrm{Temp}_i(a)$ because they conflict with an external neighbour, and the set S_3 of vertices put into $\mathrm{Temp}_i(a)$ because they conflict with an edge to an external neighbour. We prove that for $i \in \{1,2,3\}$, $\mathbf{Pr}\left(|S_i| > 100\zeta^{-1}a\right) \leq \frac{\Delta^{-10}}{3}$. The desired result follows.

Having fixed the colour assignment outside of D_i, we consider any vertex $v \in D_i$ with outdegree at most a. Clearly, v lies in Big_j for at most $a/\Delta^{7/8}$ ornery sets D_j, so by Lemma 18.2(c,d), there are fewer than $\frac{2a}{\Delta^{3/4}} \log^7 \Delta$ colours whose assignment to v would cause v to be added to S_1. Furthermore there are fewer than 2Δ vertices in D_i. So, by Corollary 18.11, for

$k \leq \frac{\zeta\Delta}{200}$, the probability that there are at least k vertices in S_1 is at most $\binom{2\Delta}{k}(\frac{2a\log^7\Delta}{\Delta^{3/4}})^k(\frac{2}{\zeta\Delta})^k$. Setting $k = 100\zeta^{-1}a$ yields the desired bound on the probability S_1 is large.

To bound the size of S_2, we actually consider the number s_4 of partition classes of D_i which contain a vertex of S_2. We note that $|S_2| \leq 2s_4$. If $s_4 = k$ then we can choose a set F of k edges from the set $D_i(a) = \{v \in D_i : |\text{Out}_v| \leq a\}$ to $V - D_i$ so that: (i) no partition class of D_i is incident to more than one edge of F, and (ii) for every edge e of F, the endpoint of e in D_i is assigned the colour assigned to the other endpoint of e. Since there are at most $2a\Delta$ edges out of $D_i(a)$, Corollary 18.11 implies that for $k \leq \frac{\zeta\Delta}{200}$, the probability that s_4 is at least k is at most $\binom{2a\Delta}{k}(\frac{2}{\zeta\Delta})^k$. Setting $k = 50\zeta^{-1}a$ yields the desired bound on the probability S_2 is large.

To obtain a bound on the probability that S_3 is large, we proceed similarly. The details are left to the fastidious reader.

Our bound on the probability of B_i falls into two parts. An expected value computation and a concentration result.

18.12 *For any ornery D_i and colour c, the number $n_{i,c}$ of vertices in D_i having an external neighbour outside* Big_i *assigned colour c in this phase has expected value at most* $\log^8\Delta$.

Proof For any vertex $v \in D_i$, the probability that v has an external neighbour assigned colour c in this phase is at most $|\text{Out}_v| \times \frac{2}{\zeta\Delta}$ by Lemma 18.9. Therefore, the expected number of vertices in D_i with such an outneighbour is at most $\sum_{v\in D_i} |\text{Out}_v| \times \frac{2}{\zeta\Delta}$ which, by Lemma 18.1(c), is less than $\log^8\Delta$. □

Now, $n_{i,c}$ is determined by the choice for each D_j of which if any vertex of D_j is assigned c. These choices are independent and each affects the value of $n_{i,c}$ by at most $\Delta^{\frac{7}{8}}$. Furthermore to certify that $n_{i,c}$ is at most k we need only specify at most k such choices. Thus, a straightforward application of Talagrand's Inequality now yields the desired bound on $\mathbf{Pr}(B_i)$.

18.4 Phase III: The Temporary Colours

In this phase, we change the colours of all temporarily coloured vertices other than those in S'. As we said earlier, the ornery sets pose the most difficulties. The first step of this phase is to deal with the kernels of the ornery sets. The second step is to recolour the remaining temporarily coloured vertices, which we will do greedily.

We do not require Restrictions 1, 2, 3 to hold during Phase III or IV. For example, it is permissible to assign a colour from Overused_i to a partition class in D_i.

We warn the reader that, while the idea behind Step 1 is fairly simple, the details of the analysis are somewhat involved. Step 2, on the other hand, will just be a simple analysis of a greedy procedure.

18.4.1 Step 1: The Kernels of the Ornery Sets

For each ornery D_i we will recolour all vertices in $\text{Temp}_i \cap K_i$, as well as a few other vertices of K_i coloured in Phase II, such that the following properties hold.

(Q3.1) We create no new external reject edges.
(Q3.2) No $v \in K_i$ has the same colour as an external neighbour.
(Q3.3) The reject degree of any vertex increases by at most 2.

For each ornery set D_i, and each $v \in \text{Temp}_i \cap K_i$, we will swap the colour of v with that of a vertex in $K_i - \text{Temp}_i$ coloured in Phase II, and so we refer to v as a *swapping vertex*. If $\{u, v\}$ form a partition class of size 2, and if $u \in K_i$, then u will also get the colour that is swapped onto v, and so we refer to $\{u, v\}$ as a *swapping pair*. But if $u \notin K_i$, then its large outdegree may make it difficult to find a colour that is suitable for both v and u. So u will not receive the same colour as v and $\{u, v\}$ is not a swapping pair. Instead, we place u into Temp_i so that we don't have to worry about the possibility of u conflicting with the vertex with which v swaps. This increases the size of Temp_i. However, we will also delete v from Temp_i at the end of this phase since it is being recoloured. So because u has larger reject degree than v,

18.13 *(P2.1) will continue to hold for all i and $ai \leq 10^{-6}\zeta^2\Delta$.*

For any swapping vertex v, we must select the vertex with which v swaps carefully, to avoid creating any conflicts. We define Swappable_v to be the set of vertices $v' \in D_i$ with the following properties:

(a) v' forms a partition class of size 1 in CP_i, which was uncoloured in Phase I and is now truly coloured,
(b) $v' \in K_i$,
(c) no vertices in $\text{Out}_{v'}$, nor any external edges from v' have colour c, and
(d) no vertices in Out_v, nor any external edges from v have the colour currently assigned to v'.

For each swapping pair $w = \{v_1, v_2\}$, we set $\text{Swappable}_w = \text{Swappable}_{v_1} \cap \text{Swappable}_{v_2}$. The reader should note that Swappable_{v_1} and Swappable_{v_2} differ in very few vertices; thus intuitively, treating these two vertices as a pair will not create significant extra difficulties.

Note that swapping the colour of any one swapping vertex/pair w with that of some $w' \in \text{Swappable}_w$ will preserve properties (Q3.1), (Q3.2) and (Q3.3). However, we will perform all swaps, over all dense sets, simultaneously and an infortuitous combination might lead to a violation of (Q3.2) or (Q3.3).

In order to deal with this concern, we will actually choose several possible candidates w' for w to select from, and show that each swapping class will be able to select from amongst its candidates in such a way as to avoid conflicts.

So for any swapping vertex/pair w of colour c, we will select a subset $\text{Candidate}_w \subseteq \text{Swappable}_w$ of size 20, whose elements we refer to as *candidates*. We want each Candidate_w to contain at least one good candidate, where "good" is defined so that, regardless of which candidates the other swapping vertices/parts choose, swapping the colour of w with that of a good candidate will neither create any conflicts, nor increase any internal reject degree which is increased by another swap. It is easy (if tedious) to formulate this condition precisely as follows:

(Q3.4) For each swapping vertex/pair w, there is some $w' \in \text{Candidate}_w$ of colour c' such that

(a) w' is not a candidate for any other swapping vertex/pair;
(b) w' is not an external neighbour of any member of Candidate_x for any swapping vertex/pair x (in some other dense set), which also has colour c;
(c) w' is not an external neighbour of any swapping vertex/pair x, which has a candidate of colour c;
(d) for every swapping vertex/pair x which has an external neighbour in w, no member of Candidate_x has colour c';
(e) w has no external neighbour which is a candidate of some swapping vertex/pair x of colour c'; and
(f) there is no vertex u, swapping vertex/pair w_1 of colour c_1, and candidate w_1' of w_1 with colour c_1', all in the same dense set as w and w', and with either
   (i) the edge (u, w') present and coloured c, and the edge (u, w_1') present and coloured c_1;
   (ii) the edge (u, w') present and coloured c, and an edge from u to w_1 present and coloured c_1';
   (iii) an edge from u to w present and coloured c', and the edge (u, w_1') present and coloured c_1; *or*
   (iv) an edge from u to w present and coloured c', and an edge from u to w_1 present and coloured c_1'.

For each swapping vertex/pair w, we will swap the colour on w with that on the candidate referred to in (Q3.4). (Q3.4) obviously enforces properties (Q3.1) and (Q3.2). It also enforces (Q3.3) since by (Q3.4(f)) no reject degree is affected by more than one swap, and each swap involves only 2 colours. It only remains to be shown that we can in fact construct the sets Candidate_w satisfying (Q3.4).

Lemma 18.14 *We can choose the sets* Candidate_w *for all swappable vertices/pairs w, such that (Q3.4) holds.*

To choose Candidate_w, we will simply select 20 members of Swappable_w at random. We will use the Local Lemma to prove that (Q3.4) holds for every w with positive probability. The first step is to show that Swappable_w is large:

Lemma 18.15 *For each swapping vertex/pair w, $|\text{Swappable}_w| \geq \frac{\Delta}{3\log^{11}\Delta}$.*

Proof We denote the colour of w by c.

Claim 1: There are at least $\frac{\Delta}{2\log^{11}\Delta}$ vertices in K_i which were truly coloured in Phase II and are neither incident to any external edge of colour c, nor adjacent to any vertex of $G - D_i$ of colour c.

Proof:

Case 1 $c \in \text{Oftenused}_i$.

By Restriction 3, no vertex of Big_i is truly coloured c. Further, by Restriction 1, $c \notin \text{Overused}_i$. Thus, by Properties (P1.6), (P1.4), and P(2.2), there are at least $\frac{\Delta}{\log^{11}\Delta} - 2\Delta^{\frac{31}{32}}$ vertices of D_i which were not coloured in Phase I and are neither adjacent to an external neighbour truly coloured c, nor joined to an external neighbour by an edge of colour c. By Lemma 18.1(b) all but $\log^5\Delta$ of these vertices are in K_i. Furthermore, since each vertex in K_i has external degree at most $\log^6\Delta$, (P2.1) implies all but fewer than $\log^7\Delta$ of these kernel vertices are truly coloured in Phase II. Thus we have at least $\frac{\Delta}{\log^{11}\Delta} - 2\Delta^{\frac{31}{32}} - \log^5\Delta - \log^7\Delta > \frac{\Delta}{2\log^{11}\Delta}$ vertices which were truly coloured in Phase II and are neither incident to any edge of $E(D_i, G - D_i)$ of colour c nor adjacent to any vertex of $G - v$ of colour c.

Case 2 $c \notin \text{Oftenused}_i$.

At most one partition class within Big_i has colour c, and by (P1.5), there are at least $2\Delta/\log\Delta$ uncoloured vertices in D_i which are not adjacent to that class. Since $c \notin \text{Oftenused}_i$, at most $\Delta/\log\Delta$ of these vertices have an external edge of colour c. We can proceed as in Case 1 to show that at least $\frac{2\Delta}{\log\Delta} - \frac{\Delta}{\log\Delta} - 2\Delta^{31/32} - \log^5\Delta - \log^7\Delta > \frac{\Delta}{2\log^{11}\Delta}$ of these vertices are as required by Claim 1. □

Since D_i is ornery, $|\mathcal{C}_i| \geq \Delta - \log^4\Delta$, and by Lemma 18.1(a), $|D_i| \leq \Delta + \log^5\Delta$. Thus, the number of vertices in D_i which do not lie in singleton partition classes is at most $3\log^5\Delta$. If w is a swapping vertex, then since $w \in K_i$, $|\text{Out}_w| \leq \log^6\Delta$ and so at most $4\log^6\Delta$ vertices of D_i have a colour appearing on an external neighbour of w or an external edge from w. If w is a swapping pair, then both vertices in w are in K_i and so at most $8\log^6\Delta$ vertices of D_i have such a colour. Also, again by Property (P2.1), $|\text{Temp}_i \cap K_i| \leq \log^7\Delta$.

Therefore, by Claim 1, $|\text{Swappable}_w| \geq \frac{\Delta}{2\log^{11}\Delta} - 3\log^5\Delta - 8\log^6\Delta - \log^7\Delta \geq \frac{\Delta}{3\log^{11}\Delta}$. □

With Lemma 18.15 in hand, we can proceed with our proof of the main lemma of this subsection:

Proof of Lemma 18.14 For each swapping vertex/pair w, we define $A2.4(w)$ to be the event that (Q3.4) fails to hold for w. Letting G' be the graph obtained by adding edges to G to make each dense set a clique, it is easy to see that that for any w, $A2.4(w)$ is mutually independent of the set $\{A2.4(w_1) : \text{dist}_{G'}(w, w_1) \geq 7\}$. Since G' has maximum degree less than 2Δ, it suffices to show that $\mathbf{Pr}(A2.4(w))$ is less than Δ^{-8}.

To this end, consider any swapping vertex/pair w of colour c. We will consider the number of members of Swappable_w which would make bad candidates, i.e. those that would not meet the criteria of condition (Q3.4). They fall into the following six (random) subsets of Swappable_w.

Bad_1 $=\{w'|w'$ is a candidate for another swapping vertex/pair $x\}$
Bad_2 $=\{w'|w'$ is an external neighbour of a candidate for another swapping vertex/pair x, of colour $c\}$
Bad_3 $=\{w'|w'$ is an external neighbour of another swapping vertex/pair x, which has a candidate of colour $c\}$
Bad_4 $=\{w'|w'$ has the same colour as a candidate for another swapping vertex/pair x, which has an external neighbour in $w\}$
Bad_5 $=\{w'|w'$ has the same colour as a swapping vertex/pair x, which has a candidate with an external neighbour in $w\}$
Bad_6 $=\{w'|$ there is some vertex u, and another swapping vertex/pair w_1 of colour c_1 and with candidate w_1', violating condition (Q3.4)(f)$\}$

We set $\text{Bad} = \cup_{i=1}^{6}\text{Bad}_i$.

Our proof will have two steps. We first randomly select the candidates of all swapping vertices/pairs other than w, and show that with high probability, $|\text{Bad}| \leq \log^{20}\Delta$. Next, we randomly select the candidates for w, showing that if $|\text{Bad}| \leq \log^{20}\Delta$ then the probability of choosing no good candidates is small.

The second of these steps is easy. By Lemma 18.14, we are choosing 20 candidates from at least $\frac{\Delta}{\log^{11}\Delta}$ vertices. So if $|\text{Bad}| < \log^{20}\Delta$, then the probability of choosing only bad candidates is less than

$$\left(\frac{\log^{20}\Delta}{\frac{\Delta}{3\log^{11}\Delta}}\right)^{20} < \Delta^{-10}.$$

So it only remains to bound the size of Bad.

Since there are at most $\log^7\Delta$ swapping vertices/pairs in D_i, each having 20 candidates, we have $|\text{Bad}_1| \leq 20\log^7\Delta$.

Since w has at most $2\log^6\Delta$ external neighbours, each having at most 20 candidates, $|\text{Bad}_4| \leq 40\log^6\Delta$.

Since D_i has at most $\log^7\Delta$ swapping vertices/pairs, each having at most 20 candidates, $|\text{Bad}_6| \leq 9 \times 20\log^7\Delta$.

To bound the size of Bad_2, we recall that by Lemma 18.1(c), $\sum_{v\in D_i}|\text{Out}_v| \leq \Delta\log^7\Delta$. Lemma 18.15 implies that the probability that a particular vertex is a candidate for a swapping vertex/pair of colour c is at most $\frac{60\log^{11}\Delta}{\Delta}$. Thus, the expected number of vertices in D_i having external neighbours which are candidates for such a vertex/pair is at most $\sum_{v\in D_i}|\mathrm{Out}_v| \times \frac{60\log^{11}\Delta}{\Delta} \leq 60\log^{18}\Delta$. Now, the exact value of Bad_2 is determined by the independent choices of the candidates for all the swappable pairs/vertices in G. Furthermore, since each such candidate lies in the kernel of another ornery set, it has at most $\log^6\Delta$ neighbours in D_i. Therefore, changing one of these choices can affect the value of $|\mathrm{Bad}_2|$ by at most $\log^6\Delta$. If $|\mathrm{Bad}_2| \geq s$ then there is a set of at most s choices which certify this, namely the choices of an external neighbour of each member of Bad_2 to be a candidate for a swapping vertex/pair of colour c. So, applying Talagrand's Inequality yields: $\mathbf{Pr}(|\mathrm{Bad}_2| > \log^{19}\Delta) \leq \Delta^{-10}$.

To bound the size of Bad_3, we once again use the fact that $\sum_{v\in D_i}|\mathrm{Out}_v| \leq \Delta\log^7\Delta$. Lemma 18.15 implies that the probability that a swapping pair/vertex chooses a candidate of colour c is at most $\frac{60\log^{11}\Delta}{\Delta}$. Thus, the expected number of external neighbours of D_i which choose such candidates is at most $60\log^{18}\Delta$. The rest of the analysis is the same as that for Bad_2, showing that $\mathbf{Pr}(|\mathrm{Bad}_3| > \log^{19}\Delta) < \Delta^{-10}$.

To bound the size of Bad_5 we use the fact that w has at most $2\log^6\Delta$ external neighbours. By property (P2.1) no dense set has more than $300\zeta^{-1}\log^6\Delta$ swapping pairs/vertices, and so each neighbour is a candidate for an expected number of at most $300\zeta^{-1}\log^6\Delta \times \frac{3\log^{11}\Delta}{\Delta} < \frac{\log^{18}\Delta}{\Delta}$ swapping pairs/vertices. Therefore, the expected size of Bad_5 is $o(1)$ and the probability that it is larger than $\log\Delta$ is less than Δ^{-10}, by a similar analysis to that in the preceding paragraphs.

Therefore, the probability that $|\mathrm{Bad}| > \log^{20}\Delta$ is less than $3\Delta^{-10}$ and so $\mathbf{Pr}(A2.4(w)) < 4\Delta^{-10}$. □

18.4.2 Step 2: The Remaining Temporary Colours

The remaining temporarily coloured vertices are relatively straightforward to deal with. We wish to colour all such vertices such that (i) we create no new external reject edges, (ii) there are no adjacent $v \in D_i$, $u \in D_j$, $j < i$, with the same colour, and (iii) the reject degree of any vertex increases by at most $C_3 - 2$. Thus, by Property (Q3.3), no vertex has its reject degree increase by more than C_3 during Phase III.

We deal with the sets D_i in sequence. Within each D_i, we colour the vertices one at a time.

For a given D_i, we first recolour all the temporary vertices with external degree at most $\log^3\Delta$. If there are any such vertices, then D_i is not ornery, for otherwise they would lie in the kernel K_i, and would have been recoloured in Step 1. Furthermore, by (18.13), $|\mathrm{Temp}_i(\log^3\Delta)| \leq 300\zeta^{-1}\log^3\Delta$.

Because D_i is not ornery, it has fewer than $\Delta + 1 - \log^4 \Delta$ partition classes, and each partition class has at most one colour. Thus there are at least $\log^4 \Delta$ colours not already used on the vertices of D_i, and we will use these new colours to recolour $\text{Temp}_i(\log^3 \Delta)$ in a greedy manner.

We will not allow a vertex to receive the same colour as a neighbour, an external edge, or an edge joining it to a neighbour whose reject degree has already increased by at least $C_3 - 3$ during this step. When we come to v, it has at most

(a) $300\zeta^{-1} \log^3 \Delta$ new colours forbidden because they have already been used in D_i,
(b) $2 \log^3 \Delta$ new colours forbidden because they appear on an external neighbour, or an external edge of v, and
(c) $300\zeta^{-1} \log^3 \Delta/(C_3 - 3)$ new colours forbidden because of vertices in D_i whose reject degrees have already been increased by $C_3 - 3$ during Step 2.

Since we have $\log^4 \Delta$ new colours to choose from, we will always be successful.

We then recolour the rest of Temp_i in non-decreasing order of external degree. When we come to colour a vertex v with external degree a, there will, by Property (P1.2), be at least $5\epsilon a$ colours available which do not appear on any neighbours of v, or any external edges of v. By (18.13), $|\text{Temp}_i(a)| \leq 300\zeta^{-1}a$ for each $a \leq 10^{-6}\zeta^2\Delta$, and for each $a > 10^{-6}\zeta^2\Delta$, $|\text{Temp}_i(a)| \leq |D_i| < 2\Delta < 2 \times 10^6\zeta^{-2}a$. Thus, there are at most $2 \times 10^6\zeta^{-2}a/(C_3 - 3)$ colours forbidden because of vertices in D_i whose reject degrees have already been increased by $C_3 - 3$ during Step 2. For C_3 sufficiently large in terms of $1/\epsilon$ (and hence, in terms of $1/\zeta$), this is less than $5\epsilon a$, so we will always be successful in finding a suitable colour to assign.

18.5 Phase IV – Finishing the Sparse Vertices

At this point, we have a partial proper colouring of G, such that:

(a) every vertex in $G - (S \cup S')$ is coloured,
(b) the reject degree of each vertex is at most $C_1 + C_2 + C_3$, and
(c) for each uncoloured v, $N(v)$ has at least $\frac{\epsilon}{4\times 10^7}\Delta$ colours which appear truly twice (i.e. P(1.2) holds).

By Lemma 17.8, we can now complete the colouring of G without increasing any reject degree by more than C_4, where $C_4 = C(\epsilon/4 \times 10^7)$ from Lemma 17.8. This yields a vertex colouring of G such that every reject degree is at most $C - 3$, and so we can recolour the reject edges with $C - 1$ new colours, thus obtaining our desired total colouring.

18.6 The Ornery Set Lemmas

We close this chapter with a few of the promised details regarding ornery sets.

Proof of Lemma 18.1:

In the proof of Lemma 15.12, we considered the clique C consisting of the singleton colour classes of CP_i and the matching M of $\overline{D_i}$ formed by the remaining two-vertex partition classes of CP_i. We actually proved that for any D_i satisfying $|CP_i| \geq \Delta - d$, we have $|C| + \frac{3}{2}|M| \leq \Delta + 1$ by considering the number of edges between C and $V(M)$. In particular, this inequality holds for the ornery dense sets. Since $|C| + |M| = |CP_i| \geq \Delta + 1 - \log^4 \Delta$, this implies $|M| \leq 2\log^4 \Delta$ and so $|D_i| = |C| + 2|M| \leq \Delta + 1 + \log^4 \Delta$. This proves (a). Furthermore, these inequalities imply $|C| \geq \Delta - 3\log^4 \Delta$. So, every vertex in C has at most $3\log^4 \Delta + 1$ external neighbours and so is in the kernel. Thus $|D_i - K_i| \leq 2|M| \leq 4\log^4 \Delta$ which proves (b). Finally, we see that $|E(D_i, V - D_i)| \leq (3\log^4 \Delta + 1)|C| + 2|M|\Delta$ which, by our bound on $|M|$, is at most $O(\Delta \log^4 \Delta) < \Delta \log^7 \Delta$, thus proving (c).

Proof of Lemma 18.2

Part (b) is a simple corollary of part (a), since D_i is ornery and so $|CP_i| \geq \Delta + 1 - \log^4 \Delta$.

Parts (c) and (d) follow immediately from the fact that each ornery S_i has at most $\Delta \log^7 \Delta$ external edges, by Lemma 18.1(c).

So the main work is to prove part (a):

In Lemma 15.12, we actually showed that for $|M| \geq \Delta - d$, and thus for ornery sets, the number of C to M edges is at least $\frac{3}{2}|C||M|$. This implies that the number of edges out of D_i from C is at most

$$|C|(\Delta - (|C| - 1)) - \frac{3}{2}|C||M| = |C|(\Delta - |C| - |M| + 1) - \frac{1}{2}|C||M|.$$

It also implies that the number of edges out of D_i from $V(M)$ is at most

$$2|M|\Delta - \frac{3}{2}|C||M| = \frac{1}{2}|M||C| + 2|M|(\Delta - |C|).$$

By our bounds on the size of C and M above this is at most

$$\frac{1}{2}|M||C| + 12\log^8 \Delta.$$

So, the total number of edges between D_i and $V - D_i$ is at most

$$\begin{aligned} & |C|(\Delta - |C| - |M| + 1) + 12\log^8 \Delta \\ = \; & |C|(\Delta + 1 - |CP_i|) + 12\log^8 \Delta \\ < \; & \left(\Delta - \frac{\Delta}{\log^{10} \Delta}\right)(\Delta + 1 - |CP_i|) \end{aligned}$$

$$\begin{aligned}
&+\left(\frac{\Delta}{\log^{10}\Delta}+1\right)(\Delta+1-|CP_i|)+12\log^8\Delta \\
\leq\ &\left(\Delta-\frac{\Delta}{\log^{10}\Delta}\right)(\Delta+1-|CP_i|)+\frac{\Delta}{\log^5\Delta}+12\log^8\Delta \\
<\ &\left(\Delta-\frac{\Delta}{\log^{10}\Delta}\right)(\Delta+2-|CP_i|)\,.
\end{aligned}$$

So, $|\text{Overused}_i| < \Delta + 2 - |CP_i|$, which is the desired result. □

Part VII

Sharpening our Tools

In the next two chapters we take a closer look at two of the most important probabilistic tools we have been using: The Local Lemma and Talagrand's Inequality.

19. Generalizations of the Local Lemma

As we have seen, the Local Lemma allows us to use a local analysis to obtain a global result. From this perspective, a drawback of the version presented in Chap. 4 is that it requires *global* bounds p and d. These global bounds can make it difficult to apply the Local Lemma if, for example, the probabilities of the bad events vary widely. In this chapter we will discuss a few useful generalizations of the Local Lemma, each of which incorporates varying probabilities of the bad events. The simplest of these is:

The Asymmetric Local Lemma *Consider a set $\mathcal{E} = \{A_1, \ldots, A_n\}$ of (typically bad) events such that each A_i is mutually independent of $\mathcal{E} - (\mathcal{D}_i \cup A_i)$, for some $\mathcal{D}_i \subseteq \mathcal{E}$. If for each $1 \leq i \leq n$*

(a) $\mathbf{Pr}(A_i) \leq \frac{1}{4}$, *and*
(b) $\sum_{A_j \in \mathcal{D}_i} \mathbf{Pr}(A_j) \leq \frac{1}{4}$,

then with positive probability, none of the events in $\mathcal{E}$ occur.

Note that the Local Lemma in its simplest form as presented in Chap. 4 is clearly a special case of the Asymmetric Local Lemma.

In our first generalization, we allow for some variance in the sets $\mathcal{D}_i$ – they can contain a few high probability events, or many low probability events, so long as the sum of the probabilities is at most $\frac{1}{4}$. In our next generalization, we allow even more variance. Essentially, we allow the sum of the probabilities over $\mathcal{D}_i$ to vary as a function of $\mathbf{Pr}(A_i)$. The tradeoff is that as $\mathbf{Pr}(A_i)$ drops exponentially, the sum can increase linearly. It is not quite as clean as this, since our sum actually involves terms which are larger than the probabilities of events in $\mathcal{D}_i$. Nevertheless, this version can be quite useful.

The Weighted Local Lemma *Consider a set $\mathcal{E} = \{A_1, \ldots, A_n\}$ of (typically bad) events such that each A_i is mutually independent of $\mathcal{E} - (\mathcal{D}_i \cup A_i)$, for some $\mathcal{D}_i \subseteq \mathcal{E}$. If we have integers $t_1, \ldots, t_n \geq 1$ and a real $0 \leq p \leq \frac{1}{4}$ such that for each $1 \leq i \leq n$*

(a) $\mathbf{Pr}(A_i) \leq p^{t_i}$; *and*
(b) $\sum_{A_j \in \mathcal{D}_i} (2p)^{t_j} \leq \frac{t_i}{2}$

then with positive probability, none of the events in $\mathcal{E}$ occur.

Roughly speaking, the Weighted Local Lemma works well when each event A_i has a "size", corresponding to t_i, such that $\mathbf{Pr}(A_i)$ is exponentially small in its size, and the number of events with which A_i is dependent is linear in its size. For example, in the next section, we see an application to hypergraph colouring where t_i corresponds to the size of the edge represented by A_i.

Remarks

1. Replacing (b) by $\sum_{A_j \in \mathcal{D}_i} p^{t_j} \leq \frac{1}{4}$ yields an immediate corollary of the Asymmetric Local Lemma. Replacing (b) by $\sum_{A_j \in \mathcal{D}_i} p^{t_j} \leq \frac{t_i}{2}$, yields an untrue statement (see Exercise 19.1). In order to permit the RHS of (b) to grow with t_i we need to increase the terms in that sum from p^{t_j} to $(2p)^{t_j}$.
2. As the reader will see upon reading the proof of the Weighted Local Lemma, the constant terms in the statement can be adjusted somewhat if needed.

Our third generalization is the most powerful (as well as the most unwieldy):

The General Local Lemma *Consider a set $\mathcal{E} = \{A_1, \ldots, A_n\}$ of (typically bad) events such that each A_i is mutually independent of $\mathcal{E} - (\mathcal{D}_i \cup A_i)$, for some $\mathcal{D}_i \subseteq \mathcal{E}$. If we have reals $x_1, \ldots, x_n \in [0, 1)$ such that for each $1 \leq i \leq n$*

$$\mathbf{Pr}(A_i) \leq x_i \prod_{A_j \in \mathcal{D}_i} (1 - x_j)$$

then the probability that none of the events in $\mathcal{E}$ occur is at least $\prod_{i=1}^{n}(1 - x_i) > 0$.

In the next few sections, we give some applications of these generalizations. Following that, we will present a proof of the General Local Lemma and show that it implies all of the other versions. Finally, we will discuss one last generalization: The Lopsided Local Lemma, which will be used in Chap. 23.

19.1 Non-Uniform Hypergraph Colouring

When we introduced the Local Lemma in Chap. 4, our first application was to a problem concerning 2-colouring hypergraphs. In this section, we present some applications of our new versions of the Local Lemma to similar problems. These examples provide a good opportunity to compare the various versions.

Theorem 19.1 *If $\mathcal{H}$ is a hypergraph with minimum edge size at least 3, such that each edge of $\mathcal{H}$ meets at most a_i other edges of size i, where*

$$\sum a_i 2^{-i} \le \frac{1}{8},$$

then $\mathcal{H}$ is 2-colourable.

Proof The proof is nearly identical to that of Theorem 4.2. We assign a uniformly random colour to each vertex (where, of course, these choices are made independently), and for each edge e, we define A_e to be the event that e is monochromatic.

If e has size i, then $\mathbf{Pr}(A_e) = 2^{-(i-1)}$. Furthermore, setting $\mathcal{D}_e = \{A_f : f \cap e \neq \emptyset\}$, we have that each A_e is mutually independent of all other events not in $\mathcal{D}_e$. Therefore, the result follows from the Asymmetric Local Lemma. □

Remark It is worth noting that in order to apply the simplest form of the Local Lemma, we would have required the much more restrictive condition that $\sum a_i 2^{-k} \le \frac{1}{8}$, where k is the smallest edge size in the graph.

Our next application will be to colouring hypergraphs for which we have a bound on the number of edges that each *vertex* lies in, rather than on the number of edges that each *edge* meets.

Theorem 19.2 *If $\mathcal{H}$ is a hypergraph with minimum edge size at least 3, such that each vertex of $\mathcal{H}$ lies in at most Δ_i edges of size i, where*

$$\sum_i \Delta_i 2^{-i/2} \le \frac{1}{6\sqrt{2}},$$

then $\mathcal{H}$ is 2-colourable.

Proof Again, we assign a uniformly random colour to each vertex (where, of course, these choices are made independently), and for each edge e, we define A_e and $\mathcal{D}_e$ as in the previous proof.

We will apply the Weighted Local Lemma. The obvious choice for the "size" of A_e would be the number of vertices in e. Clearly, this choice yields that $\mathbf{Pr}(A_e)$ is exponentially small in its size, and the number of other events on which A_e is dependent is linear in its size. Thus, the Weighted Local Lemma seems promising.

Since $\mathbf{Pr}(A_e) = \left(\frac{1}{2}\right)^{|e|-1}$, the most natural thing to do would be to set $t_e = |e| - 1$, and $p = 1/2$. Of course, this creates problems with the "$2p$" terms in condition (b). To avoid these problems, we scale the size a little, setting $t_e = \frac{1}{2}(|e| - 1)$ and $p = \frac{1}{4}$. We still have that for each hyperedge e, $\mathbf{Pr}(A_e) = p^{t_e}$ and furthermore, we now have:

$$\sum_{A_f \in \mathcal{D}_e} (2p)^{t_f} \le \sum_j |e| \Delta_j \left(\frac{1}{2}\right)^{\frac{1}{2}(j-1)}$$

$$= \sqrt{2}|e| \times \sum_j \Delta_j 2^{-j/2}$$
$$\leq \frac{|e|}{6}$$
$$\leq \frac{t_e}{2},$$

where the last inequality uses the fact that $|e| \geq 3$. Therefore our result follows from the Weighted Local Lemma . □

Remark It is worth noting that if we only have a bound on the number of edges that each vertex lies in, and if the maximum edge size of the hypergraph is not bounded by a constant, then we are unable to bound the number of edges of size i that each *edge* intersects, and this makes the Asymmetric Local Lemma difficult to apply.

19.2 More Frugal Colouring

Recall from Chap. 9 that a proper vertex-colouring of a graph is said to be *β-frugal*, if for each vertex v and colour c, the number of times that c appears in the neighbourhood of v is at most β.

Consider any constant $\beta \geq 1$. Alon (see [80]) has shown that for each Δ, there exist graphs with maximum degree Δ for which the number of colours required for a β-frugal colouring is at least of order $\Delta^{1+\frac{1}{\beta}}$. We prove here that this is best possible as shown by Hind, Molloy and Reed [80].

Theorem 19.3 *If G has maximum degree $\Delta \geq \beta^\beta$ then G has a β-frugal proper vertex colouring using at most $16\Delta^{1+\frac{1}{\beta}}$ colours.*

Proof For $\beta = 1$ this is easy. We are simply trying to find a proper vertex colouring of the square of G, i.e. the graph obtained from G by adding an edge between any two vertices of distance 2 in G. It is straightforward to show that this graph has maximum degree at most Δ^2 and so it can be properly $(\Delta^2 + 1)$-coloured.

For $\beta \geq 2$, we use the Asymmetric Local Lemma. Set $C = 16\Delta^{1+\frac{1}{\beta}}$. We assign to each vertex of G a uniformly random colour from $\{1, \ldots, C\}$. For each edge uv we define the Type A event $A_{u,v}$ to be the event that u, v both receive the same colour. For each $\{u_1, \ldots, u_{\beta+1}\}$ all in the neighbourhood of one vertex, we define the Type B-event $B_{u_1,\ldots,u_{\beta+1}}$ to be the event that $u_1, \ldots, u_{\beta+1}$ all receive the same colour. Note that if none of these events hold, then our random colouring is β-frugal.

The probability of any Type A event is at most $1/C$, and the probability of any Type B event is at most $1/C^\beta$. By the Mutual Independence Principle, each event is mutually independent of all events with which it does not have any common vertices, which is all but at most $(\beta+1)\Delta$ Type A events and $(\beta+1)\Delta\binom{\Delta}{\beta}$ Type B events. Now,

$$\begin{aligned}(\beta+1)\Delta \times \frac{1}{C} + (\beta+1)\Delta\binom{\Delta}{\beta} \times \frac{1}{C^\beta} &< \frac{(\beta+1)\Delta}{C} + \frac{(\beta+1)\Delta^{\beta+1}}{\beta! C^\beta} \\ &= \frac{\beta+1}{16\Delta^{\frac{1}{\beta}}} + \frac{\beta+1}{\beta! 16^\beta} \\ &< \frac{1}{4}\end{aligned}$$

for $\Delta \geq \beta^\beta$. The proof now follows from the Asymmetric Local Lemma. □

It is instructive to note here that if we had tried to use the Local Lemma in its simplest form, we would have had to take $p = 1/C$ and $d = (\beta+1)\Delta\binom{\Delta}{\beta}$. Thus pd would have been much bigger than 1 for large Δ and so that form of the Local Lemma would not have applied.

19.2.1 Acyclic Edge Colouring

A proper edge colouring of a graph is said to be *acyclic* if the union of any two colour classes is a forest. This concept was introduced by Grünbaum [72]; see [85] for a more thorough discussion of acyclic colourings. The following result was proven by Alon, McDiarmid and Reed [8].

Theorem 19.4 *If G has maximum degree Δ then G has an acyclic proper edge colouring using at most 9Δ colours.*

The constant 9 can be easily improved. In fact, Alon, Sudakov and Zaks [9] have conjectured that "9Δ" can be replaced by "$\Delta + 2$". Reed and Sudakov in [135] prove that it can be replaced by $\Delta + o(\Delta)$.

Proof Set $C = 9\Delta$. We assign to each edge of G a uniformly random colour from $\{1, \ldots, C\}$. For each pair of incident edges e, f, we define the Type 1 event $A_{e,f}$ to be the event that e, f both receive the same colour. For each $2k$-cycle C, we define the Type k event A_C to be the event that the edges of C are properly 2-coloured. The probability of each Type 1 event is $1/C$ and the probability of each Type k event, $k \geq 2$, is less than $1/C^{2(k-1)}$. In order to apply any version of the Local Lemma, we must bound the number of events that each event is dependent on.

Claim: For each $k \geq 2$, no edge lies in more than Δ^{2k-2} $2k$-cycles.

Proof Consider any edge u, v. There are at most Δ^{2k-2} paths of the form $u, x_1, x_2, \ldots, x_{2k-2}$. Therefore, there are at most Δ^{2k-2} cycles of the form $u, x_1, x_2, \ldots, x_{2k-2}, v, u$.

By the Mutual Independence Principle, each event is mutually independent of the set of events with which it does not share any common edges. For any $k \geq 1$, any Type k event involves $2k$ edges, and each edge belongs to at most 2Δ Type 1 events and, by our Claim, at most $\Delta^{2(\ell-1)}$ Type ℓ events for $\ell \geq 2$. Therefore, each Type k event is mutually independent of all but at most $4k\Delta$ Type 1 events and $2k\Delta^{2(\ell-1)}$ Type ℓ events, $\ell \geq 2$.

We set $p = 1/C$. For each Type 1 event E, we set $t_E = 1$, and for each Type k event E, $k \geq 2$, we set $t_E = 2(k-1)$. As we shall see, this satisfies the conditions of the Weighted Local Lemma.

(a) If E is Type 1 then $\mathbf{Pr}(E) = 1/C = 1/C^{t_E}$. If E is Type k, for $k \geq 2$, then $\mathbf{Pr}(E) = 1/C^{2(k-1)} = p^{t_E}$.

(b) If E is Type k for any $k \geq 1$ then

$$\begin{aligned}\sum_{A_j \in \mathcal{D}_i} (2p)^{t_i} &\leq 4k\Delta \times \frac{2}{C} + \sum_{\ell \geq 2} \left(2k\Delta^{2(\ell-1)}\right) \times \left(\frac{2}{C}\right)^{2\ell-2} \\ &= \frac{8k}{9} + 2k \times \sum_{\ell \geq 2} \left(\frac{2}{9}\right)^{2\ell-2} \\ &= \frac{8}{9}k + \frac{8}{77}k < k \leq t_E.\end{aligned}$$

Therefore, the proof follows from the Weighted Local Lemma. □

Remark Note the similarities between this proof and the proof of Theorem 19.2. In both cases, there was no upper bound on the number of events of any one type that another event might interact with, which is why we needed to use the Weighted Local Lemma.

19.3 Proofs

Proof of the General Local Lemma. Consider any set S of our events, along with any event $A_i \notin S$. We will prove by induction on the size of S that

19.5 $\mathbf{Pr}(A_i | \cap_{A_j \in S} \overline{A_j}) \leq x_i$.

Upon proving (19.5), the General Local Lemma follows immediately as

$$\begin{aligned}\mathbf{Pr}\left(\overline{A_1} \cap \ldots \cap \overline{A_n}\right) &= \mathbf{Pr}(\overline{A_1}) \times \mathbf{Pr}(\overline{A_2} | \overline{A_1}) \times \mathbf{Pr}(\overline{A_3} | \overline{A_1} \cap \overline{A_2}) \\ &\qquad \times \ldots \times \mathbf{Pr}(\overline{A_n} | \overline{A_1} \cap \ldots \cap \overline{A_{n-1}}) \\ &\geq \prod_{i=1}^{n} (1 - x_i) > 0.\end{aligned}$$

Since $\mathbf{Pr}(A_i) \leq x_i$, (19.5) holds for $|S| = 0$. Now, consider any particular A_i, S, and suppose that (19.5) holds for every event and every smaller set. We will need the following consequence of the hypotheses of the lemma:

19.6 *(19.5) holds if S is disjoint from D_i.*

Now, let $S_1 = S \cap \mathcal{D}_i$ and let $S_2 = S - S_1$. Applying (19.6), we see that we can assume that S_1 is non-empty. Using the fact that $\mathbf{Pr}(X|Y) = \mathbf{Pr}(X \cap Y)/\mathbf{Pr}(Y)$ three times, we obtain $\mathbf{Pr}(A|B \cap C) = \mathbf{Pr}(A \cap B|C)/\mathbf{Pr}(B|C)$. Therefore

$$\mathbf{Pr}(A_i| \cap_{A_j \in S} \overline{A_j}) = \frac{\mathbf{Pr}(A_i \cap (\cap_{A_j \in S_1} \overline{A_j})| \cap_{A_j \in S_2} \overline{A_j})}{\mathbf{Pr}(\cap_{A_j \in S_1} \overline{A_j}| \cap_{A_j \in S_2} \overline{A_j})}.$$

Clearly, the numerator is at most $\mathbf{Pr}(A_i| \cap_{A_j \in S_2} \overline{A_j})$. So, applying (19.6) again, we have:

19.7 *the numerator is at most $\mathbf{Pr}(A_i)$.*

For convenience of notation, relabel the events of S_1 as $B_1, \ldots, B_t$. The denominator is equal to

$$\begin{aligned}\mathbf{Pr}(\overline{B_1}| \cap_{A_j \in S_2} \overline{A_j}) \quad \times \quad & \mathbf{Pr}(\overline{B_2}|\overline{B_1} \cap (\cap_{A_j \in S_2} \overline{A_j})) \\ \times \ldots \times \; & \mathbf{Pr}(\overline{B_t}|\overline{B_1} \cap \ldots \cap \overline{B_{t-1}} \cap (\cap_{A_j \in S_2} \overline{A_j})).\end{aligned}$$

Thus, by our induction hypothesis,

$$\mathbf{Pr}(A_i| \cap_{A_j \in S} \overline{A_j}) \leq \frac{\mathbf{Pr}(A_i)}{\prod_{A_j \in S_1}(1 - x_j)} \leq \frac{\mathbf{Pr}(A_i)}{\prod_{A_j \in \mathcal{D}_i}(1 - x_j)} \leq x_i,$$

as required. □

Proof of the Asymmetric Local Lemma. Set $x_i = 2\mathbf{Pr}(A_i)$ for each i. Since $\mathbf{Pr}(A_i) \leq \frac{1}{4}$, we have $x_i \leq \frac{1}{2}$. Setting $\alpha = 2 \ln 2$, we will use the fact that $(1 - x) \geq \mathrm{e}^{-\alpha\, x}$ for $x \leq \frac{1}{2}$.

$$\begin{aligned} x_i \prod_{A_j \in \mathcal{D}_i} (1 - x_j) &\geq x_i \prod_{A_j \in \mathcal{D}_i} \mathrm{e}^{-\alpha\, x_i} \\ &\geq 2\mathbf{Pr}(A_i) \times \exp\left(-\alpha \sum_{A_j \in \mathcal{D}_i} 2\mathbf{Pr}(A_j)\right) \\ &\geq 2\mathbf{Pr}(A_i) \times \exp^{-\alpha/2} \\ &= \mathbf{Pr}(A_i) \end{aligned}$$

Therefore the Asymmetric Local Lemma follows from the General Local Lemma (and thus, so does the simple form of the Local Lemma). □

Proof of the Weighted Local Lemma. Set $x_i = (2p)^{t_i}$ for each i. Since $p \leq \frac{1}{4}$ and $t_i \geq 1$, we again have $x_i \leq \frac{1}{2}$. Again we set $\alpha = 2 \ln 2$ and make use of the fact that $(1 - x) \geq \mathrm{e}^{-\alpha x}$ for $x \leq \frac{1}{2}$.

$$\begin{aligned} x_i \prod_{A_j \in \mathcal{D}_i} (1 - x_j) &= (2p)^{t_i} \prod_{A_j \in \mathcal{D}_i} \left(1 - (2p)^{t_j}\right) \\ &\geq (2p)^{t_i} \times \exp\left(-\alpha \sum_{A_j \in \mathcal{D}_i} (2p)^{t_j}\right) \\ &\geq (2p)^{t_i} \times \exp(-\alpha\, t_i/2) \\ &= p^{t_i} \\ &\geq \mathbf{Pr}(A_i). \end{aligned}$$

Therefore the Weighted Local Lemma follows from the General Local Lemma.

□

19.4 The Lopsided Local Lemma

The astute reader may have noticed that in our proof of the General Local Lemma, we did not require the full power of the independence properties guaranteed in its hypotheses. All we needed was for (19.6) to hold, and this might have happened even if A_2 was not quite mutually independent of the events in S_2. By replacing our usual independence condition by a condition which is essentially the same as (19.6) we get a slightly stronger tool:

The Lopsided Local Lemma [44]: *Consider a set $\mathcal{E}$ of (typically bad) events such that for each $A \in \mathcal{E}$, there is a set $D(A)$ of at most d other events, such that for all $S \subset \mathcal{E} - (A \cup D(A))$ we have*

$$\mathbf{Pr}(A | \cap_{A_j \in S} \overline{A_j}) \leq p.$$

If $4pd \leq 1$ then with positive probability, none of the events in $\mathcal{E}$ occur.

Our independence condition simply says that conditioning on no bad events outside of $D(A)$ occurring does not increase the chances of A occurring.

Remark Similar modifications can be made to each of our other versions of the Local Lemma.

We close this chapter with the following illustration of the Lopsided Local Lemma, due to McDiarmid [113], which is an improvement of our first application of the Local Lemma, Theorem 4.2.

Theorem 19.8 *If $\mathcal{H}$ is a hypergraph such that each hyperedge has size at least k and intersects at most 2^{k-2} other hyperedges, then $\mathcal{H}$ is 2-colourable.*

Proof The proof follows along the same lines as that of Theorem 4.2. This time, we define our bad events as follows: For each edge e, we R_e, B_e are the events that e is monochromatically Red, Blue respectively. $D(R_e) = \{B_f : f \cap e \neq \emptyset\}$ and $D(B_e) = \{R_f : f \cap e \neq \emptyset\}$. Intuitively, it seems clear that we satisfy the first condition of the Lopsided Local Lemma, since if f intersects e then conditioning on f not being monochromatically Red will decrease the probability of e being monochromatically Red. We leave a formal proof as Exercise 19.2. Since each event has probability $\frac{1}{2^k}$ and we are taking $d = 2^{k-2}$, we also satisfy the second condition, and so the Theorem follows. □

Theorem 4.2 implies that every hypergraph with minimum edge size k and maximum degree k is 2-colourable for $k \geq 10$. Theorem 19.8 implies that "10" can be replaced by "8". Using Remark 4.1 to make our calculations more precise, we can get tighter versions of these two theorems and they yield $k \geq 9$ and $k \geq 7$, respectively. Thomassen [151], using very different techniques, improved this result further to $k \geq 4$. It is not true for $k \leq 3$.

Exercises

Exercise 19.1 Show that there exists a set of events $\mathcal{E} = \{A_1, \ldots, A_n\}$ with each A_i is mutually independent of $\mathcal{E} - (\mathcal{D}_i \cup A_i)$, for some $\mathcal{D}_i \subseteq \mathcal{E}$, along with integers $t_1, \ldots, t_n \geq 1$ and a real $0 \leq p \leq \frac{1}{4}$ such that for each $1 \leq i \leq n$

(a) $\mathbf{Pr}(A_i) \leq p^{t_i}$,
(b) $\sum_{A_j \in \mathcal{D}_i} p^{t_j} \leq \frac{t_i}{2}$,

and $\mathbf{Pr}(\overline{A}_1 \cap \ldots \cap \overline{A}_n) = 0$.

Exercise 19.2 Complete the proof of Theorem 19.8.

20. A Closer Look at Talagrand's Inequality

When presenting Talagrand's Inequality in Chap. 10, we sacrificed power for simplicity. The original inequality provided by Talagrand is much more general than those we stated, but it is somewhat unwieldy. In this chapter, we will see Talagrand's original inequality, and we will show how to derive from it the weaker inequalities of Chap. 10. In order to give the reader a better idea of how the full inequality can be used, we will present a few other weakenings that can be derived from it, each one a generalization of those that we already know.

We do not include a proof of Talagrand's Inequality, as this proof is presented (very well) elsewhere, for example in [114, 146], and in the original paper [148].

20.1 The Original Inequality

Consider any n independent random trials $T_1, \dots, T_n$, and let $\mathcal{A}$ be the set of all the possible sequences of n outcomes of those trials. For any subset $A \subseteq \mathcal{A}$, and any real ℓ, we define $A_\ell \subseteq \mathcal{A}$ to be the subset of sequences which are within a distance ℓ of some sequence in A with regards to an unusual measure. In particular, we say that $x = (x_1, \dots, x_n) \in A_\ell$ if for every set of reals $b_1, \dots, b_n$, there exists at least one $y = (y_1, \dots, y_n) \in A$ such that

$$\sum_{x_i \neq y_i} b_i < \ell \left(\sum_{i=1}^{n} b_i^2 \right)^{1/2}.$$

Setting each $b_i = 1$ (or in fact, setting each $b_i = c$ for any constant $c > 0$), we see that if $y \in A_\ell$ then there is an $x \in A$ such that x and y differ on at most $\ell\sqrt{n}$ trials. Furthermore, if $y \in A_\ell$, then no matter how we weight the trials with b_i's, there will be an $x \in A$ such that the total weight of the trials on which x and y differ is small.

Talagrand's Inequality: For any n independent trials $T_1, \dots, T_m$, any set $A \subseteq \mathcal{A}$ and any real ℓ,

$$\mathbf{Pr}(A) \times \mathbf{Pr}(\overline{A}_\ell) \leq \mathrm{e}^{-\ell^2/4}.$$

By again considering $b_i = c$ for all i, we see that if an event A holds with reasonably high probability then with very high probability the random outcome will not differ in many trials from at least one event in A.

Suppose that X is a real-valued variable determined by our sequence of trials such that changing the outcome of a trial can affect X by at most c. Suppose further that for some interval R of the real line, X lies in R with reasonably high probability (say, at least $\frac{1}{2}$). Then Talagrand's Inequality implies that with very high probability (i.e. $1 - 2e^{-\ell^2/4}$), X will not differ in more than $l\sqrt{n}$ trials from some outcome yielding a value in R and so X will take a value that is very close to R (i.e. within $cl\sqrt{n}$). With a bit of work, these ideas can be fleshed out to obtain the Simple Concentration Bound from Chap. 10.

As we shall see now, by considering only slightly more general sequences $\{b_i\}$, namely each $b_i \in \{0, c\}$, we will obtain Talagrand's Inequality I from Chap. 10.

Talagrand's Inequality I *Let X be a non-negative random variable, not identically 0, which is determined by n independent trials $T_1, \dots, T_n$, and satisfying the following for some $c, r > 0$:*

1. *changing the outcome of any one trial can affect X by at most c, and*
2. *for any s, if $X \geq s$ then there is a set of at most rs trials whose outcomes certify that $X \geq s$,*

then for any $0 \leq t \leq \mathbf{Med}(X)$,

$$\mathbf{Pr}(|X - \mathbf{Med}(X)| > t) \leq 4e^{-\frac{t^2}{8c^2 r \mathbf{Med}(X)}}.$$

Proof Define $A = \{x : X(x) \leq \mathbf{Med}(X)\}$, and $C = \{y : X(y) > \mathbf{Med}(X)+t\}$. We first prove that $C \subseteq \overline{A_\ell}$, where $\ell = t/c\sqrt{r(\mathbf{Med}(X)+t)}$. So consider some $x \in C$. Let I be the set of indices of the at most $r(\mathbf{Med}(X)+t)$ trials whose outcomes certify that $X(x) \geq \mathbf{Med}(X)+t$. For each i, we set $b_i = c$ if $i \in I$ and $b_i = 0$ if $i \notin I$. Note that

$$\sum_{i=1}^{m} b_i^2 = c^2|I| \leq rc^2(\mathbf{Med}(X)+t) = (t/\ell)^2.$$

Now consider any $y \in A$. Define y' to be the outcome which agrees with x on all indices of I, and on y on all other indices. Since I certifies that $X(x) \geq \mathbf{Med}(X)+t$, we also have $X(y') \geq \mathbf{Med}(X)+t$. Since y and y' differ only on trials in I on which x and y differ, and since changing the outcome of any one $T_i \in I$ can affect X by at most $b_i = c$, we have that $X(y) \geq X(y') - \sum_{x_i \neq y_i} b_i \geq \mathbf{Med}(X) + t - \sum_{x_i \neq y_i} b_i$. Thus,

$$\sum_{x_i \neq y_i} b_i > t \geq \ell \sqrt{\sum_{i=1}^{m} b_i^2}.$$

Since this is true for every $y \in A$, we have $x \notin A_\ell$.

Therefore $C \subseteq \overline{A_\ell}$. Now, $\mathbf{Pr}(A) \geq \frac{1}{2}$, by the definition of median. So by Talagrand's Inequality:

$$\mathbf{Pr}(C) \leq \mathbf{Pr}(\overline{A_\ell}) \leq 2\mathrm{e}^{-\ell^2/4} < 2\mathrm{e}^{-\frac{t^2}{4c^2r(\mathbf{Med}(X)+t)}}.$$

This is the first side of our concentration bound.

Next, set $C^{'} = \{x : X(x) \geq \mathbf{Med}(X)\}$, $A^{'} = \{y : X(y) < \mathbf{Med}(X) - t\}$ and $\ell = t/c\sqrt{r\mathbf{Med}(X)}$. By a nearly identical argument, we obtain that $C^{'} \subseteq \overline{A^{'}_\ell}$, and so $\mathbf{Pr}(\overline{A^{'}_\ell}) \geq \frac{1}{2}$ by the definition of median. So, applying Talagrand's Inequality, we have:

$$\mathbf{Pr}(A^{'}) \leq 2\mathrm{e}^{-\frac{t^2}{4c^2r\mathbf{Med}(X)}}.$$

Therefore

$$\begin{aligned}\mathbf{Pr}(|X - \mathbf{Med}(X)| > t) &\leq \mathbf{Pr}(C \cup A') \\ &\leq 4\mathrm{e}^{-\frac{t^2}{4c^2r(\mathbf{Med}(X)+t)}} \\ &\leq 4\mathrm{e}^{-\frac{t^2}{8c^2r\mathbf{Med}(X)}}\end{aligned}$$

as required, since $t \leq \mathbf{Med}(X)$. □

To obtain Talagrand's Inequality II, we first prove Fact 10.1; we restate both of these below.

Fact 20.1 *Under the conditions of Talagrand's Inequality I,* $|\mathbf{E}(X) - \mathbf{Med}(X)| \leq 40c\sqrt{r\mathbf{E}(X)}$.

Proof First, observe that $\mathbf{E}(X) - \mathbf{Med}(X) = \mathbf{E}(X - \mathbf{Med}(X))$. So, since the absolute value of a sum is less than the sum of the absolute values of its terms, we obtain: $|\mathbf{E}(X) - \mathbf{Med}(X)| = |\mathbf{E}(X - \mathbf{Med}(X))| \leq \mathbf{E}(|X - \mathbf{Med}(X)|)$. We will bound this latter term by partitioning the positive real line into the intervals $I_i = (i \times c\sqrt{r\mathbf{Med}(X)}, (i+1) \times c\sqrt{r\mathbf{Med}(X)}]$, defined for each integer $i \geq 0$. Clearly, $\mathbf{E}(|X - \mathbf{Med}(X)|)$ is at most the sum over all I_i of the maximum value in I_i times the probability that $|X - \mathbf{Med}(X)| \in I_i$. We bound this latter probability by the probability that $|X - \mathbf{Med}(X)|$ is greater than the left endpoint of the interval.

$$\begin{aligned}\mathbf{E}(|X-\mathbf{Med}(X)|) &\le \sum_{i\ge 0}(i+1)c\sqrt{r\mathbf{Med}(X)} \times \mathbf{Pr}(|X-\mathbf{Med}(X)| \\ &> ic\sqrt{r\mathbf{Med}(X)}) \\ &< c\sqrt{r\mathbf{Med}(X)} \times \sum_{i\ge 0}(i+1)4\mathrm{e}^{-\frac{i^2}{8}} \\ &< 28c\sqrt{r\mathbf{Med}(X)}.\end{aligned}$$

Since $X \ge 0$, we have $\mathbf{E}(X) \ge \frac{1}{2}\mathbf{Med}(X)$, and so $28c\sqrt{r\mathbf{Med}(X)} \le \sqrt{2} \times 28c\sqrt{r\mathbf{E}(X)} < 40c\sqrt{r\mathbf{E}(X)}$ as required. □

Talagrand's Inequality II *Let X be a non-negative random variable, not identically 0, which is determined by n independent trials $T_1, \dots, T_n$, and satisfying the following for some $c, r > 0$:*

1. *changing the outcome of any one trial can affect X by at most c, and*
2. *for any s, if $X \ge s$ then there is a set of at most rs trials whose outcomes* certify *that $X \ge s$,*

then for any $0 \le t \le \mathbf{E}(X)$,

$$\mathbf{Pr}(|X-\mathbf{E}(X)| > t + 60c\sqrt{r\mathbf{E}(X)}) \le 4\mathrm{e}^{-\frac{t^2}{8c^2r\mathbf{E}(X)}}.$$

Proof Applying Fact 20.1, then Talagrand's Inequality I, and then Fact 20.1 again, we obtain:

$$\begin{aligned}&\mathbf{Pr}(|X-\mathbf{E}(X)| > t + 60c\sqrt{r\mathbf{E}(X)}) \\ &< \mathbf{Pr}(|X-\mathbf{Med}(X)| > t + 20c\sqrt{r\mathbf{E}(X)}) \\ &\le 4\mathrm{e}^{-(t+20c\sqrt{r\mathbf{E}(X)})^2/8c^2r\mathbf{Med}(X)} \\ &\le 4\mathrm{e}^{-(t+20c\sqrt{r\mathbf{E}(X)})^2/8c^2r(\mathbf{E}(X)+40c\sqrt{r\mathbf{E}(X)})}.\end{aligned}$$

So it suffices to show that $(t+20c\sqrt{r\mathbf{E}(X)})^2/8c^2r(\mathbf{E}(X)+40c\sqrt{r\mathbf{E}(X)})$ is at least as big as $t^2/(8c^r\mathbf{E}(X))$. This is true since

$$\begin{aligned}&(t+20c\sqrt{r\mathbf{E}(X)})^2 \times \mathbf{E}(X) - t^2(\mathbf{E}(X)+40c\sqrt{r\mathbf{E}(X)}) \\ &> 40tc\sqrt{r\mathbf{E}(X)} \times \mathbf{E}(X) - 40t^2c\sqrt{r\mathbf{E}(X)} \\ &\ge 0,\end{aligned}$$

as $t \le \mathbf{E}(X)$. □

20.2 More Versions

In this section, we further illustrate the power of Talagrand's Inequality by presenting a few more consequences of it. We leave the proofs as exercises since they are variations of the proofs in the previous section.

When we used Talagrand's original inequality to prove Talagrand's Inequality I, we used the certificate I and the bound c to prove that, for our choice of b_i, if $X(y)$ is much smaller than $X(x)$ then $\sum_{x_i \neq y_i} b_i$ must be large in terms of $(\sum b_i^2)^{1/2}$.

The key to the proof was to choose an I certifying that $X(x) \geq s$ and set b_i to be c if $i \in I$ and 0 otherwise. This immediately yields:

20.2 $\sum b_i^2 \leq c \times |I|$.

Using the fact that changing the outcome of any one trial can affect X by at most c, we obtained:

20.3 $X(y) \geq s - c \times |\{i \in I : x_i \neq y_i\}| \geq s - \sum_{x_i \neq y_i} b_i$.

These two bounds were the main parts of the proof.

Our next inequality is more suitable when the amounts by which changing the outcome of each trial can cause X to decrease vary significantly. Here, instead of conditions 1 and 2 from Talagrand's Inequality I, we simply require (i) a reworking of (20.3), which uses a general sequence $\{b_i\}$ rather than setting each $b_i = c$ or 0, and (ii) a bound on $\sum b_i^2$ instead of (20.2) (since we no longer have the bound $b_i \leq c$).

Talagrand's Inequality III: *Let X be a non-negative random variable, not identically 0, which is determined by n independent trials $T_1, \ldots, T_n$. Let r be an arbitrary positive constant. Suppose that for every outcome $x = (x_1, \ldots, x_n)$ of the trials, there exists a list of non-negative weights $b_1, \ldots, b_n$ such that:*

1. *$\sum b_i^2 \leq r \times X(x)$; and*
2. *for any outcome y, we have $X(y) \geq X(x) - \sum_{x_i \neq y_i} b_i$,*

then for any $0 \leq t \leq \mathbf{E}(X)$,

$$\mathbf{Pr}(|X - \mathbf{E}(X)| > t + 60\sqrt{r\mathbf{E}(X)}) \leq 4e^{-\frac{t^2}{8r\mathbf{E}(X)}}.$$

Enforcing that $\sum b_i^2 \leq r \times X(x)$ can be difficult when $X(x)$ is very small. So instead, it is often convenient to replace this by an absolute bound D (which is usually taken to be of the same order as $\mathbf{E}(X)$):

Talagrand's Inequality IV: *Let X be a non-negative random variable, not identically 0, which is determined by n independent trials $T_1, \ldots, T_n$. Suppose that for every outcome $x = (x_1, \ldots, x_n)$ of the trials, there exists a list of non-negative weights $b_1, \ldots, b_n$ such that:*

1. *$\sum b_i^2 \leq D$; and*
2. *for any outcome y, we have $X(y) \geq X(x) - \sum_{x_i \neq y_i} b_i$,*

then for any $0 \leq t \leq \mathbf{E}(X)$,

$$\mathbf{Pr}(|X - \mathbf{E}(X)| > t + 60\sqrt{D}) \leq 4e^{-\frac{t^2}{8D}}.$$

Our final generalization is that our conditions do not have to hold for *every* outcome X. They only have to hold with moderately high probability, say, at least $\frac{3}{4}$:

Talagrand's Inequality V: *Let X be a non-negative random variable, not identically 0, which is determined by n independent trials $T_1, \ldots, T_n$. Let F be the event that for the outcome $x = (x_1, \ldots, x_n)$ of the trials, there exists a list of non-negative weights $b_1, \ldots, b_n$ such that:*

1. *$\sum b_i^2 \leq D$; and*
2. *for any outcome y, we have $X(y) \geq X(x) - \sum_{x_i \neq y_i} b_i$,*

then for any $0 \leq t \leq \mathbf{E}(X)$,

$$\mathbf{Pr}(|X - \mathbf{E}(X)| > t + 60\sqrt{D}) \leq 4e^{-\frac{t^2}{8D}} + 2\mathbf{Pr}(\overline{F}).$$

In Exercise 20.2 you will show that Talagrand's Inequality IV simplifies the proof of Lemma 10.4.

Exercises

Exercise 20.1 Prove Talagrand's Inequalities III, IV, V.
Hints: For III, consider the function $f(\alpha) = \alpha - t\sqrt{\alpha/M}$, and show that if $\alpha \geq M$ then $\phi(\alpha) \geq f(M)$. For V, consider the set $C = F \cap \{x : X(x) \geq \mathbf{Med}(X)\}$.

Exercise 20.2 Use Talagrand's Inequality IV to simplify the proof of Lemma 10.4 by analyzing X_v directly instead of focusing on AT_v and Del_v.

Part VIII

Colour Assignment via Fractional Colouring

In this part of the book, we introduce a new technique for assigning colours to the vertices of a graph when applying the naive colouring procedure. In particular, rather than picking the colours independently at random and resolving conflicts by uncolouring, we will avoid conflicts by choosing the vertices to be assigned each colour from a probability distribution on the stable sets. We will prove that we can still perform a local analysis and apply the Local Lemma, even though our choice of colour assignments is made globally.

In order to do so we need to show that, for the distributions we use, the colour assigned to a vertex is essentially independent of the assignments in distant parts of the graph.

Chapter 23 presents the most difficult theorem proved using this technique: Kahn's proof that the list chromatic index is assymptotically equal to the fractional chromatic index. Chapter 22 presents an analysis of the probability distributions that Kahn uses in his proof, the so-called hardcore distributions. Chapter 21 presents some results with the same flavour which are much easier to prove.

21. Finding Fractional Colourings and Large Stable Sets

In this preliminary chapter, we introduce the notion of a fractional colouring. We then present some results on finding large stable sets and fractional colourings of certain special graphs. These results give some of the flavour of the approach taken in the remainder of this part of the book. Although the approach described in the next two chapters is similar, the technical complications will increase dramatically.

21.1 Fractional Colouring

A *fractional vertex colouring* is a set $\{S_1, \ldots, S_l\}$ of stable sets and corresponding positive real weights, $\{w_1, .., w_l\}$ such that the sum of the weights of the stable sets containing each vertex is one. I.e. $\forall v \in V, \; \sum_{\{S_i : v \in S_i\}} w_i = 1$. As an example, consider C_5, the chordless cycle on five vertices. There are exactly five stable sets of size 2 in C_5. Furthermore, each vertex of C_5 is in exactly two of these stable sets. Thus, assigning a weight of $\frac{1}{2}$ to each of these five stable sets yields a fractional vertex colouring of G.

Of course, each stable set weight in a fractional vertex colouring will be at most one. We note that a vertex colouring is simply a fractional vertex colouring in which every weight is an integer (to be precise, each weight is 1). Letting $\mathcal{S} = \mathcal{S}(G)$ be the set of stable sets in G, we note that we can also describe a fractional vertex colouring by an assignment of a non-negative real weight w_S to each stable set S in $\mathcal{S}(G)$ so that the sum of the weights of the stable sets containing each vertex is 1.

A fractional vertex colouring is a *fractional vertex c-colouring* if $\sum_{S \in \mathcal{S}(G)} w_S = c$. The *fractional chromatic number* of G, denoted $\chi_v^*(G)$, is the minimum c such that G has a fractional vertex c-colouring. Since every colouring is a fractional colouring, $\chi_v^*(G) \leq \chi(G)$. To see that this inequality can be strict, we note that $\chi(C_5) = 3$ but that the fractional colouring given above shows that $\chi_v^*(C_5) \leq \frac{5}{2}$.

We define fractional edge colourings and the fractional chromatic index, $\chi_e^*(G)$ similarly. In the same vein, we define fractional total colourings and the fractional total chromatic number, $\chi_T^*(G)$. Of course, here we need to assign weights to the set of matchings of G, $\mathcal{M}(G)$, and the set of total stable sets of G, $\mathcal{T}(G)$.

We note that if G has n vertices then $\chi_v^*(G) \geq \frac{n}{\alpha(G)}$ (recall that $\alpha(G)$ is the size of a largest stable set in G), because for any fractional colouring we have:

$$n = \sum_{v \in V} \sum_{\{S \in \mathcal{S}: v \in S\}} w_S = \sum_{S \in \mathcal{S}} \sum_{v \in S} w_S \leq \sum_{S \in \mathcal{S}} \alpha w_S.$$

Thus for example we see that $\chi_v^*(C_5) \geq \frac{5}{2}$ and hence the fractional colouring given above is optimal and $\chi_v^*(C_5) = \frac{5}{2}$.

Since the fractional chromatic number does not decrease as we add vertices and edges, we see that $\chi_v^*(G) \geq \max\left\{\frac{|V(H)|}{\alpha(H)} | H \subseteq G\right\}$.

In fact, as the reader can easily verify, a similar argument shows that $\chi_v^*(G)$ is at least the solution to:

21.1 max $\sum_{v \in V} x(v)$

s.t. $\forall S \in \mathcal{S}(G), \quad \sum_{v \in S} x(v) \leq 1$

over all non-negative weightings x on V.

We obtain the first bound given above on $\chi_v^*(G)$ from (21.1) by setting each $x(v) = \frac{1}{\alpha(G)}$ and our second by setting each $x(v)$ either to 0 or $\frac{1}{\alpha(H)}$ depending on whether or not v is in H.

Now, it follows immediately from LP duality that the fractional chromatic number is in fact equal to the maximum in (21.1), (see e.g. [140]). Despite this pleasing result, it is still NP-complete to compute the fractional chromatic number of a graph. In fact, recent results [13] show that it is difficult to approximate the fractional chromatic number of graphs with n vertices to within a factor of n^ϵ for some small positive ϵ.

The seminal result on fractional edge colourings, due to Edmonds [37], is:

Theorem 21.2 *The fractional chromatic index of G is the maximum of Δ and* $\max\left\{\frac{2|E(H)|}{|V(H)|-1} : H \subseteq G, \; |V(H)| \; odd\right\}$.

We now restate this result in terms of weightings of the line graphs as in (21.1). Theorem 21.2 says that a non-negative weighting of maximum total weight in which no matching has weight exceeding one can be obtained in one of the following two simple ways:

(i) assigning 1 to every edge incident to one particular vertex, or
(ii) for an appropriate odd cardinality subgraph H, assigning the weight $\frac{2}{|V(H)|-1}$ to each of the edges of H.

Remark Clearly, for any $H \subseteq G$ s.t. the maximum matching in H has size k, assigning the weight $\frac{1}{k}$ to each edge of H yields a feasible solution to (21.1) in the line graph of G. Thus, if the weighting of maximum weight is as in (ii) then the corresponding H contains a matching of size $\frac{|V(H)|-1}{2}$ (the largest value possible).

Using this characterization, Padberg and Rao [127] were able to obtain a polynomial time algorithm for computing the fractional chromatic index of a graph and indeed an optimal fractional edge colouring.

The complexity of computing the fractional total chromatic number of a graph remains unclear. See [94] for some partial results.

We close this section by presenting Edmonds' result in its more well-known garb, as a characterization of the matching polytope, and then explaining why the two results are equivalent.

Let G be a graph with m edges. Fix an enumeration $e_1, \ldots, e_m$ of $E(G)$. A is the *incidence vector* of a matching M if $M = \{e_i | x_i = 1\}$. We use x_M to denote the incidence vector of M. The *matching polytope* for G, denoted $\mathcal{MP}(G)$ is the convex hull of the incidence vectors of the matchings in $\mathcal{R}^m$. Equivalently, $y \in \mathcal{R}^m$ is in $\mathcal{MP}(G)$ if and only if it can be expressed as a convex combination of incidence vectors of matchings (i.e. $y = \sum_{M \in \mathcal{M}(G)} a_M x_m$ where the a_M are non-negative and sum to 1). Edmonds proved:

Theorem 21.3 *A non-negative vector x is in $\mathcal{MP}(G)$ if and only if:*

(a) for each vertex v, $\sum_{e_i \ni v} x_i \leq 1$, and
(b) for each subgraph H which has an odd number of vertices, $\sum_{e_i \in E(H)} x_i \leq \frac{|V(H)|-1}{2}$.

Remark Since (a) and (b) hold for every incidence vector of a matching, they clearly hold for every vector in the matching polytope. The difficult part of the theorem is to prove the converse.

Remark Edmonds was interested in finding the maximum weight matching in a graph whose edges had been assigned weights. It is well known and easy to see that the maximum of a linear function on a convex polytope is achieved at a vertex. Thus, finding the weight of a maximum weight matching amounts to computing the maximum of a linear function on $\mathcal{MP}(G)$. Edmonds first developed a polynomial-time algorithm for finding maximum weight matchings, and then used it to obtain the polyhedral characterization of Theorem 21.3 [37].

We observe that given a fractional c edge colouring, we can demonstrate that $(\frac{1}{c}, .., \frac{1}{c})$ is in $\mathcal{MP}(G)$ by setting $a_M = \frac{w_M}{c}$ for each matching M. Conversely, if we have some a_M which demonstrate that $(\frac{1}{c}, .., \frac{1}{c})$ is in $\mathcal{MP}(G)$ then we can obtain a fractional c edge colouring by setting $w_M = a_M c$. Thus, we have:

21.4 *G has a fractional c edge colouring if and only if $(\frac{1}{c}, \ldots, \frac{1}{c})$ is in $\mathcal{MP}(G)$.*

So, Theorem 21.2 follows from Theorem 21.3. In fact, both theorems hold for graphs with multiple edges, and so we can obtain the reverse implication by taking multiple copies of edges and scaling. We omit the details as we do not need the result.

21.2 Finding Large Stable Sets in Triangle-Free Graphs

In Chap. 13, we showed that there is a constant c such that every triangle-free graph has chromatic number at most $\frac{c\Delta}{\log \Delta}$. A corollary of this theorem is that there is a positive ϵ $(= \frac{1}{c})$ such that every triangle free graph contains a stable set with $\epsilon\frac{|V(G)|\log \Delta}{\Delta}$ vertices. This result had been proved earlier by Ajtai, Komlos and Szemeredi [1]. The proof was simplified and the constant improved by Shearer [143] and Alon [5].
In this section, we show:

Theorem 21.5 *Every triangle-free graph G contains a stable set with at least $\frac{|V(G)|\log \Delta}{8\Delta}$ vertices.*

We do so by computing the average size of a uniformly chosen random stable set. The result and proof technique are both due to Shearer [143], although our exposition is closer to that given in [5].

Proof of Theorem 21.5 We can assume that $\Delta \geq 2^8$, as otherwise we need only use a colour class in a Δ colouring. We let $n = |V(G)|$. We choose a uniformly random stable set S and bound $\mathbf{E}(|S|)$. We show that $\mathbf{E}(|S|) \geq \frac{n \log \Delta}{8\Delta}$, which proves the theorem.

By Linearity of Expectation: $\mathbf{E}(|S|) = \sum_{v\in V}\mathbf{Pr}(v \in S)$. We show that for every vertex v,

Claim 21.6 $\mathbf{Pr}(v \in S) + \frac{1}{\Delta}\mathbf{E}(|S \cap N(v)|) \geq \frac{\log \Delta}{4\Delta}$.

This implies that

$$\sum_{v\in V}\mathbf{Pr}(v \in S) + \frac{1}{\Delta}\sum_{v\in V}\mathbf{E}(|S \cap N(v)|) \geq \frac{n\log \Delta}{4\Delta}.$$

Since $\mathbf{E}(|S \cap N(v)|) = \sum_{w\in N(v)}\mathbf{Pr}(w \in S)$, and each vertex w is in $d(w)$ neighbourhoods, this yields:

$$\sum_{v\in V}\left(1 + \frac{d(v)}{\Delta}\right)\mathbf{Pr}(v \in S) \geq \frac{n\log \Delta}{4\Delta}.$$

Since $\frac{d(v)}{\Delta} \leq 1$, we have $\sum_{v\in V}\mathbf{Pr}(v \in S) \geq \frac{n\log \Delta}{8\Delta}$, which is the desired result.

Thus, to prove Theorem 21.5, it remains only to prove Claim 21.6. To do this, we condition on the intersection of S with $G - v - N(v)$. In particular, letting S' be the random stable set $S - v - N(v)$, we show that for each $R \in \mathcal{S}(G - v - N(v))$:

$$\mathbf{Pr}(v \in S | S' = R) + \frac{1}{\Delta}\mathbf{E}(|S \cap N(v)||S' = R) \geq \frac{\log \Delta}{4\Delta}.$$

Since

$$\mathbf{Pr}(v \in S) = \sum_{R \in \mathcal{S}(V - N(v) - v)} \mathbf{Pr}(v \in S | S' = R)\mathbf{Pr}(S' = R)$$

and

$$\mathbf{E}(|N(v) \cap S|) = \sum_{R \in \mathcal{S}(V - N(v) - v)} \mathbf{E}(|N(v) \cap S||S' = R)\mathbf{Pr}(S' = R),$$

summing over all such R yields Claim 21.6.

So, we fix a stable set R of $G - v - N(v)$, and condition on the event that $S' = R$. Note that this implies that $S - S' \subseteq (v \cup N(v)) - N(R)$. We set $W = (v \cup N(v)) - N(R)$. Since we are choosing S from the uniform distribution on $\mathcal{S}(G)$, for any two stable sets W_1 and W_2 contained in W we have:

$$\mathbf{Pr}(S = R + W_1) = \mathbf{Pr}(S = R + W_2).$$

Thus,

$$\mathbf{Pr}(S = R + W_1 | S' = R) = \mathbf{Pr}(S = R + W_2 | S' = R).$$

This implies that if S'' is a uniformly chosen random stable set in W, $\mathbf{Pr}(v \in S | S' = R) = \mathbf{Pr}(v \in S'')$ and $\mathbf{E}(|S \cap N(v)||S' = R) = \mathbf{E}(|S'' \cap N(v)|)$. So, we need only prove:

$$\mathbf{Pr}(v \in S'') + \frac{1}{\Delta}\mathbf{E}(|S'' \cap N(v)|) \geq \frac{\log \Delta}{4\Delta}.$$

Now, W consists of v along with a stable set W' all of whose vertices are adjacent to v. Thus, W contains $2^{|W'|} + 1$ stable sets, one of which is v and the rest of which are the subsets of W'. So, $\mathbf{Pr}(v \in S'') = \frac{1}{2^{|W'|}+1}$ and $\mathbf{E}(|S'' \cap N(v)|) = \frac{|W'|}{2}\frac{2^{|W'|}}{2^{|W'|}+1}$. Now, if $|W'| \leq \log \Delta - \log\log \Delta + 1$ then $\mathbf{Pr}(v \in S'')$ exceeds $\frac{\log \Delta}{4\Delta}$. On the other hand, if $|W'| \geq \log \Delta - \log\log \Delta + 1$, then $\mathbf{E}(|S'' \cap N(v)|) \geq \frac{\log \Delta - \log\log \Delta}{2}$. Since for $\Delta \geq 2^8$, we have $\log\log \Delta < \frac{\log \Delta}{2}$, the result follows. □

21.3 Fractionally, $\chi \leq \frac{\omega+\Delta+1}{2}$

Recall that Conjecture 16.2 states that if a graph has maximum clique size ω and maximum degree Δ then its chromatic number is at most $\lceil \frac{\omega+\Delta+1}{2} \rceil$. In this section, we prove a fractional version of this conjecture:

Theorem 21.7 *For every graph G, we have $\chi_v^*(G) \leq \frac{\omega(G)+\Delta(G)+1}{2}$.*

Note that, since $\chi^*(v) \geq \frac{n}{\alpha(G)}$, this theorem implies that every graph G contains a stable set of size $\frac{2n}{\omega+\Delta+1}$. This weaker statement was proven independently and much earlier by Fajtlowicz [48] (see also [49]).

Proof of Theorem 21.7. Our approach is to consider choosing stable sets where each *maximum* stable set is equally likely. The key to the proof is the following lemma, whose proof is similar to that of Claim 21.6 of the previous section.

Lemma 21.8 *If S is a randomly chosen uniform maximum stable set then for every vertex v of G, we have:*

$$\mathbf{Pr}(|S \cap N(v)| = 1) \leq (\omega - 1)\mathbf{Pr}(v \in S).$$

Proof We again condition on our choice of $S' = S - v - N(v)$. We show that for each stable set $R \in \mathcal{S}(G - v - N(v))$, which can be extended to a maximum stable set by adding vertices of $v + N(v)$, we have:

21.9 $\mathbf{Pr}(|S \cap N(v)| = 1 | S' = R) \leq (\omega - 1)\mathbf{Pr}(v \in S | S' = R)$.

As in the last section, summing over the possible choices for R yields the desired result.

So, fix such a stable set R in $V - N(v) - v$ and let $W = N(v) + v - N(R)$. By our choice of R, we have $\alpha(G) = |R| + \alpha(W)$. I.e., any maximum stable set U with $U - v - N(v) = R$ must contain a maximum stable set of W. Thus, if W is not a clique then we are done because $\mathbf{Pr}(|N(v) \cap S| = 1 | S' = R) = 0$. If W is a clique then $|R| = \alpha(G) - 1$, and S is equally likely to be $R + w$ for each vertex of W. So, we are done because $\mathbf{Pr}(v \in S | S' = R) = \frac{1}{|W|}$ and: $\mathbf{Pr}(|N(v) \cap S| = 1 | S' = R) = \frac{|W|-1}{|W|} \leq \frac{\omega-1}{|W|}$. □

We note that since S is a maximum stable set, we have $\mathbf{Pr}(v \in S) = \mathbf{Pr}(|S \cap N(v)| = 0)$. So, since for any non-negative integer valued random variable X, $E(X) \geq 2 - \mathbf{Pr}(X = 1) - 2\mathbf{Pr}(X = 0)$, Lemma 21.8 implies the following analogue of Claim 21.6:

21.10 $\mathbf{E}(|S \cap N(v)|) \geq 2 - (\omega + 1)\mathbf{Pr}(v \in S)$.

Summing (21.10) over all $v \in V$, we obtain $\Delta\mathbf{E}(|S|) \geq 2n - (\omega+1)\mathbf{E}(|S|)$. I.e.,

21.11 $\mathbf{E}(|S|) \ge \frac{2n}{\omega+\Delta+1}$,

which implies Fajtlowicz's result.

Now, suppose that every vertex is equally likely to be in S. That is, there is some p such that $\forall v \in V$, $\mathbf{Pr}(v \in S) = p$. If this were the case then we would be done as follows. Applying (21.11), we obtain $p \ge \frac{2}{\omega+\Delta+1}$. So, setting $w_R = \frac{1}{p}\mathbf{Pr}(S = R)$ for each maximum stable set R in G yields the desired fractional colouring of G using total weight $\frac{1}{p} \le \frac{\omega+\Delta+1}{2}$.

Of course, some vertices may be more likely to appear in S than others, and so the argument above does not always apply. For example, if G is $K_{1,\Delta}$ and v is the vertex of degree Δ in G then $\mathbf{Pr}(v \in S) = 0$ whilst for every other vertex w of G, $\mathbf{Pr}(w \in S) = 1$. In such situations, if we set $w_R = \frac{\omega+\Delta+1}{2}\mathbf{Pr}(S = R)$ for each stable set $R \in \mathcal{S}$ then, by (21.11), the sum over all of the vertices of the weight of the stable sets containing them is at least n. However, we might not have a fractional colouring as for some vertices the total weight of the stable sets in which they appear may be less than one, whilst for other vertices this value may exceed one. So, in the general situation, we will need to use a more sophisticated (iterative) approach to construct the desired fractional colouring.

We will once again begin by assigning each maximum stable set of G the same (positive) weight. However, we will choose this weight so that the weight of the stable sets containing each vertex is at most one. Avoiding waste in this way will allow us to complete the fractional colouring in subsequent iterations.

In iteration 1, we set $w = \frac{1}{\max\{\mathbf{Pr}(v \in S) | v \in V\}}$ and set $w_R = w \cdot \mathbf{Pr}(S = R)$ for each maximum stable set R. Now, for every vertex v, the sum of the weights of the stable sets containing v is at most one, and for at least one vertex this inequality is tight. We delete any such vertices and continue. In each iteration we increase the weight on each maximum stable set in the graph induced by the remaining vertices by the same amount. We bound this increase so as to ensure that the total weight of the stable sets containing a vertex never exceeds one. The formal description of the procedure follows.

1. Set $w_S = 0$, $\forall S \in \mathcal{S}(G)$. Set $G_0 = G$. Set $i = 0$.
   Set $T = 0$ (T stands for Total weight used).
   For each $v \in V(G)$, set $wo_v = 0$ (wo stands for weight on).
2. If $V(G_i) = \emptyset$ or $T = \frac{\omega+\Delta+1}{2}$ then stop.
3. For each vertex v of G_i, let $p_i(v)$ be the probability that v is in a uniformly chosen random maximum stable set of G_i. Set $low = \min\{\frac{1-wo_v}{p_i(v)} | v \in V(G_i)\}$. Set $val_i = \min(low, \frac{\omega+\Delta+1}{2} - T)$.
4. Let $\mathcal{S}_i$ be the set of maximum stable sets of G_i. For each stable set in $\mathcal{S}_i$, increase w_S by $\frac{val_i}{|\mathcal{S}_i|}$. For each vertex v of G_i, increase wo_v by $p_i(v)val_i$. Increase T by val_i.

5. Let G_{i+1} be the graph induced by those vertices v which satisfy $wo_v < 1$. Increment i and go to Step 2.

We show now that this procedure terminates. Our choice of val_i ensures that T never exceeds $\frac{\omega+\Delta+1}{2}$. Our choice of val_i thus ensures that if the ith iteration is not the last then $V(G_{i+1})$ is strictly contained in $V(G_i)$, as every vertex of G_i minimizing $\frac{1-wo_v}{p_i(v)}$ will not be in G_{i+1}. Thus, the algorithm must terminate.

We claim that at the end of this procedure, the w_S yield the desired fractional vertex colouring. We note that it is easy to verify by induction that at the end of each iteration, $\forall v \in V, wo_v = \sum_{\{S\in\mathcal{S}|v\in S\}} w_S$ and $T = \sum_{S\in\mathcal{S}} w_S$. Furthermore, our choice of *low* and val_i ensure that no wo_v ever exceeds 1. Thus, if we stop because $V(G_i) = \emptyset$ then we have the desired fractional colouring. It remains only to show that if we stop because $T = \frac{\omega+\Delta+1}{2}$ then each $wo_v = 1$ and we still have the desired fractional colouring.

To this end, assume the contrary and let v be a vertex of G with $wo_v < 1$ when we complete the iterative process. For each vertex u and iteration i, we let $a_i(u)$ be the amount by which wo_u was augmented in iteration i, i.e. $a_i(u) = val_i p_i(u)$. By (21.10), since v is in each G_i, we have:

21.12 *For each i, $\sum_{u\in N(v)} a_i(u) \geq 2(val_i) - (\omega+1)a_i(v)$.*

Summing over the iterations, we obtain:

21.13 $\sum_{u\in N(v)} wo_u \geq 2T - (\omega+1)wo_v > \omega + \Delta + 1 - (\omega+1) = \Delta$.

But for each u in $N(v)$, $wo_u \leq 1$, so we obtain a contradiction. □

Exercises

Exercise 21.1 Show that in fact the fractional chromatic number of G is at most the maximum over all vertices of $\frac{|N_v(G)|+\omega_v}{2}$ where ω_v is the maximum size of a clique containing v (McDiarmid, unpublished).

22. Hard-Core Distributions on Matchings

In the preceding chapter, we analyzed two distributions on stable sets. In the first, each stable set was equally likely. In the second, each maximum stable set was equally likely. Our analyses allowed us to find large stable sets and fractional colourings using few colours. In both cases, the analysis involved showing that certain properties held in the neighbourhood of a vertex, regardless of what the stable set looked like further away from the vertex. This makes these two distributions attractive candidates for use in conjunction with the Local Lemma, for their local analysis leads to global results.

In this chapter, we present a special type of probability distribution, the hard-core distributions, of which these two are (essentially) examples. We then discuss a specific hard-core distribution which we will use in the next chapter to construct edge colourings of multigraphs. Our analysis focuses on showing that certain properties hold in the neighbourhood of a vertex, regardless of any conditioning on what the stable set looks like far away from the vertex. We hope that the results of the last chapter have whetted the reader's appetite for this more extensive excursion into the realm of hard-core distributions.

22.1 Hard-Core Distributions

A probability distribution p on the stable sets of G is *hard-core* if it is obtained by associating a positive real $\lambda(v)$ to each vertex of G so that that the probability that we pick a stable set S is proportional to $\prod_{v\in S}\lambda(v)$. I.e., setting $\lambda(S)=\prod_{v\in S}\lambda(v)$ we have:

$$p(S)=\frac{\lambda(S)}{\sum_{T\in\mathcal{S}(G)}\lambda(T)}.$$

We call the $\lambda(v)$ the *activities* of p.

The use of the name hard-core for such distributions arose in statistical physics, see e.g. [68]. They have also proven important in other contexts, see e.g. [103, 130], although under different names.

We note that if we set each $\lambda(v)=1$ then we obtain the uniform distribution as a hard-core distribution. Also, if we let each $\lambda(v)=N$ where N goes to infinity, then we approach the uniform distribution on the maximum stable sets, as the probability of not picking a maximum stable set goes to zero. Thus, the distributions of the last chapter are (essentially) hard-core distributions where the $\lambda(v)$ are all equal to some fixed λ. In these distributions, all the stable sets of a given size are equally likely and our choice of λ indicates whether we prefer larger or smaller stable sets, and to what extent.

For a given probability distribution on the stable sets of a graph, we refer to the probability that a particular vertex v is in the random stable set as the *marginal* of p at v and denote it $f_p(v)$. The vector f_p is called the *marginals of* p .

All this notation carries through for probability distributions on matchings in multigraphs in the natural way.

The following three observations are fundamental, the second in particular will be used frequently without being referenced.

22.1 *If M is chosen according to a hard-core distribution on the matchings of a multigraph G then:*

$$\forall e = xy, \ f_p(e) = \lambda(e)\mathbf{Pr}(x \notin V(M), y \notin V(M)).$$

Proof For each matching L containing e, $p(L) = \lambda(e)p(L - e)$. □

22.2 *If M is chosen according to a hard-core distribution on the matchings of a multigraph G then:*

$$\forall e, \ f_p(e) \leq \lambda(e).$$

22.3 *If M is chosen according to a hard-core distribution on the matchings of a multigraph G then: $\forall x \in V(G)$*

$$\mathbf{Pr}(x \in V(M)) = \sum_{xy \in E(G)} \lambda(xy)\mathbf{Pr}(x \notin V(M), y \notin V(M)).$$

The following lemma shows that we can generate a matching drawn from a hard-core distribution one edge at a time.

Lemma 22.4 *Suppose $e = xy$ is an edge of a multigraph G and M is chosen according to a hard-core distribution on the matchings of G. Suppose M_1 (resp. M_2) is chosen from the matchings of $G - e$ (resp. $G - x - y$) using the hard-core distribution with the same activities. Then:*

(a) for any matching N in $G - e$, $\mathbf{Pr}(M_1 = N) = \mathbf{Pr}(M = N | e \notin M)$;
(b) for any matching N in $G - x - y$, $\mathbf{Pr}(M_2 = N) = \mathbf{Pr}(M = N + e | e \in M)$.

Proof Let $W_1 = \sum_{M \in \mathcal{M}(G-e)} \lambda(M)$. Let $W_2 = \sum_{M \in \mathcal{M}(G-x-y)} \lambda(M)$. Let $W_3 = \sum_{M \in \mathcal{M}(G)} \lambda(M)$. Then $W_3 = W_1 + \lambda(e)W_2$. Furthermore, for

$N \in \mathcal{M}(G-e)$, we have: $\mathbf{Pr}(M_1 = N) = \frac{\lambda(N)}{W_1}$, $\mathbf{Pr}(M = N) = \frac{\lambda(N)}{W_3}$, and $\mathbf{Pr}(e \notin M) = \frac{W_1}{W_3}$. Thus, (a) holds. Similarly, for $N \in \mathcal{M}(G-x-y)$ we have: $\mathbf{Pr}(M_2 = N) = \frac{\lambda(N)}{W_2}$, $\mathbf{Pr}(M = N+e) = \frac{\lambda(N)\lambda(e)}{W_3}$, and $\mathbf{Pr}(e \in M) = \frac{\lambda(e)W_2}{W_3}$. Thus, (b) holds. □

22.2 Hard-Core Distributions from Fractional Colourings

In this chapter, we prove the existence of hard-core distributions on the matchings of G with certain desirable approximate independence properties. More precisely, we consider the question: what conditions on a vector x (indexed by the edges of a graph) are sufficient to ensure that there is a hard-core distribution with marginals x for which the probability a particular edge is in the random matching is not significantly affected by the choices made in distant parts of the graph? (Recall that the marginal on an edge is the probability this edge is in the random matching.) To begin, we consider an easier related question: for which non-negative vectors x does a hard-core distribution p exist with marginals x?

Our first remark is that a probability distribution is nothing but a convex combination of matchings (we set $p(M) = a_M$). Thus, Edmonds' characterization of the Matching Polytope (Theorem 21.3) can be restated as:

Theorem 22.5 *There is a probability distribution p with marginals x if and only if:*

(a) $\forall v \in V,\ \sum_{e \ni v} x_e \leq 1,$ *and*
(b) $\forall H \subseteq G, |V(H)| \text{ odd},\ \sum_{e \in E(H)} x_e \leq \frac{|V(H)|-1}{2}.$

Surprisingly, to ensure that there is a *hard-core* distribution with given marginals, we do not need to strengthen these conditions to any great extent. Specifically, Lee [103] and independently Rabinovich, Sinclair and Widgerson [130] showed:

Theorem 22.6 *There is a hard-core probability distribution p with marginals x if and only if the inequalities (a) and (b) above are strict.*

This result can be obtained via the classical technique of Lagrange multipliers. Since this technique is not a topic of the book, we omit the proof. Applying (21.4), we deduce from this result:

Theorem 22.7 *There is a hard-core probability distribution p with marginals $(\frac{1}{c}, \ldots, \frac{1}{c})$ if and only if there is a fractional c'-edge colouring of G with $c' < c$, i.e. if and only if $\chi_e^* < c$.*

It turns out that to ensure that we can choose a hard-core distribution p with marginals x such that the probability that a particular edge is in the random matching is not significantly affected by our choices in distant parts of the graph, we need only strengthen the conditions in Theorem 22.6 slightly.

Definition For $0 < t < 1$, we say that a non-negative vector x is in $t\mathcal{MP}(G)$ if

(a) $\forall v \in V, \ \sum_{v \in e} x_e \leq t$, and
(b) $\forall H \subseteq G, |V(H)|$ odd, $\sum_{e \in E(H)} x_e \leq t\frac{|V(H)|-1}{2}$.

Theorem 22.6 states that for any $\delta > 0$, if x is in $(1-\delta)\mathcal{MP}(G)$ then there is a hard-core distribution whose marginals are x. Actually, for any such hard-core distribution, the probability that a particular edge is in the random matching is not significantly affected by our choices in distant parts of the graph. The only catch is that our definition of distant depends on δ. We now make this rather vague statement more precise.

We saw in applying the hard-core distributions of the last chapter that there seemed to be some degree of independence between what the random stable set looked like in one neighbourhood of the graph, and what it looked like elsewhere. For example, in studying triangle-free graphs, we obtained a lower bound on $\mathbf{Pr}(v \in S) + \Delta^{-1}\mathbf{E}(|S \cap N(v)|)$ which held independently of our choice of $S - v - N(v)$. We will need very strong independence properties of this type. In particular, we will use hard-core distributions such that for a random matching M chosen according to the distribution and for any two far apart edges e and f: $\mathbf{Pr}(e \in M | f \in M) \approx \mathbf{Pr}(e \in M)$. (I.e. $\mathbf{Pr}(e \in M, f \in M) \approx \mathbf{Pr}(e \in M)\mathbf{Pr}(f \in M)$.)

This will not be the case for an arbitrary choice of graph and activities. Consider, for example, a chordless cycle C with $2n$ edges and a hard-core distribution in which $\lambda(e) = 2^{3n}$ for each edge e. Since C has fewer than 2^{2n} matchings, we can easily compute that M is almost always one of the two perfect matchings on C. In fact, M is a perfect matching with probability exceeding $1 - \frac{1}{2^n}$. Thus, for each edge e of C, $\mathbf{Pr}(e \in M) \approx \frac{1}{2}$. Moreover, for any two edges e and f of C, no matter how far apart, we have either $\mathbf{Pr}(e \in M | f \in M) \approx 1$ or $\mathbf{Pr}(e \in M | f \in M) \approx 0$. Thus, this hard-core distribution does not have the desired independence properties.

The hard-core distribution of the last paragraph is badly behaved because the associated λ are too large. It turns out that the hard-core distributions we will use have bounded λ and this fact will provide the independence properties we need. Specifically, we will show:

Lemma 22.8 [92] *$\forall 0 < \delta < 1, \quad \exists \delta' > 0$ such that if p is a hard-core distribution with marginals in $(1-\delta)\mathcal{MP}(G)$ and M is a matching chosen according to p then $\forall x, y \in V(G), \mathbf{Pr}(x \notin V(M), y \notin V(M)) > \delta'$.*

Applying (22.1), we can then obtain:

Corollary 22.9 $\forall 0 < \delta < 1$, $\exists K$ *such that if* p *is a hard-core distribution with marginals in* $(1-\delta)\mathcal{MP}(G)$ *then*

1. $\forall e \in E(G), \lambda(e) < Kf_p(e)$, *and*
2. $\forall x \in V(G)$ *we have:* $\sum_{e \ni x} \lambda(e) < K$.

Proof Set $K = \frac{1}{\delta'}$. By (22.1), for an edge e with endpoints x and y, $f_p(e) = \lambda(e)\mathbf{Pr}(x \notin V(M), y \notin V(M))$. So, by Lemma 22.8, $\lambda(e) < Kf_p(e)$. Since $\sum_{e \ni x} f_p(e) = \mathbf{Pr}(x \in V(M)) < 1$, this implies the desired results. □

Applying this corollary, Kahn and Kayll [92] obtained the following strong result about the independence properties of hard-core distributions on matchings.

Definition Suppose that we are choosing a random matching M from some probability distribution. For a vertex v we say that an event Q is *t-distant* from v if it is completely determined by the choice of all the matching edges at distance t or greater from v. We say that an event is t-distant from an edge e if it is t-distant from both ends of e.

Lemma 22.10 *Fix* $K > 0$ *and an* ϵ *strictly between* 1 *and* 0. *Let* $t = t(\epsilon) = 8(K+1)^2\epsilon^{-1}+2$. *Consider a multigraph* G *and hard-core distribution* p *whose activities satisfy:* $(*)\forall x \in V(G)$, $\sum_{f \ni x} \lambda(f) < K$. *If we choose a matching* M *according to* p *then the following is true:*

> *For any edge* e *and event* Q *which is* t*-distant from* e,
> $(1-\epsilon)\mathbf{Pr}(e \in M) \leq \mathbf{Pr}(e \in M|Q) \leq (1+\epsilon)\mathbf{Pr}(e \in M)$.

Their results also imply:

Corollary 22.11 *For any* $0 < \delta < 1$, *there exists* $K = K(\delta)$ *such that for any multigraph* G *with fractional chromatic index* c, *there is a hard-core distribution* p *with marginals* $(\frac{1-\delta}{c}, \ldots, \frac{1-\delta}{c})$, *such that* $\forall e \in E(G), \lambda(e) \leq \frac{K}{c}$ *and for all* $v \in V(G)$, $\sum_{f \ni x} \lambda(f) \leq K$.

Proof By Theorem 22.7 there exists a hard-core distribution with the given marginals. By (21.4), the vector $(\frac{1}{c}, \ldots, \frac{1}{c})$ is in $\mathcal{MP}(G)$ and so $(\frac{1-\delta}{c}, \ldots, \frac{1-\delta}{c})$ is in $(1-\delta)\mathcal{MP}(G)$. The result follows by Corollary 22.9. □

In the next chapter, we will combine these last two approximate independence results with the Lopsided Local Lemma to obtain results on edge-colouring multigraphs.

The rest of this chapter is devoted to proving Lemma 22.8 and Lemma 22.10. The key to these results is showing that hard-core distributions are fixed points of a certain map on probability distributions over $\mathcal{M}(G)$ which we do in the next section.

Thus far in the book, we have tended not to focus on the proofs of our probabilistic tools, particularly when these proofs are long, focussing instead on presenting applications of these tools. We break with this tradition here, partly because the proofs are combinatorial in nature, partly because we present modified versions of the original proof which some readers may find easier to digest.

The reader who wishes that we had continued to spare him the grubby details, can skip immediately to the next chapter. He will not miss anything that is required in the remainder of the book.

22.3 The Mating Map

In analyzing hard-core distributions we will often generate a matching from a pair of matchings. We say that two (ordered) pairs of matchings (M_1, M_2) and (M_3, M_4) are *compatible* if $M_1 \cap M_2 = M_3 \cap M_4$ and $M_1 \cup M_2 = M_3 \cup M_4$. We note that compatability is an equivalence relation. For each pair (M_1, M_2) of matchings, each component of $M_1 \cup M_2$ is a path or a cycle. Furthermore, $M_1 \cap M_2$ consists of the union of some of the single edge components of $M_1 \cup M_2$.

We say that a component of $M_1 \cup M_2$ is *interesting* if it is not an edge of $M_1 \cap M_2$. We note that each interesting component U of $M_1 \cup M_2$ has precisely one partition into two matchings, namely: $\{M_1 \cap U, M_2 \cap U\}$. It follows that if $M_1 \cup M_2$ has l interesting components then there are precisely 2^l pairs of matchings compatible with (M_1, M_2).

For any pair (M_1, M_2) of matchings, we let $\phi(M_1, M_2)$ be a uniformly chosen pair of matchings (M_1^*, M_2^*) compatible with (M_1, M_2). We define $\theta(M_1, M_2)$ to be M_1^*. Clearly, $\theta(M_1, M_2)$ can be generated from $M_1 \cup M_2$ by setting for each component U of $M_1 \cup M_2$: $\theta(M_1, M_2) \cap U = M_1 \cap U$ with probability $\frac{1}{2}$, and $\theta(M_1, M_2) \cap U = M_2 \cap U$ with probability $\frac{1}{2}$, with these choices made independently.

For a given probability distribution p, we can generate a random matching M by choosing M_1 and M_2 independently according to p, and then setting $M = \theta(M_1, M_2)$. This generates a new probability distribution on $\mathcal{M}(G)$. We obtain the *mating map* Υ by letting Υ_p be the new distribution obtained upon applying this procedure to p. We use Υ for Υ_p when the choice of p is clear. Our analysis of hard-core distributions relies on the following result:

Lemma 22.12 *Every hard-core distribution is a fixed point of the mating map.*

Proof We use g_p or simply g to denote the probability distribution on pairs of matchings obtained by choosing independent random M_1, M_2 according to p, and taking the random pair of matchings $\phi(M_1, M_2)$. Let p be a hard-core distribution. We actually show

22.13 *For all $M_1', M_2' \in \mathcal{M}(G)$, $g(M_1', M_2') = p(M_1')p(M_2')$.*

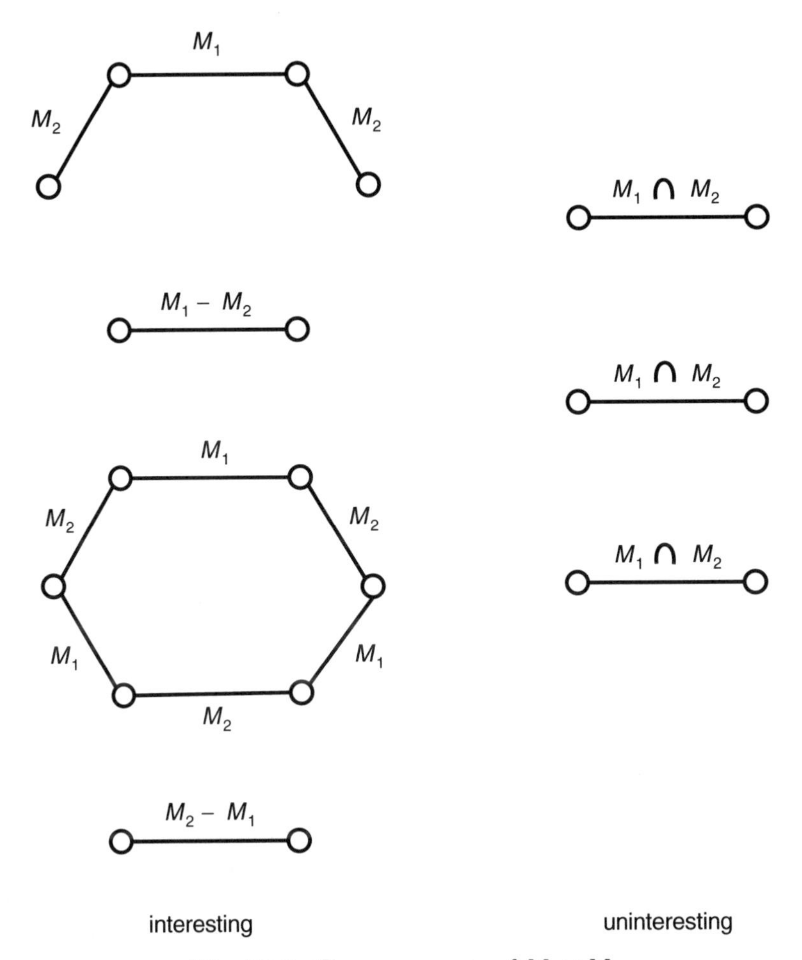

Fig. 22.1. The components of $M_1 \cup M_2$

This implies: $\Upsilon(M) = \sum_{M' \in \mathcal{M}(G)} g(M, M') = \sum_{M' \in \mathcal{M}(G)} p(M)p(M') = p(M)$, as required. It remains only to prove (22.13).

So, fix a pair (M_1', M_2') of matchings. Let $\mathcal{S} = \mathcal{S}(M_1', M_2')$ be the equivalence class of pairs of matchings compatible with (M_1', M_2'). Let l be the number of interesting components of $M_1' \cup M_2'$. Then $|\mathcal{S}| = 2^l$. So, for any $(N_1, N_2) \in \mathcal{S}$, we have:

22.14 $\mathbf{Pr}(\phi(N_1, N_2) = (M_1', M_2')) = 2^{-l}$.

Furthermore, setting $\mathcal{W} = \sum_{M \in \mathcal{M}(G)} \lambda(M)$, for such a pair we have:

$$\begin{aligned} p(N_1)p(N_2) &= \frac{\lambda(N_1)}{\mathcal{W}} \frac{\lambda(N_2)}{\mathcal{W}} \\ &= \frac{\lambda(N_1 \cup N_2)}{\mathcal{W}} \frac{\lambda(N_1 \cap N_2)}{\mathcal{W}} \end{aligned}$$

$$= \frac{\lambda(M_1' \cup M_2')}{\mathcal{W}} \frac{\lambda(M_1' \cap M_2')}{\mathcal{W}}$$
$$= \frac{\lambda(M_1')}{\mathcal{W}} \frac{\lambda(M_2')}{\mathcal{W}} = p(M_1')p(M_2').$$

Thus,

$$\begin{aligned} g(M_1', M_2') &= \sum_{(N_1,N_2)\in\mathcal{S}} \mathbf{Pr}\left((M_1, M_2) = (N_1, N_2)\right) \\ &\qquad \times \mathbf{Pr}\left(\phi(N_1, N_2) = (M_1', M_2')\right) \\ &= \sum_{(N_1,N_2)\in\mathcal{S}} p(N_1)p(N_2)2^{-l} \\ &= \sum_{(N_1,N_2)\in\mathcal{S}} p(M_1')p(M_2')2^{-l} \\ &= |\mathcal{S}|p(M_1')p(M_2')2^{-l} \\ &= p(M_1')p(M_2'). \end{aligned}$$

□

22.4 An Independence Result

We now restrict our attention to some positive $\delta < 1$ and a hard-core distribution p with marginals in $(1-\delta)\mathcal{MP}(G)$. Using the seemingly innocuous fact that p is a fixed point of the mating map, we will give a straightforward description of how conditioning on some vertex not being on the random matching generated according to p affects the probability that other vertices are on this matching. We then apply this result to prove Lemma 22.8.

Notation: For a matching N and vertex x, we use $x \prec N$ to mean x is an endpoint of an edge of N. For $Z \subseteq V(G)$, we use $Z \cap N$ for $\{x | x \in Z, x \prec N\}$. M always refers to a random matching generated according to p, and M_1, M_2 to a random pair of matchings generated independently according to p.

For any such pair M_1, M_2 we will think of $M_1 \cup M_2$ as a spanning subgraph and thus $M_1 \cup M_2$ may have singleton (i.e. single vertex) components. For each component U of $M_1 \cup M_2$, we set $\theta(M_1, M_2) \cap U$ to be either $M_1 \cap U$ or $M_2 \cap U$, where each possibility is equally likely. Of course, if U is a singleton component, it makes no difference which choice we make, since $M_1 \cap U = M_2 \cap U$. But it will be convenient for us to distinguish between the choice of $M_1 \cap U$ or $M_2 \cap U$, even when U is a singleton.

We use d_{xy} to denote $\mathbf{Pr}(x \not\prec M, y \not\prec M)$.

To begin, we relate d_{xy} to $\mathbf{Pr}(x \not\prec M)\mathbf{Pr}(y \not\prec M)$ which would be its value if the events $x \prec M$ and $y \prec M$ were independent. This result, though simple, is the key to the analysis:

Definition For a pair $\{x, y\}$ of vertices of G, we let $Ev(x, y)$ be the event that x and y are the endpoints of an even path which is a component of $M_1 \cup M_2$ (recall that the length of a path is the number of edges it contains). We let $Od(x, y)$ be the event that x and y are the endpoints of an odd path which is a component of $M_1 \cup M_2$.

Lemma 22.15 *For all $x, y \in V(G)$ with $x \neq y$, we have:*

$$d_{xy} = \mathbf{Pr}(x \not\prec M)\mathbf{Pr}(y \not\prec M) + \frac{1}{2}\mathbf{Pr}(Od(x, y)) - \frac{1}{2}\mathbf{Pr}(Ev(x, y)).$$

Proof We generate the random matching M by first generating M_1,M_2 and then setting $M = \theta(M_1, M_2)$.

We let U_x be the component of $M_1 \cup M_2$ containing x and U_y be the component containing y. We let $A_1(x)$ be the event that $x \not\prec M_1$ and $M \cap U_x = M_1 \cap U_x$. We let $A_2(x)$ be the event that $x \not\prec M_2$ and $M \cap U_x = M_2 \cap U_x$. We recall that, by our remarks above, these events are well-defined and disjoint even if U_x is a singleton. We define $A_1(y)$ and $A_2(y)$ similarly. Then,

$$\begin{aligned} d_{xy} &= \mathbf{Pr}(A_1(x), A_1(y)) + \mathbf{Pr}(A_1(x), A_2(y)) + \mathbf{Pr}(A_2(x), A_1(y)) \\ &\quad + \mathbf{Pr}(A_2(x), A_2(y)) \\ &= 2\mathbf{Pr}(A_1(x), A_1(y)) + 2\mathbf{Pr}(A_1(x), A_2(y)), \quad \text{by symmetry.} \end{aligned}$$

Now

$$\begin{aligned} \mathbf{Pr}(A_1(x), A_1(y)) &= \frac{1}{4}\mathbf{Pr}(x \not\prec M_1, y \not\prec M_1, U_x \neq U_y) \\ &\quad + \frac{1}{2}\mathbf{Pr}(x \not\prec M_1, y \not\prec M_1, U_x = U_y), \\ &= \frac{1}{4}\mathbf{Pr}(x \not\prec M_1, y \not\prec M_1) \\ &\quad + \frac{1}{4}\mathbf{Pr}(x \not\prec M_1, y \not\prec M_1, U_x = U_y), \end{aligned}$$

and

$$\begin{aligned} \mathbf{Pr}(Od(x, y)) &= \mathbf{Pr}(x \not\prec M_1, y \not\prec M_1, U_x = U_y) \\ &\quad + \mathbf{Pr}(x \not\prec M_2, y \not\prec M_2, U_x = U_y) \\ &= 2\mathbf{Pr}(x \not\prec M_1, y \not\prec M_1, U_x = U_y), \quad \text{by symmetry.} \end{aligned}$$

So, we obtain:

$$\mathbf{Pr}(A_1(x), A_1(y)) = \frac{1}{4}\mathbf{Pr}(x \not\prec M_1, y \not\prec M_1) + \frac{1}{8}\mathbf{Pr}(Od(x, y)).$$

Similarly, we have:

$$\mathbf{Pr}(A_1(x), A_2(y)) = \frac{1}{4}\mathbf{Pr}(x \not\sim M_1, y \not\sim M_2) - \frac{1}{8}\mathbf{Pr}(Ev(x,y)).$$

So,

$$\begin{aligned} d_{xy} = & \frac{1}{2}\mathbf{Pr}(x \not\sim M_1, y \not\sim M_1) + \frac{1}{2}\mathbf{Pr}(x \not\sim M_1, y \not\sim M_2) \\ & + \frac{1}{4}\mathbf{Pr}(Od(x,y)) - \frac{1}{4}\mathbf{Pr}(Ev(x,y)). \end{aligned}$$

Since M, M_1 and M_2 all have the same distribution, we obtain:

$$d_{xy} = \frac{1}{2}d_{xy} + \frac{1}{2}\mathbf{Pr}(x \not\sim M)\mathbf{Pr}(y \not\sim M) + \frac{1}{4}\mathbf{Pr}(Od(x,y)) - \frac{1}{4}\mathbf{Pr}(Ev(x,y)).$$

Rearranging this equation yields the desired result. □

Now, the proof of Lemma 22.8 falls into two parts. We show:

(i) if some d_{xy} is very small then there is a small set S such that for all $u, v \in S$, d_{uv} is small whilst for all $u \in S, v \notin S$, d_{uv} is big.
(ii) there is no set S as in (i).

Combining these two results yields a lower bound on min $\{d_{xy}\}$ whose precise value depends on the definition of "very small" in (i). Both parts of the proof require us to apply Lemma 22.15. We shall also need the following weak triangle inequality, which holds for all d.

Lemma 22.16 *Suppose $d_{xy} < d$ and $d_{xz} < d$. Then $d_{yz} < 8\delta^{-1}d$.*

Proof Consider our random pair of matchings (M_1, M_2). Let E be the event that $y \not\sim M_1, z \not\sim M_1$ and $x \not\sim M_2$. Note that $\mathbf{Pr}(x \not\sim \theta(M_1, M_2), y \not\sim \theta(M_1, M_2) | (U_x \neq U_y) \cap E) \geq \frac{1}{4}$ and $\mathbf{Pr}(x \not\sim \theta(M_1, M_2), z \not\sim \theta(M_1, M_2) | (U_x \neq U_z) \cap E) \geq \frac{1}{4}$. Furthermore, if E holds then either $U_x \neq U_y$ or $U_z \neq U_y$. Combining these two facts we obtain:

$$\begin{aligned} & \mathbf{Pr}(x \not\sim \theta(M_1, M_2), y \not\sim \theta(M_1, M_2) | E) \\ + & \mathbf{Pr}(x \not\sim \theta(M_1, M_2), z \not\sim \theta(M_1, M_2) | E) \geq \frac{1}{4}. \end{aligned}$$

This implies:

22.17 $d_{xy} + d_{xz} \geq \frac{1}{4}\mathbf{Pr}(E)$.

Now, since p has marginals in $(1-\delta)\mathcal{MP}(G)$, $\mathbf{Pr}(x \not\sim M) \geq \delta$. So, as M_1, M_2 and M have the same distribution, we obtain:

$$\mathbf{Pr}(E) = d_{yz}\mathbf{Pr}(x \not\sim M) \geq \delta d_{yz}$$

So $d_{xy} + d_{xz} \geq \frac{\delta}{4} d_{yz}$, and the result follows. □

To prove Lemma 22.8, we need only prove it for $\delta < \frac{1}{10}$, since if the statement of the lemma holds for some value of δ it holds for all larger values of δ with the same value of δ'. So we assume $\delta < \frac{1}{10}$ in what follows. We can now establish item (i) preceding the statement of Lemma 22.16:

Lemma 22.18 *There is a $\delta' > 0$ such that if there exist x and y with $d_{xy} < \delta'$ then there exists a set S with $2 \leq |S| \leq \delta^{-2} + 1$ and $\delta^* > 0$ such that: $\forall u, v \in S, d_{uv} < \frac{\delta^{12}(\delta^*)^2}{8000}$, and $\forall u \in S, v \notin S, d_{uv} > \delta^*$.*

Proof Set $\delta_0 = \frac{\delta^2}{2}$ and recursively define for $1 \leq i \leq \lceil \delta^{-2} \rceil$, $\delta_i = 10^{-7}\delta_{i-1}^2\delta^{15}$. Set $\delta' = \delta_{\lceil \delta^{-2} \rceil}$.

Suppose that there exists a pair of vertices $\{x, y\}$ with $d_{xy} < \delta'$. Set $S_x = \{z : d_{xz} < \frac{\delta^2}{2}\}$. We need:

22.19 $|S_x| < \delta^{-2}$.

Proof Since $f_p \in (1-\delta)\mathcal{MP}(G)$, by definition $\mathbf{Pr}(x \not\sim M) \geq \delta$, so $\forall z \in V(G) : \mathbf{Pr}(x \not\sim M)\mathbf{Pr}(z \not\sim M) \geq \delta^2$. Thus, by Lemma 22.15 $\forall z \in S_x : \mathbf{Pr}(Ev(x, z)) > \delta^2$. Since the events $Ev(x, z_1)$ and $Ev(x, z_2)$ are disjoint events for $z_1 \neq z_2$, the result follows. □

Since S_x has fewer than δ^{-2} elements, there is an i with $0 \leq i < \delta^{-2}$ such that there is no z with $d_{xz} \in [\delta_{i+1}, \delta_i)$. That is, for some i with $0 \leq i < \delta^{-2}$, we have:

(i) $d_{xy} < \delta' < 10^{-7}\delta_i^2\delta^{15}$
(ii) $\nexists z$ such that $10^{-7}\delta_i^2\delta^{15} \leq d_{xz} < \delta_i$

For some such i, we let $S = x \cup \{z : d_{xz} < 10^{-7}\delta_i^2\delta^{15}\}$. For any $u, v \in S$, applying Lemma 22.16, to x, u, v yields that: $d_{uv} < 10^{-6}\delta_i^2\delta^{14}$. Now for any $u \in S$ and $v \notin S$, as (ii) holds, applying Lemma 22.16, to u, x, v yields: $d_{uv} \geq \frac{\delta}{8}\delta_i$. By (i), $\{x, y\} \subseteq S$ and so $|S| \geq 2$. Since $S \subseteq S_x + x$, $|S| \leq \delta^{-2} + 1$. So, we obtain the desired result with $\delta^* = \frac{\delta}{9}\delta_i$. □

By this result, to complete the proof of Lemma 22.8, we need only show:

Lemma 22.20 *There is no set S with $2 \leq |S| \leq \delta^{-2} + 1$ and $\delta^* > 0$ such that: $\forall u, v \in S, d_{uv} < \frac{\delta^{12}(\delta^*)^2}{8000}$, and $\forall u \in S, v \notin S, d_{uv} > \delta^*$.*

Proof We assume the contrary and choose an S and δ^* for which the lemma fails. We first prove:

22.21 $\exists$ *distinct* $y_1, y_2 \in S$ *s.t.* $\mathbf{Pr}(y_1 \not\sim (M \cap E(S)), y_2 \not\sim (M \cap E(S)) \geq \delta^6$.

Proof We assume the contrary and define the random set $S' = \{z | z \in S, z \not\sim M \cap E(S)\}$. By assumption,

$$\mathbf{Pr}(|S'| \geq 2) < \binom{\lfloor \delta^{-2} \rfloor + 1}{2} \delta^6 < \delta^2, \text{ as } \delta < \frac{1}{10}.$$

We note that $|S|$ and $|S'|$ have the same parity.

Thus, if $|S|$ is even then, $\mathbf{Pr}(|S'| > 0) = \mathbf{Pr}(|S'| \geq 2) < \delta^2$, and so: $\mathbf{E}(|S'|) < \delta^2 |S|$. But for each $z \in V(G)$, $\mathbf{Pr}(z \not\prec M) \geq \delta$ and summing over all $z \in S$ we see that $\mathbf{E}(|S'|) \geq \delta |S|$, a contradiction.

If $|S|$ is odd, then our bound on $\mathbf{Pr}(|S'| \geq 2)$ yields: $\mathbf{E}(|S'|) < 1 + \delta^2(|S| - 1)$. So, $\mathbf{E}(|M \cap E(S)|) = \frac{|S| - \mathbf{E}(|S'|)}{2} > \frac{|S|-1}{2}(1 - \delta^2)$. But since p has marginals in $(1-\delta)\mathcal{MP}(G)$, $E(|M \cap E(S)|) = \sum_{e \in E(S)} f_p(e) \leq \frac{|S|-1}{2}(1-\delta)$, which yields a contradiction. □

Now if $y \not\prec M \cap E(S)$ then either $y \not\prec M$ or $y \prec M - E(S)$. Further, by assumption, $\forall y_1, y_2 \in S$, $d_{y_1, y_2} < \frac{\delta^{12}(\delta^*)^2}{8000} < \frac{\delta^6}{4}$. So, by (22.21) and the fact that $\frac{\delta^6}{4} + \frac{2\delta^{10}\delta^*}{160} + \frac{2\delta^6}{3} \leq \delta^6$, either:

$$\exists y_1, y_2 \in S, \ s.t. \ \mathbf{Pr}(y_1 \not\prec M, y_2 \prec M - E(S)) > \frac{\delta^{10}\delta^*}{160},$$

or

$$\exists y_1, y_2 \in S, \ s.t. \ \mathbf{Pr}(y_1 \prec M - E(S), y_2 \prec M - E(S)) > \frac{2\delta^6}{3}.$$

Case 1: $\exists y_1, y_2 \in S, \ s.t. \ \mathbf{Pr}(y_1 \not\prec M, y_2 \prec M - E(S)) > \frac{\delta^{10}\delta^*}{160}$.

We choose such a pair (y_1, y_2). We focus on the event A that

$$y_1 \not\prec M_1, y_2 \not\prec (M_1 \cap E(S)),$$
$$V(M_1) \cap (S - y_1 - y_2) = S - y_1 - y_2, y_2 \not\prec M_2.$$

Using the fact that for all $w \in S$, $d_{y_1 w} \leq \frac{1}{8000}\delta^{12}(\delta^*)^2$ and summing over all of the at most $\lfloor \delta^{-2} \rfloor$ w in $S - y_1 - y_2$ we see:

$$\mathbf{Pr}(y_1 \not\prec M, V(M) \cap (S - y_1 - y_2) \neq S - y_1 - y_2) < \frac{\delta^{-2}}{8000}\delta^{12}(\delta^*)^2.$$

So, we obtain:

22.22 $\mathbf{Pr}(y_1 \not\prec M, y_2 \prec M - E(S), V(M) \cap (S - y_1 - y_2) = S - y_1 - y_2)$
$> \frac{1}{160}\delta^{10}\delta^* - \frac{\delta^{-2}}{8000}\delta^{12}(\delta^*)^2 \geq \frac{1}{200}\delta^{10}\delta^*.$

Since $\mathbf{Pr}(y_2 \not\prec M_2) > \delta$, (22.22) implies

$$\mathbf{Pr}(A) \geq \frac{1}{200}\delta^{11}\delta^*.$$

We partition the pairs of matchings in A into equivalence classes where (M_1^*, M_2^*) and (M_1', M_2') are in the same equivalence class if $M_2^* = M_2'$ and

M_1^* either is M_1' or differs from M_1' only in the choice of the edge incident to y_2. (I.e. all of the at most two edges in the symmetric difference of M_1^* and M_1' contain y_2.) We note that for a given equivalence class C, there is at most one pair of matchings $N(C) = (N_1, N_2)$ in C, such that y_1 and y_2 are in the same component of $N_1 \cup N_2$. Furthermore, there exists a pair of matchings in C, $L(C) = (L_1, L_2)$ such that $y_2 \not\prec L_1$ and $y_2 \not\prec L_2$. Note that $N_1 = L_1 \cup e$ for some edge $e = y_2 w$ with $w \notin S$ and $N_2 = L_2$. Thus $\lambda(N_1)\lambda(N_2) = \lambda(L_1)\lambda(L_2)\lambda(e)$.

Since $y_2 \in S$ and $w \notin S$, we have: $\mathbf{Pr}(y \not\prec M, w \not\prec M) \geq \delta^*$. By (22.1), it follows that $\lambda(e) \leq (\delta^*)^{-1}$.

Next, we note that, letting Q be the event that $(M_1, M_2) = (N_1, N_2)$, we have $\mathbf{Pr}(Q|(M_1, M_2) \in C) \leq \mathbf{Pr}(Q|(M_1, M_2) \in \{(N_1, N_2), (L_1, L_2)\}) = \frac{\lambda(e)}{\lambda(e)+1} \leq \frac{(\delta^*)^{-1}}{1+(\delta^*)^{-1}}$. Moreover, $\mathbf{Pr}(y_1 \not\prec M, y_2 \not\prec M|(M_1, M_2) \in C)) \geq \frac{1}{4}(1 - \mathbf{Pr}(Q|(M_1, M_2) \in C) \geq \frac{1}{4+4(\delta^*)^{-1}} \geq \frac{\delta^*}{8}$.

Summing over the equivalence classes in A, we see:

$$\mathbf{Pr}(y_1 \not\prec M, y_2 \not\prec M|(M_1, M_2) \in A) \geq \frac{\delta^*}{8}.$$

So,

$$d_{y_1 y_2} \geq \frac{\delta^*}{8}\mathbf{Pr}(A) > \frac{(\delta^*)^2 \delta^{12}}{8000}.$$

But, this contradicts our choice of S.

Case 2: Case 1 does not hold. In this case, we know

$$\exists y_1, y_2 \in S,\ s.t.\ \mathbf{Pr}(y_1 \prec M - E(S), y_2 \prec M - E(S)) > \frac{2\delta^6}{3}.$$

We choose such a pair (y_1, y_2). We focus on the event B that

$$y_1 \prec M_1 - E(S), y_2 \prec M_1 - E(S),$$
$$V(M_1) \cap (S - y_1 - y_2) = S - y_1 - y_2, y_2 \not\prec M_2.$$

Using the fact that Case 1 does not hold, and summing over all pairs (w, y_1) with w in $S - y_1 - y_2$ we see:

$$\mathbf{Pr}(y_1 \prec M - E(S), y_2 \prec M - E(S), V(M) \cap (S - y_1 - y_2) = S - y_1 - y_2)$$
$$> \frac{2\delta^6}{3} - \frac{\lfloor \delta^{-2} \rfloor}{160}\delta^{10}\delta^* > \frac{\delta^6}{2}.$$

Since $\mathbf{Pr}(y_2 \not\prec M_2) > \delta$, this implies $\mathbf{Pr}(B) \geq \frac{\delta^7}{2}$. We can partition B into equivalence classes, just as we partioned A, and mimicing the reasoning in Case 1, show that for each equivalence class $\mathcal{C}$:

$$\mathbf{Pr}(y_1 \prec M - E(S), y_2 \not\prec M|(M_1, M_2) \in \mathcal{C}) \geq \frac{\delta^*}{16}.$$

So,

$$\mathbf{Pr}(y_1 \prec M - E(S) \cap y_2 \not\prec M) \geq \frac{\delta^*}{16}\mathbf{Pr}(B) > \frac{(\delta^*)\delta^8}{160},$$

which contradicts the fact that Case 1 does not hold.

Remark The 8 from the analogous inequalites in Case 1 is replaced by a 16 here because there may be two matchings in each equivalence class for which $U_{y_1} = U_{y_2}$ rather than just one. □

22.5 More Independence Results

In this section, we prove Lemma 22.10. Thus, we no longer assume that $f_p \in (1-\delta)\mathcal{MP}(G)$, simply that for some K and all x, we have: $\sum_{f \ni x} \lambda(f) \leq K$.

The idea of the proof is quite simple: we consider some fixed edge e of G and, by analyzing the mating map, bound the probability that a random pair (M_1, M_2) of matchings is bad in the sense that (i) $e \in M_1 \cup M_2$, and (ii) the component, U_e, of $M_1 \cup M_2$ containing e intersects the set of edges t-distant from e. This will be enough to imply the result, for if we consider a fixed pair M_1, M_2 which is not bad then the event "$e \in \theta(M_1, M_2)$" will be independent of the choice of the edges of $\theta(M_1, M_2)$ which are t-distant from e.

To show that the probability of (M_1, M_2) being bad is very small, we compare the set B of bad pairs with the set B' of pairs which are not bad but are obtained from a bad pair by deleting one edge in one of the matchings.

Note that if (M_1, M_2) is a bad pair then there are at least one and at most two paths P of U_e satisfying: one endpoint of P is on e, the other endpoint of P is t-distant from e, and no internal vertex of P is t-distant from e or on e. Deleting one edge from each such path yields a pair in B'. Thus, there are at least $t-1$ pairs of B' corresponding in this fashion to each pair of B.

Remark Note that although every nearly bad pair is obtained from a bad pair by deleting one edge, there are bad pairs from which we must delete two edges to obtain a nearly bad pair. We highlight this asymmetry between the definition of nearly bad pair and the correspondence between bad pairs and nearly bad pairs to simplify the reader's task.

Of course, there may be many pairs in B corresponding to each pair in B'. In fact, for some graphs, B is much larger than B'. However, every pair in B corresponding to a specific pair (L_1, L_2) in B' arises by adding up to two edges each of which is incident to one of the two vertices which are the endpoints of the component of $L_1 \cup L_2$ containing e. This allows us to use the fact that $\sum_{f \ni y} \lambda_f \leq K$ for each y in V, to show that the probability that (M_1, M_2) is bad is much smaller than the probability it is in B'. Forthwith the details.

Proof of Lemma 22.10 We let $G(e,d)$ be the graph induced by the vertices at distance at least d from e. Recall that $t = 8(K+1)^2\epsilon^{-1}+2$. and consider a fixed matching N in $E(G(e,t-1))$. A matching L is *consistent* (with N) if $L \cap E(G(e,t-1)) = N$. Let *Cons* be the event that M is consistent. We shall prove that, for every edge e, letting A_e be the event that M contains e, we have:

22.23 $(1-\epsilon)\mathbf{Pr}(A_e) < \mathbf{Pr}(A_e|Cons) < (1+\epsilon)\mathbf{Pr}(A_e)$.

Now, by definition, for each event Q which is t-distant from e there is a subset $\mathcal{M}(Q)$ of $\mathcal{M}(G(e,t-1))$ such that

$$Q = \cup_{L\in\mathcal{M}(Q)} M \text{ is consistent with } L.$$

Summing (22.23) over all the matchings in $\mathcal{M}(Q)$ yields Lemma 22.10.

So, we need only prove (22.23). In doing so, we essentially follow the outline sketched above. However, since we are focusing on a particular matching N of $G(e,t-1)$, our definition of bad must be modified slightly. To wit, we say a pair of matchings (L_1,L_2) is *bad* if L_1 is consistent, $L_1\cup L_2$ contains e, and the component of $L_1\cup L_2$ containing e intersects $E(G(e,t-1))$. We let $(M_1',M_2') = \phi(M_1,M_2)$. Thus, M, M_1, M_2, M_1', M_2' all have the same distribution and (M_1,M_2) and (M_1',M_2') have the same (joint) distribution. Let *Bad* be the event that (M_1',M_2') is bad. We show:

22.24 $|\mathbf{Pr}(Cons\cap A_e) - \mathbf{Pr}(Cons)\mathbf{Pr}(A_e)| \le 2\mathbf{Pr}(Bad)$.

22.25 $\mathbf{Pr}(Bad) < \frac{\epsilon}{2}\mathbf{Pr}(Cons)\mathbf{Pr}(A_e)$.

Combining these results we have:

$$\begin{aligned} &|\mathbf{Pr}(A_e|Cons) - \mathbf{Pr}(A_e)| \\ &= \left|\frac{\mathbf{Pr}(Cons\cap A_e)}{\mathbf{Pr}(Cons)} - \mathbf{Pr}(A_e)\right| \le \frac{2\mathbf{Pr}(Bad)}{\mathbf{Pr}(Cons)} < \epsilon\mathbf{Pr}(A_e). \end{aligned}$$

This proves (22.23); it remains only to prove (22.24) and (22.25).
In doing so, we will use:

22.26 $\mathbf{Pr}(Cons\cap A_e) - \mathbf{Pr}(Cons)\mathbf{Pr}(A_e)$
$= \mathbf{Pr}(Cons\cap A_e)\mathbf{Pr}(\overline{A_e}) - \mathbf{Pr}(Cons\cap\overline{A_e})\mathbf{Pr}(A_e)$.

Proof

$$\begin{aligned} &\mathbf{Pr}(Cons\cap A_e)\mathbf{Pr}(\overline{A_e}) - \mathbf{Pr}(Cons\cap\overline{A_e})\mathbf{Pr}(A_e) \\ =\ &\mathbf{Pr}(Cons\cap A_e) - \mathbf{Pr}(Cons\cap A_e)\mathbf{Pr}(A_e) - \mathbf{Pr}(Cons\cap\overline{A_e})\mathbf{Pr}(A_e) \\ =\ &\mathbf{Pr}(Cons\cap A_e) - \mathbf{Pr}(Cons)\mathbf{Pr}(A_e). \end{aligned}$$

□

Proof of (22.24). Using the fact that p fixes the joint distribution on matchings, we have:

$$\mathbf{Pr}(e \notin M_1', M_1' \text{ is consistent}, e \in M_2') = \mathbf{Pr}(Cons \cap \overline{A_e})\mathbf{Pr}(A_e),$$

and

$$\mathbf{Pr}(e \in M_1', M_1' \text{ is consistent}, e \notin M_2') = \mathbf{Pr}(Cons \cap A_e)\mathbf{Pr}(\overline{A_e}).$$

For M_1', M_2' with $e \in M_1' \cup M_2'$, we let U_e be the component of $M_1' \cup M_2'$ containing e, otherwise U_e is not defined. We consider generating a new pair of matchings (L_1, L_2) by swapping our choice of $M_1' \cap U_e$ and $M_2' \cap U_e$ if U_e exists. Clearly, this new pair of matchings has the same distribution as (M_1', M_2') (as we can make our choice on U_e by choosing one of $M_1 \cup e$ or $M_2 \cup e$ not to be $M_1' \cup e$). Furthermore, if neither the new pair nor the old pair is bad then

$$e \notin M_1', M_1' \text{ is consistent}, e \in M_2'$$

if and only if

$$e \in L_1, L_1 \text{ is consistent}, e \notin L_2.$$

Combining this fact with the two equations above, we obtain:

$$|\mathbf{Pr}(Cons \cap \overline{A_e})\mathbf{Pr}(A_e) - \mathbf{Pr}(Cons \cap A_e)\mathbf{Pr}(\overline{A_e})| \le 2\mathbf{Pr}(Bad).$$

By (22.26), this is the desired result. □

Proof of (22.25). We say a pair of matchings (L_1, L_2), which is not bad, is *nearly bad* if L_1 is consistent, $e \in L_1 \cup L_2$, and there is a bad pair of matchings (L_1', L_2') from which we can obtain (L_1, L_2) by deleting an edge $f \neq e$ from the component of $L_1' \cup L_2'$ containing e. Note that this implies that the component of $L_1 \cup L_2$ containing e is a path rather than a cycle.

For any bad pair (L_1', L_2') of matchings, we say a path of $L_1' \cup L_2'$ is *critical* if one of its endpoints is on e, one of its endpoints is in $G(e, t-1)$, and none of its internal vertices are in $G(e, t-1)$ or on e. Clearly, $L_1' \cup L_2'$ contains either one or two critical paths and these paths are disjoint. Furthermore, a critical path contains at least $t-2$ edges and if we delete exactly one edge from each critical path in a bad pair of matchings, we obtain a nearly bad pair of matchings.

We consider a bipartite graph F whose edges link the bad pairs of matchings to the nearly bad pairs. Specifically, a bad pair (L_1', L_2') is linked to a nearly bad pair (L_1, L_2) if we can obtain (L_1, L_2) from (L_1', L_2') by deleting exactly one edge from each critical path of $L_1' \cup L_2'$. The remarks of the last paragraph imply that every bad pair has at least $t-2$ neighbours in F.

We consider a nearly bad pair (L_1, L_2) and let x_1 and x_2 be the endpoints of the component of $L_1 \cup L_2$ containing e. We let $\mathcal{S} = \mathcal{S}(L_1, L_2)$ be the set of bad pairs of matchings adjacent to (L_1, L_2) in F. Thus any pair in $\mathcal{S}$ is obtained from (L_1, L_2) by adding an edge adjacent to x_1, an edge adjacent to x_2, or two edges, one adjacent to x_1 and the other to x_2. So,

$$\begin{aligned}\sum_{(L_1',L_2')\in\mathcal{S}} \lambda(L_1')\lambda(L_2') \leq & \sum_{f_1\ni x_1} \lambda(f_1)\lambda(L_1)\lambda(L_2) \\ & + \sum_{f_2\ni x_2} \lambda(f_2)\lambda(L_1)\lambda(L_2) \\ & + \sum_{f_1\ni x_1}\sum_{f_2\ni x_2} \lambda(f_1)\lambda(f_2)\lambda(L_1)\lambda(L_2).\end{aligned}$$

I.e.,

$$\begin{aligned}&\sum_{(L_1',L_2')\in\mathcal{S}} \lambda(L_1')\lambda(L_2') \\ &\leq \left(1+\sum_{f_1\ni x_1}\lambda(f_1)\right)\left(1+\sum_{f_2\ni x_2}\lambda(f_2)\right)\lambda(L_1)\lambda(L_2) \\ &\leq (K+1)^2\lambda(L_1)\lambda(L_2).\end{aligned}$$

On the other hand, as noted above, each bad pair is adjacent in F to at least $t-2$ nearly bad pairs. So, reversing the order of a double summation, we obtain:

$$\sum_{\substack{(L_1,L_2)\\ \text{nearly bad}}} \sum_{(L_1',L_2')\in\mathcal{S}(L_1,L_2)} \lambda(L_1')\lambda(L_2') \geq (t-2)\times \sum_{\substack{(L_1',L_2')\\ \text{bad}}} \lambda(L_1')\lambda(L_2').$$

Combining these results we see:

$$\sum_{\substack{(L_1',L_2')\\ \text{bad}}} \lambda(L_1')\lambda(L_2') \leq \frac{(K+1)^2}{t-2} \sum_{\substack{(L_1,L_2)\\ \text{nearly bad}}} \lambda(L_1)\lambda(L_2).$$

Now, for every nearly bad pair of matchings (L_1, L_2), L_1 is consistent, by definition. Furthermore, e is in precisely one of L_1 or L_2 because otherwise, however we add an edge to one of the matchings, e remains a component of the union of the two new matchings and hence we have not constructed a bad pair. So, since $t = 8(K+1)^2\epsilon^{-1} + 2$, the last inequality implies:

$$\begin{aligned}\mathbf{Pr}(Bad) &\leq \frac{\epsilon}{8}(\mathbf{Pr}(Cons\cap A_e)\mathbf{Pr}(\overline{A_e}) + \mathbf{Pr}(Cons\cap \overline{A_e})\mathbf{Pr}(A_e)) \\ &\leq \frac{\epsilon}{8}(2\mathbf{Pr}(Cons\cap A_e)\mathbf{Pr}(\overline{A_e}) \\ &\quad +(\mathbf{Pr}(Cons\cap \overline{A_e})\mathbf{Pr}(A_e) - \mathbf{Pr}(Cons\cap A_e)\mathbf{Pr}(\overline{A_e}))).\end{aligned}$$

Since $\mathbf{Pr}(\overline{A_e}) \leq 1$, applying (22.26), we obtain:

$$\begin{aligned}\mathbf{Pr}(Bad) &\leq \frac{\epsilon}{8}(2\mathbf{Pr}(Cons \cap A_e) + (\mathbf{Pr}(Cons \cap A_e) - \mathbf{Pr}(Cons)\mathbf{Pr}(A_e))) \\ &\leq \frac{\epsilon}{8}(2\mathbf{Pr}(Cons)\mathbf{Pr}(A_e) + |\mathbf{Pr}(Cons \cap A_e) - \mathbf{Pr}(Cons)\mathbf{Pr}(A_e)|)\end{aligned}$$

So, (22.24) yields:

$$\mathbf{Pr}(Bad) \leq \frac{\epsilon}{8}(2\mathbf{Pr}(Cons)\mathbf{Pr}(A_e) + 2\mathbf{Pr}(Bad)).$$

I.e.

$$\mathbf{Pr}(Bad) \leq (1 - \frac{\epsilon}{4})^{-1}\frac{\epsilon}{4}\mathbf{Pr}(Cons)\mathbf{Pr}(A_e)$$

$$< \frac{\epsilon}{2}\mathbf{Pr}(Cons)\mathbf{Pr}(A_e).$$

□

23. The Asymptotics of Edge Colouring Multigraphs

As we mentioned in Chap. 1, one of the most celebrated conjectures concerning edge colouring is the Goldberg–Seymour Conjecture which states that for any multigraph G: $\chi_e(G) \le \max(\Delta+1, \lceil \chi_e^*(G) \rceil)$. In this chapter, we present Kahn's proof of:

Theorem 23.1 [90] $\chi_e(G) \le (1+o(1))\chi_e^*(G)$.

We then discuss Kahn's proof of the analogous result for list colouring:

Theorem 23.2 [91] $\chi_e^l(G) \le (1+o(1))\chi_e^*(G)$.

In both proofs, we use a variant of our naive colouring procedure. The key new ingredient in the proof of Theorem 23.1 is a new method of assigning colours. For each colour c, we choose a matching M_c from some hard-core distribution on $\mathcal{M}(G)$ and assign the colour c to the edges in M_c. By assigning each colour exclusively to the edges of one matching, we avoid conflicting colour assignments and the resulting uncolourings. This new assignment process is also an important ingredient in the proof of Theorem 23.2, which requires a number of other ingenious new ideas.

23.1 Assigning the Colours

As we saw in Chap. 22, if we scale a fractional c-colouring by dividing each matching's weight by c then we obtain a probability distribution p on the matchings of G, under which we expect each edge to be in a random matching with probability $\frac{1}{c}$. In both proofs, we shall assign colours to the edges of G using such a probability distribution.

At first sight, this may appear to be a significant departure from our standard approach of randomly assigning each vertex a colour independently and then uncolouring vertices involved in conflicts. However, further reflection shows that the two approaches are, in fact, quite similar. To illustrate, we recall our approach to colouring triangle-free graphs.

For each vertex v in the triangle-free graph, and colour γ, we assigned v the colour γ with a probability $p_\gamma(v)$, independently of the colour assignments to the other vertices. Thus, for each set T, the probability $q_\gamma(T)$ of the set of vertices assigned γ being T, was $\prod_{v\in T} p_\gamma(v)\prod_{v\notin T}(1-p_\gamma(v))$. Now, for any stable set S, we let $\mathcal{F}(S)$ be the set of all T such that $S=\{v\in T|\ \nexists u\in T\ s.t.\ uv\in E(G)\}$. Then the probability of S being the set of vertices assigned γ and not uncoloured due to conflicts was $p^*_\gamma(S)=\sum_{T\in\mathcal{F}(S)} q_\gamma(T)$. So, in this earlier proof, we were implicitly assigning each colour γ using a probability distribution p^*_γ on the stable sets.

The difference in our new approach is that we explicitly impose the condition that the set of edges to which we assign γ is a matching from the outset, rather than dealing with this problem in a second phase by uncolouring edges in conflicts. With the previous approach, the colour assignments in the first phase were completely independent, so we were in a position to apply the simplest version of the Local Lemma in our analysis. We will still need to apply the Local Lemma to obtain global results via a local analysis under the new procedure. In order to do so, we need to know that for the probability distribution we choose from, events in distant parts of the graph have a limited effect on the probability that a particular edge is in our random matching. It is for this reason that we will restrict ourselves to hard-core distributions. For such distributions, the *approximate* independence results of the last chapter imply that we can use the Local Lemma in performing our analysis. However, because these results yield only approximate independence we will need to apply the Lopsided Local Lemma rather than the simpler version we have used previously. Before presenting our two proofs, we recall the approximate independence results we will use.

Remark Note that, in our procedure for colouring triangle-free graphs, if we let $Z=\prod_{v\in V}(1-p_\gamma(v))$ and $\lambda_\gamma(v)=\frac{p_\gamma(v)}{1-p_\gamma(v)}$ then $q_\gamma(v)=(\prod_{v\in T}\lambda_\gamma(v))Z$. I.e., the probability that we pick a set T is proportional to $\lambda_\gamma(T)=\prod_{v\in T}\lambda_\gamma(v)$. So, in the triangle-free case we assigned colours according to a hard-core distribution over the subsets of V and then dealt with conflicts by uncolouring the vertices involved. The only difference in our new approach is that we deal with conflicts by restricting our attention to matchings, i.e. restricting our random choice under the hard-core distribution to these conflict-free sets of edges.

23.1.1 Hard-Core Distributions and Approximate Independence

We will use the following results developed in the previous chapter:

Corollary 22.11 *For any $\delta>0$, there exists a $K=K(\delta)$ such that for any multigraph G with fractional chromatic number c there is a hard-core distribution p with marginals $(\frac{1-\delta}{c},\ldots,\frac{1-\delta}{c})$ such that $\forall e\in E(G), \lambda(e)\le\frac{K}{c}$ and hence $\forall v\in V(G),\quad \sum_{e\ni v}\lambda(e)\le K$.*

Lemma 22.10 *Consider any $K>0$, multigraph G, and hard-core distribution p such that (*) $\forall v\in V(G),\quad \sum_{e\ni v}\lambda(e)\le K$. If we choose* *a matching M according to p then, for any ϵ strictly between* 1 *and* 0*, setting* $t = 8(K+1)^2\epsilon^{-1}+2$*, we have:*

For any edge e and event Q which is t-distant from e, we have:

$$(1-\epsilon)\mathbf{Pr}(e \in M) \leq \mathbf{Pr}(e \in M|Q) \leq (1+\epsilon)\mathbf{Pr}(e \in M).$$

In the proofs, we will repeatedly apply Corollary 22.11 to a graph F to obtain a hard-core distribution p with marginals $(\frac{1-\delta}{\chi_e^*(F)}, \dots, \frac{1-\delta}{\chi_e^*(F)})$ for some small $\delta > 0$ such that for every edge e of G, $\sum_{e \ni v} \lambda(e) \leq K$, for some constant $K = K(\delta)$. We will then be able to apply Lemma 22.10 to this distribution to prove that whether or not a given edge is in a random matching chosen from this distribution is essentially independent of distant events.

23.2 The Chromatic Index

We now prove Theorem 23.1. As discussed above, we will assign colours to edges using a hard-core probability distribution on the matchings where each edge is in the random matching with probability $\frac{1}{c}$ for some $c \approx \chi_e^*(G)$. Now, if we choose c matchings independently from this distribution then the expected number of colours assigned to each edge is one. However, we would be unlikely to obtain a colouring, as some edges would be assigned many colours and others none. In fact we would expect to leave roughly $\frac{|E|}{e}$ edges uncoloured. So, instead, for some N much less than c we will delete a set of N matchings chosen using our probability distribution. We want to show that when we delete these matchings we reduce the fractional chromatic index by about N. In order to do so, we will need to choose N reasonably large for if N is too small, then the probability that the random matchings behave as desired locally is too small. Forthwith the details.

Proof of Theorem 23.1 We show that for all $\epsilon > 0$ there is a $\chi_0 = \chi_0(\epsilon)$ such that in any multigraph whose fractional chromatic index exceeds χ_0 we can choose a set of $N > 0$ matchings whose removal reduces χ_e^* by at least $(1+\epsilon)^{-1}N$. Using the fact that $\chi_e \leq 2\Delta \leq 2\chi_e^*$ to resolve the base case in which $\chi_e^* \leq \chi_0$, we can therefore obtain by induction that every multigraph has an edge colouring using at most $(1+\epsilon)\chi_e^* + \chi_0$ colours. Since this bound holds for all $\epsilon > 0$, Theorem 23.1 follows.

Specifically, we prove:

Lemma 23.3 *$\forall \epsilon > 0$, $\exists \chi_0$ s.t. if $\chi_e^*(G) \geq \chi_0$ then we can find $N = \lfloor \chi_e^*(G)^{\frac{3}{4}} \rfloor$ matchings in G whose deletion leaves a graph G' with $\chi_e^*(G') \leq \chi_e^*(G) - (1+\epsilon)^{-1}N$.*

Proof Clearly we need only prove the lemma for ϵ less than $\frac{1}{10}$ as if it holds for ϵ then it holds for all $\epsilon' > \epsilon$. So, fix an ϵ strictly between 0 and $\frac{1}{10}$. We define $\chi_0(\epsilon)$ so that it satisfies certain implicit inequalities scattered throughout the proof. Fix a graph G. Let $N = \lfloor \chi_e^*(G)^{\frac{3}{4}} \rfloor$ and let $c^* = \chi_e^* - (1+\epsilon)^{-1}N$. Define $\delta = \frac{\epsilon}{4}$. Take a hard-core distribution p and corresponding K as in Corollary 22.11 for this value of δ. Set $t = 8(K+1)^2(\frac{\epsilon}{4})^{-1} + 2$. We will delete N random matchings chosen according to p and show that with positive probability the resulting graph G' has fractional chromatic index at most c^*.

Because of Edmonds' characterization of the matching polytope, the fractional chromatic index of G' is this low precisely if

23.4 $\forall v: \ d_{G'}(v) \leq c^*$

and

23.5 $\forall H \subseteq G'$ with $|V(H)|$ odd: $|E(H)| \leq (\frac{|V(H)|-1}{2})c^*$.

We actually show that with positive probability:

23.6 $\forall v: \ d_{G'}(v) \leq c^* - \frac{\epsilon}{4}N$

and

23.7 $\forall$ odd connected $H \subseteq G'$ with $|V(H)| \leq \frac{\Delta}{\frac{\epsilon}{4}N}$ we have: $|E(H)| \leq \left(\frac{|V(H)|-1}{2}\right)c^*$.

Clearly, (23.6) implies (23.4). We claim that (23.6) and (23.7) imply (23.5). To prove our claim, we note first that if (23.6) holds then by summing degrees we have that for subgraphs F with an even number of vertices, $|E(F)| < \left(\frac{|V(F)|}{2}\right) c^*$. Further, any odd subgraph H can be split into a component H' with an odd number of vertices, and a subgraph F with an even number of vertices. These two remarks imply that given (23.6), to prove (23.5) it is enough to prove it for connected H. Further, by again summing degrees, we see that if (23.6) holds then (23.5) can only fail for H with fewer than $\frac{\Delta}{\frac{\epsilon}{4}N}$ vertices. This proves our claim.

So, we let A_v be the event that (23.6) fails for v. For each subset H of V which induces a connected subgraph of G, has an odd number of vertices and satisfies $|H| \leq \frac{\Delta}{\frac{\epsilon}{4}N}$ we let A_H be the event that (23.7) fails for the subgraph of G' induced by H. We will use the Lopsided Local Lemma to show that with positive probability none of these bad events hold thereby proving (23.6), (23.7), the lemma and the theorem.

To apply the Lopsided Local Lemma, we need to perform a local analysis conditioned on the outcome of distant events. To do so, we need to introduce some notation.

So, for each vertex v, we let $S_{<t}(v)$ be the set of vertices within distance t of v, and we let $S^*(v)$ be the set of events indexed by a vertex of $S_{<t}(v)$ or a set H intersecting $S_{<t}(v)$. For each set H for which we have defined A_H we let $S^*(H)$ be the union of $S^*(v)$ over all v in H. Clearly, every $S_{<t}(v)$ has at most Δ^t elements. Further, since $c^* \leq 2\Delta$, we have: $\frac{\Delta}{\frac{\epsilon}{4}N} \leq \Delta^{\frac{1}{3}}$. So, every u is

in at most $\Delta^{\Delta^{\frac{1}{3}}}$ sets H for which we have defined A_H. Hence every $S^*(v)$ has at most $\Delta^{t+\Delta^{\frac{1}{3}}+1}$ elements Thus, since every H for which we define $S_1^*(H)$ has fewer than Δ vertices, every $S^*(H)$ has less than $D = \Delta^{t+\Delta^{\frac{1}{3}}+2}$ elements.

We let $M_1, \ldots, M_N$ be our random matchings. For any vertex v, we let Q_v be our (random) choice of $\{M_1 - S_{<t}(v), M_2 - S_{<t}(v), \ldots, M_N - S_{<t}(v)\}$ (recall that $M - X = M \cap E(G - X)$). For any set H, for which we have defined A_H, we let Q_H be our (random) choice of $\{M_1 - S_{<t}(H), M_2 - S_{<t}(H), \ldots, M_N - S_{<t}(H)\}$.

To be able to apply the Lopsided Local Lemma to complete the proof, we need to show:

Lemma 23.8 *(a) For every v and Q_v we have: $\mathbf{Pr}(A_v|Q_v) \leq \frac{1}{eD}$, and (b) for every H for which we define A_H and Q_H we have: $\mathbf{Pr}(A_H|Q_H) \leq \frac{1}{eD}$.*

Proof We note first that by Lemma 22.10 and our choice of t, for any Q_v and for each i between 1 and N, the probability that an edge e incident to v is in M_i conditional on our choice of Q_v is at least $(1 - \frac{\epsilon}{4})\frac{1-\delta}{\chi_e^*(G)} \geq \frac{1-\frac{\epsilon}{2}}{\chi_e^*(G)}$. Since $\chi_e^*(G) \geq \Delta$, we have:

$$\mathbf{E}(d_{G'}(v)) \leq \chi_e^*(G)\left(1 - \frac{1-\frac{\epsilon}{2}}{\chi_e^*(G)}\right)^N .$$

Now, $N = o(\chi_e^*(G))$ and so:

$$\mathbf{E}(d_{G'}(v)) \leq \chi_e^*(G)\left(1 - (1+o(1))\frac{\left(1-\frac{\epsilon}{2}\right)N}{\chi_e^*(G)}\right) \leq \chi_e^*(G) - \left(1 - \frac{9\epsilon}{17}\right)N.$$

Since, $c^* = \chi_e^*(G) - (1+\epsilon)^{-1}N$ and $\epsilon \leq \frac{1}{10}$, this yields:

$$\mathbf{E}(d_{G'}(v)) \leq c^* - \left(1 - \frac{9\epsilon}{17} - (1+\epsilon)^{-1}\right)N. \leq c^* - \frac{\epsilon}{3}N.$$

As the choices of the M_i are independent and each affects the degree of v in G' by at most 1, we can apply the Simple Concentration Bound to prove (a).

The proof of (b) is similar. By Lemma 22.10, for any Q_H and for each i between 1 and N, the probability that an edge e with both endpoints in H is in M_i, conditional on our choice of Q_H, is at least $(1 - \frac{\epsilon}{4})\frac{1-\delta}{\chi_e^*(G)} \geq \frac{1-\frac{\epsilon}{2}}{\chi_e^*(G)}$. Furthermore, the number of edges of G with both endpoints in H is at most $\chi_e^*(G)\lfloor\frac{|H|-1}{2}\rfloor$. Performing calculations similar to those in the last paragraph, we obtain that the expected number of edges in the subgraph of G' induced by H is less than $\left(\frac{|H|-1}{2}\right)(c^* - \frac{\epsilon}{3}N)$. Since the choices of the M_i are independent and each affects the number of edges in H by at most $\frac{|V(H)|-1}{2}$, we can apply the Simple Concentration Bound to prove (b). □

The proof of Lemma 23.3 follows via the Lopsided Local Lemma because every event t-distant from v (resp. H) is determined by Q_v (resp. Q_H). For completeness, we provide the details.

We note that all the bad events not in $S^*(v)$ are determined by Q_v. Thus, if for some v, we choose a subset $\mathcal{F}$ of this set of events and let E be the event $(\cap_{A \in \mathcal{F}} \overline{A})$, then we can partition the possible choices for Q_v into two families $F(E)$ and $F(\overline{E})$ such that $E = \cup_{L \in F(E)}(Q_v = L)$ and $\overline{E} = \cup_{L \in F(\overline{E})}(Q_v = L)$. So, for any such E, summing the bound given by Lemma 23.8(a), over the L in $F(E)$ yields $\mathbf{Pr}(A_v|E) \le \frac{1}{eD}$. Similarly, all the bad events not in $S^*(H)$ are determined by Q_H and so for any event E obtained in a similar way by choosing a subset of these bad events, we have: $\mathbf{Pr}(A_H|E) \le \frac{1}{eD}$. The fact that there is a non-zero probability that both (23.6) and (23.7) hold now follows, via an application of the Lopsided Local Lemma. □

23.3 The List Chromatic Index

In this section, we discuss the proof of Theorem 23.2. The iterative proof technique is very similar to that we used to bound the chromatic number of triangle free graphs in Chap. 13. In each iteration, we have a list L_e of acceptable colours for each edge, and a probability distribution on the matchings in each colour class (of course, in Chap. 13, we were considering vertices and stable sets). Our distributions in each iteration are chosen so that for each edge e, the expected number of matchings containing e (or, as we expressed it in Chap. 13, the sum over all colours γ of the probability that e is assigned γ) is very close to 1. We choose a matching of each colour from the corresponding distribution, with these choices made independently. Next, for each colour γ, we activate each edge assigned γ independently with some probability α which is $o(1)$. We only retain colours assigned to activated edges. This ensures that very few edges are assigned more than one colour, which is what allows us to choose the matchings of each colour independently. We then restrict our attention to the uncoloured edges, updating the graph by deleting the coloured edges and updating the L_e by deleting any colour assigned to an edge incident to e.

To make this proof technique work, we need to show that:

(i) at the beginning of each iteration, we can choose new probability distributions so that (a) for each uncoloured edge e, we maintain the property that the expected number of random matchings containing e is very near 1, and (b) we can apply the Lopsided Local Lemma to analyze the iteration,
(ii) after some number of iterations, we can complete the colouring greedily.

As in the last section, we will choose our random matchings from a hard-core distribution. We assume that each L_e originally has C colours for some

$C \geq (1+\epsilon)\chi_e^*(G)$. For each colour class γ, we let G_γ be the subgraph of G formed by the edges for which γ is acceptable. Since $G_\gamma \subseteq G$, $\chi_e^*(G_\gamma) \leq \chi_e^*(G)$. Thus, by Corollary 22.11, we can find a hard-core distribution on the matchings in G_γ with marginals $(\frac{1}{C}, \ldots, \frac{1}{C})$ whose activity vector λ_γ satisfies: $\lambda_\gamma(e) \leq \frac{K}{C}$ for all e, where K is a constant that depends on ϵ. The choice of the marginals ensures that (i)(a) holds in the first iteration. By Lemma 22.10, the bound on the activities ensures that the distribution satisfies (*) and hence (i)(b) holds.

As we shall see, to ensure that (i) continues to hold, we will use the same activity vector λ to generate the random matching assigned colour γ throughout the process. This is a new and clever twist to our approach. Of course in each iteration we restrict our attention to the subgraph of G_γ obtained by deleting the set E^* of edges coloured (with any colour) in previous iterations and the endpoints of the edges in the set E_γ^* of edges coloured γ in previous iterations.

This technique for choosing (new) probability distributions ensures that all of our distributions satisfy (*) of Lemma 22.10, as the bound on the sum of the λ around a vertex still holds. So, (i)(b) also remains true throughout the procedure.

Unlike the λ values, the marginals on each edge will change drastically from iteration to iteration. Indeed if e is incident to an edge coloured γ then the probability it is assigned γ drops to 0. However, by Lemma 22.4, choosing a matching M of $G_\gamma - V(E_\gamma^*) - (E^* - E_\gamma^*)$ using the original λ is equivalent to choosing a random matching M' on G_γ using the hard-core distribution with the same λ but conditioned on $E_\gamma^* \in M'$ and $E^* - E_\gamma^* \notin M'$, and then setting $M = M' - E_\gamma^*$. This equivalence is what allows us to show that, for any particular uncoloured edge e, the expected number of random matchings containing e remains near 1, and hence that (i)(a) continues to hold throughout the process. The proof is discussed more fully below.

The fact that (i)(a) holds throughout implies that, in each iteration, the probability that an edge retains a colour remains near the activation probability α. This allows us to prove that the maximum degree in the uncoloured graph drops by a factor of about $1-\alpha$ in each iteration. The proof that (ii) holds is now straightforward. After $\log_{\frac{1}{1-\alpha}} 3K$ iterations, the maximum degree in the uncoloured graph will be less than $\frac{\Delta}{2K}$. Furthermore, for each e and γ, the probability that e is in the random matching of colour γ is at most $\lambda_\gamma(e) \leq \frac{K}{C}$. Since (i)(a) continues to hold, this implies there are at least $\frac{C}{K} > \frac{\Delta}{K}$ colours available on each edge and so the colouring can be completed greedily.

Note that in proving (i) and (ii) we repeatedly use the fact that though the marginals may vary, the probability that e is assigned γ is bounded above by $\lambda_\gamma(e) \leq \frac{K}{C}$, a fact the reader would do well to have in mind whilst reading the remainder of the chapter.

The proof that this approach works is long and complicated, and we present only the key ideas.

23.3.1 Analyzing an Iteration

An iteration proceeds as follows:

Step 1. We pick, for each colour γ, a matching M_γ according to a hard-core probability distribution p_γ on $\mathcal{M}(G_\gamma)$ with activities λ_γ such that for some constant K:
(a) $\forall e \in E(G), \sum_\gamma \mathbf{Pr}(e \in M_\gamma) \approx 1$,
(b) $\forall \gamma, e \in E(G), \lambda_\gamma(e) \leq \frac{K}{C}$ and hence $\forall v \in V(G), \sum_{e \ni v} \lambda_\gamma(e) \leq K$.

Step 2. For each γ, we activate each edge of M_γ independently with probability $\alpha = \frac{1}{\log \Delta(G)}$, to obtain a submatching F_γ. We colour the edges of F_γ with the colour γ and delete $V(F_\gamma)$ from G_γ. We also delete from G_γ, every edge not in M_γ which is in $F_{\gamma'}$ for some $\gamma' \neq \gamma$. For technical reasons, we do not delete edges of $(M_\gamma - F_\gamma) \cap F_{\gamma'}$ from G_γ (this may result in edges receiving more than one colour, which is not a problem).

Step 3. Note that the expected number of edges removed in Step 2 is less than $\alpha|E(G)|$ because the expected number of colours retained by an edge is very close to α but some edges retain more than one colour. As in previous chapters, we will perform an equalizing coin flip for each edge e of $G_\gamma - M_\gamma$ so that every edge not in M_γ is deleted from G_γ in either Step 2 or Step 3 with probability exactly α.

By the outcome of the iteration we mean the choices of matchings, activations, and helpful coin flips. We let G'_γ be the graph obtained after carrying out the modifications to G_γ performed in Steps 2 and 3 of the iteration. We let M'_γ be a random matching in G'_γ chosen according to the hard-core distribution with activities λ_γ. For an edge e, we define

$$L_1(e) = \sum_{G_\gamma \ni e} \mathbf{Pr}(e \in M_\gamma).$$

We need to show that we can perform our iteration in such a way that G'_γ and M'_γ satisfy the conditions which allow us to continue our iterative process.

Lemma 23.9 *We can choose an outcome R such that for every uncoloured edge e of G we have:*

$$(P1) \left| \sum_{G'_\gamma \ni e} \mathbf{Pr}(e \in M'_\gamma | R) - L_1(e) \right| \leq \frac{1}{(\log \Delta)^4},$$

and for every vertex v incident to more than $\frac{\Delta}{K}$ edges we have:
$(P2)$ the proportion of edges incident to v which are coloured is $\geq \alpha - \frac{1}{(\log \Delta)^4}$.

As sketched in the last section, this result allows us to prove Theorem 23.2. We omit the details of the proof that Lemma 23.9 implies the theorem, as this is the type of routine iterative analysis we have seen many times. Instead, in the remainder of the chapter, we focus on the proof of Lemma 23.9.

Of course, we apply the Lopsided Local Lemma. The difficult part of the proof is to show that, for each edge $e = uv$, the probability that Property (P1) holds is very close to 1, conditioned on any choice of outcomes for distant events. The proof of the analogous result for (P2) is much simpler and very similar to the proof of Lemma 23.8 from the last section.

In proving that (P1) is extremely likely to hold for e, we find it convenient to avoid conditioning on e being uncoloured. To state the precise result we will prove, we need some definitions. So, for each e and γ with $e \in G_\gamma$, if e is not incident to any vertex of F_γ we let $G^e_\gamma = G'_\gamma + e$. Otherwise, we let $G^e_\gamma = G'_\gamma$. We let M^e_γ be a random matching chosen according to the hard-core distribution on $\mathcal{M}(G^e_\gamma)$ using the activity vector λ_γ. For an outcome R we let

$$L_2(e,R) = \sum_{G^e_\gamma \ni e} \mathbf{Pr}(e \in M^e_\gamma | R).$$

We let $t' = 8(K+1)^2(\log \Delta)^{20} + 2$ and let $t = (t')^2$. We use Q to denote our random choice of the outcome of the iteration. For any edge $e = uv$, we let R_e be the (random) outcome of our iteration in $G - S_{<t}(\{u,v\})$, i.e. R_e consists of $\{M_\gamma - S_{<t}(\{u,v\}) | \gamma \text{ a colour}\}$, together with the choices of the activated edges in $G - S_{<t}(\{u,v\})$ which determine the $\{F_\gamma - S_{<t}(\{u,v\}) | \gamma \text{ a colour}\}$, and the outcomes of the equalizing coin flips for edges in in this subgraph. For a choice R^*_e for R_e, we let $Q(R^*_e)$ be a random outcome chosen conditional on $R_e = R^*_e$. We will show

23.10 *for every edge e of G and possible choice R^*_e for R_e, we have:*

$$\mathbf{Pr}\left(|L_2(e, Q(R^*_e)) - L_1(e)| > \frac{1}{(\log \Delta)^4} \right) \le e^{-\Delta^{\frac{1}{4}}}.$$

Remark This statement, at first glance, may be confusing. It concerns the random process of choosing the outcome $Q(R^*_e)$ of our iteration, given any choice R^*_e of the outcome in distant parts of the graph. The claim is that the random sum $L_2(e, Q(R^*_e))$, each of whose terms is determined by the colour assignments and coin flips for the edges near e, is highly concentrated around $L_1(e)$ which, by hypothesis, is near one.

We note that for an uncoloured edge e, $G'_\gamma = G^e_\gamma$ and so $M'_\gamma = M^e_\gamma$. Thus (23.10) implies that for each such uncoloured edge e, (P1) holds for e with probability very near one. To avoid burdening the reader with unimportant and cumbersome technical details we actually first prove the following weakening of (23.10).

23.11 *for every edge e of G, we have:*

$$\mathbf{Pr}\left(|L_2(e,Q) - L_1(e)| > \frac{1}{(\log \Delta)^4}\right) \le e^{-\Delta^{\frac{1}{4}}}.$$

We note that this result concerns an unconditioned choice of the random outcome whereas (23.10) concerns an outcome where the choices made far from e are fixed. The new ideas needed to prove Lemma 23.9. are all found in the proof of (23.11). This proof occupies the next three sections. In the final section of the chapter, we discuss strengthening it to obtain (23.10), and then combining this result with the Lopsided Local Lemma to prove Lemma 23.9.

23.3.2 Analyzing a Different Procedure

To prove (23.11), we first focus on one particular colour γ. As a prelude to analyzing the three step procedure that makes up an iteration we first analyze a similar but slightly different procedure. From now on, we refer to our first process as REAL and the new process as IDEAL. IDEAL has two steps. In the first, we choose a matching M_γ according to p_γ as in REAL. In the second, the edges of G_γ are independently activated with probability α and we obtain a new graph H_γ by deleting the vertices in the set F_γ of activated edges in M_γ and the activated edges not in M_γ.

Note that both processes make the same choice of M_γ and F_γ. In fact, there is only one difference between the way in which IDEAL constructs H_γ and the way in which REAL constructs G'_γ. It is that the edge deletions are not independent under REAL, as we determine which edges of $G_\gamma - V(F_\gamma)$ to delete by choosing the $M_{\gamma'}$ and activating some of their edges. However, as we shall argue later, the two processes are so similar, that we can use our analysis of IDEAL to prove (23.11). To do so, we let N'_γ be a random matching chosen from H_γ according to the hard-core distribution with activities λ_γ. We first study the probability that e is in N'_γ and then show this is close to the probability that it is in M'_γ.

We begin with the following result, which is the key to the whole analysis.

23.12 *Under the IDEAL process, for each colour γ and edge e,*

$$\mathbf{Pr}(e \in N'_\gamma | e \text{ } unactivated \text{ } for \text{ } \gamma) = \mathbf{Pr}(e \in M_\gamma).$$

Proof We note that every edge, in M_γ or not, is activated with probability α. So, we can perform the activation and choice of M_γ independently. In fact, we shall first decide which edges are activated and then choose M_γ. Knowing which edges are activated allows us to choose M_γ via a two step process. We can first make our choices on the activated edges and then make our choices on the unactivated ones, conditioned on our choices for the activated edges. That is we first choose the matching F_γ consisting of those activated edges of M_γ and then choose the rest of M_γ. We can choose F_γ by considering each activated edge in turn. Recursively applying Lemma 22.4, we see that the distribution of $M_\gamma - F_\gamma$ is precisely the same as the hard-core distribution obtained using the activity vector λ_γ on the graph obtained from $G_\gamma - V(F_\gamma)$ by deleting all the activated edges. That is, M_γ and $F_\gamma \cup N'_\gamma$ have exactly the same distribution. Now, the probability that $e \in F_\gamma$ is clearly $\alpha\mathbf{Pr}(e \in M_\gamma)$ so $\mathbf{Pr}(e \in N'_\gamma) = (1-\alpha)\mathbf{Pr}(e \in M_\gamma)$. Further, the probability that e is unactivated is exactly $1-\alpha$, so (23.12) follows. □

Once again, we want to avoid conditioning on e being unactivated. So for each e and γ with $e \in G_\gamma$, if e is not incident to any vertex of F_γ we let $H^e_\gamma = H_\gamma + e$. Otherwise, we let $H^e_\gamma = H_\gamma$. We let N^e_γ be a random matching chosen according to the hard-core distribution on $\mathcal{M}(H^e_\gamma)$ using the activity vector λ_γ. From (23.12), some straightforward calculations allow us to deduce:

23.13 *for each colour γ and edge e,*

$$\mathbf{Pr}(e \in M_\gamma) - \frac{K^2}{C^2} \leq \mathbf{Pr}(e \in N^e_\gamma) \leq \mathbf{Pr}(e \in M_\gamma).$$

Proof For any two matchings F and M of G_γ with $F \subseteq M$ we let $Ev(F,M)$ be the event that $F_\gamma = F$ and $M_\gamma = M$.

If $e \notin M$ then if e is unactivated $N'_\gamma = N^e_\gamma$ while if e is activated then $e \notin N'_\gamma$. So, for any two matchings F and M of G_γ with $F \subseteq M$ and $e \notin F$ we have:

$$\begin{aligned} &\mathbf{Pr}((e \in N'_\gamma) \cap Ev(F,M)) \\ =\ &\mathbf{Pr}(e \textit{ unactivated} | Ev(F,M))\mathbf{Pr}((e \in N^e_\gamma) \cap Ev(F,M)) \\ =\ &(1-\alpha)\mathbf{Pr}((e \in N^e_\gamma) \cap Ev(F,M)). \end{aligned}$$

For any two matchings F and M of G_γ with $F \subseteq M$ and $e \in F$ we have:

$$\mathbf{Pr}((e \in N'_\gamma) \cap Ev(F,M)) = \mathbf{Pr}((e \in N^e_\gamma) \cap Ev(F,M)) = 0.$$

For any two matchings F and M of G_γ with $F \subseteq M$ and $e \in M - F$ we have:

$$\mathbf{Pr}((e \in N'_\gamma) \cap Ev(F,M)) = \mathbf{Pr}((e \in N^e_\gamma) \cap Ev(F,M)).$$

Summing over all F, M with $F \subseteq M$, yields:

$$\mathbf{Pr}(e \in N'_\gamma) = (1-\alpha)\mathbf{Pr}(e \in N^e_\gamma) + \alpha\mathbf{Pr}((e \in N^e_\gamma) \cap (e \in M_\gamma - F_\gamma)).$$

In the proof of (23.12) we obtained: $\mathbf{Pr}(e \in N'_\gamma) = (1-\alpha)\mathbf{Pr}(e \in M_\gamma)$, which implies:

$$\mathbf{Pr}(e \in M_\gamma) = \mathbf{Pr}(e \in N^e_\gamma) + \frac{\alpha}{1-\alpha}\mathbf{Pr}((e \in N^e_\gamma) \cap (e \in M_\gamma - F_\gamma)).$$

Now, $\mathbf{Pr}(e \in M_\gamma - F_\gamma) \leq \mathbf{Pr}(e \in M_\gamma) \leq \lambda_\gamma(e) \leq \frac{K}{C}$. In the same vein, $\mathbf{Pr}(e \in N^e_\gamma | e \in M_\gamma - F_\gamma) \leq \lambda_\gamma(e) \leq \frac{K}{C}$. The desired result follows. □

Having dealt with the conditioning, we can now prove an analog of (23.11). For any possible choice R of M_γ and the set of activated edges for each γ in IDEAL, we let

$$L_3(e,R) = \sum_{G^*_\gamma \ni e} \mathbf{Pr}(e \in N^e_\gamma | R).$$

We let Q^* be a random choice of the M_γ and the set of activated edges for each γ. We show:

23.14 *for every edge e of G, we have:*

$$\mathbf{Pr}\left(|L_3(e,Q^*) - L_1(e)| > \frac{1}{2(\log \Delta)^4}\right) \leq \frac{1}{2}e^{-\Delta^{\frac{1}{4}}}.$$

Proof By (23.13), the expected value of the first sum differs from the second sum by at most $\frac{K^2}{C}$. So, we need only show that the first sum is highly concentrated around its expected value. However, this is easy to see. Note first that each term is between 0 and $\frac{K}{C}$, as the λ_γ are bounded by $\frac{K}{C}$. Further, since we are considering IDEAL, the term corresponding to γ depends only on the choices for G_γ, so we are considering the sum of a set of $|L_e| \leq C$ independent random variables. The Simple Concentration Bound now yields the desired result. □

It remains only to compare IDEAL with REAL. In doing so, it is important to recall that our choice of Q^* is coupled to our choice of Q in that the two procedures make the same choice of F_γ and M_γ. So, H_γ and G'_γ differ only in the choice of which edges of $G_\gamma - M_\gamma - V(F_\gamma)$ are deleted. Note that this implies that $e \in H^e_\gamma$ if and only if $e \in G^e_\gamma$.

We shall show, in the next two sections, that the probability that e is in N^e_γ and the probability that $e \in M^e_\gamma$ are essentially determined by the common choice of (F_γ,M_γ) and hence $L_2(e,Q)$ is very near $L_3(e,Q^*)$ with high probability. To this end, for any matchings F, M of G_γ with $F \subseteq M$, we let $Q(F,M)$ be a random choice of Q conditioned on $Ev(F,M)$ and let $Q^*(F,M)$ be a random choice of Q^* conditioned on $Ev(F,M)$. We shall compare $L_2(e,Q)$ and $L_3(e,Q^*)$ term by term, and show:

23.15 *for each colour γ, edge e, and matchings F, M in G_γ with $F \subseteq M$:*

$$\mathbf{Pr}\Bigg(|\mathbf{Pr}(e \in M^e_\gamma | Q(F,M)) - \mathbf{Pr}(e \in N^e_\gamma | Q^*(F,M))| > \frac{\mathbf{Pr}(e \in N^e_\gamma | Q^*(F,M))}{2(\log \Delta)^4}\Bigg) \leq e^{-\Delta^{\frac{1}{3}}}.$$

Summing over all C choices for γ, we obtain that for any (coupled) choice of Q and Q^*, we have:

23.16 *for every edge e of G, we have:*

$$\mathbf{Pr}\left(|L_2(e,Q) - L_3(e,Q^*)| > \frac{L_3(e,Q^*)}{2(\log \Delta)^4}\right) \leq e^{-\Delta^{\frac{1}{3}}}C.$$

We note that $L_3(e,Q^*)$ is concentrated near $L_1(e)$, by (23.14). Also, since $C = O(\Delta)$, we have $e^{-\Delta^{\frac{1}{3}}}C < \frac{e^{-\Delta^{\frac{1}{4}}}}{2}$, so (23.16) combined with (23.14) yields (23.11). Thus, it remains only to prove (23.15).

23.3.3 One More Tool

To compare $\mathbf{Pr}(e \in M^e_\gamma)$ with $\mathbf{Pr}(e \in N^e_\gamma)$ we need to introduce path-trees, defined by Godsil [66] who called them trees of walks.

For a vertex v in a graph H, we use $T(H,v)$, or simply T, to denote the *path-tree of H rooted at v*. The vertices of T correspond to the non-empty paths beginning at v. The root of T is $\{v\}$ and for any path $P = \{v_1 = v, e_1, v_2, e_2, \ldots, v_l, e_l, v_{l+1}\}$ the father of P is $P - e_l$, i.e. $\{v_1 = v, e_1, v_2, e_2, \ldots, v_l\}$. Note that

23.17 *If v has neighbours $u_1, \ldots, u_l$ then $T(H,v)$ is isomorphic to the tree obtained from disjoint copies of $T(H-v,u_i)$ for $i = 1, \ldots, l$ by adding a new root adjacent to the root of each of these l trees.*

We define a natural projection π from T to H by setting $\pi(\{v_1 = v, e_1, v_2, e_2, \ldots, v_l, e_l, v_{l+1}\}) = v_{l+1}$ and $\pi(P, P-e_l) = e_l$. Note that $\pi^{-1}(v) = \{v\}$ and for each edge e with endpoints u and v, $\pi^{-1}(e)$ is the edge between v and $\{v,e,u\}$, but the preimages of other vertices and edges typically have many elements. For any weighting λ on $E(H)$ we obtain a corresponding weighting on $E(T)$ by setting $\lambda(e) = \lambda(\pi(e))$, $\forall e \in E(T)$. Given a hard-core distribution p_H or simply p, on $\mathcal{M}(H)$ with an associated λ, we obtain a corresponding hard-core distribution p_T on $\mathcal{M}(T)$ by using the weighting λ on $E(T)$.

We now present the central result which allows us to consider trees rather than general graphs in the analysis of our iterative colouring procedure.

Definition As in the last chapter, for a vertex v and matching N we use $v \prec N$ to mean that v is the endpoint of an edge of N.

Lemma 23.18 *For every edge e of T incident to $\{v\}$, $p_T(e) = p_H(\pi(e))$. Hence, if M_H is chosen according to p_H and M_T is chosen according to p_T then:*

$$\mathbf{Pr}(v \not\prec M_H) = \mathbf{Pr}(\{v\} \not\prec M_T).$$

Proof We prove this statement by induction on $|V(H)|$. It is clearly true if v is incident to no edges. Hence, we can assume that the set $N(v)$ of neighbours of v is non-empty. We need the following which is obtained from (22.3) by replacing $\mathbf{Pr}(x \not\sim M \cap y \not\sim M)$ by $\mathbf{Pr}(x \not\sim M)\mathbf{Pr}(y \not\sim M|x \not\sim M)$.

23.19 *For any vertex x in a graph F and random matching M, chosen according to a hard-core probability distribution with activities λ:*

$$\mathbf{Pr}(x \not\sim M) = \left(1 + \sum_{xy \in E(F)} \lambda(xy)\mathbf{Pr}(y \not\sim M|x \not\sim M)\right)^{-1}. \quad (23.1)$$

For each child u of $\{v\}$ in T, let T_u be the subtree of T rooted at u which contains all its descendants. Let M_u be a random matching in T_u drawn from the hard-core distribution using the same λ as p_T. By (23.17) and our induction hypothesis we have that for each such u,

$$\mathbf{Pr}(u \not\sim M_u) = \mathbf{Pr}(\pi(u) \not\sim M_{H-v})$$

Combining this fact with an application of (22.4) to the set of edges of H incident to v, we obtain:

$$\mathbf{Pr}(u \not\sim M_u) = \mathbf{Pr}(\pi(u) \not\sim M_H|v \not\sim M_H).$$

By again applying (22.4), this time in T, we obtain:

$$\mathbf{Pr}(u \not\sim M_T|\{v\} \not\sim M_T) = \mathbf{Pr}(u \not\sim M_u).$$

Thus:

23.20 $\mathbf{Pr}(\pi(u) \not\sim M_H|v \not\sim M_H) = \mathbf{Pr}(u \not\sim M_T|\{v\} \not\sim M_T)$.

Further, $\lambda(\{v\}, \{v, e, u\}) = \lambda(\pi(e))$ for each child u of $\{v\}$. So, applying (23.19) with $x = v$ in G and $x = \{v\}$ in T yields $\mathbf{Pr}(v \not\sim M_H) = \mathbf{Pr}(\{v\} \not\sim M_T)$. Now, using this fact, the fact that $\lambda(\{v\}, \{v, e, u\}) = \lambda(\pi(e))$ and (23.20) we see that for every child u of $\{v\}$ in T we have: $\mathbf{Pr}(e \in M_T) = \mathbf{Pr}(\pi(e) \in M_T)$.

□

We show now that the marginals for a hard-core distribution on the matchings in a tree are indeed easy to compute. This suggests that transforming the analysis of a hard-core distribution on an arbitrary graph to an analysis of the corresponding tree using Lemma 23.18 should indeed help us.

Consider a rooted tree T, hard-core distribution p on $\mathcal{M}(T)$, corresponding λ and matching M_T chosen according to p. We can work out the probability that each edge is in M_T recursively, working our way up from the leaves, as described below.

(A) For each leaf w set $r(w) = 1$.

(B) For each non-leaf node w after having defined $r(u)$ for every u in the set $C(w)$ of children of w set $r(w) = \left(1 + \sum_{u \in C(w)} \lambda(wu) r(u)\right)^{-1}$.

Applying (23.19) as in the proof of Lemma 23.18, and working our way up from the leaves, we can show that for every node w with father x, $r(w) = \mathbf{Pr}(w \not\sim M_T | x \not\sim M_T)$. Similarly, for the root w_0 of T: $r(w_0) = \mathbf{Pr}(w_0 \not\sim M_T)$.

Thus, by Observation 22.1, for every edge $w_0 w$ of T incident to the root:

$$\mathbf{Pr}(w_0 w \in M_T) = \lambda(w_0 w) r(w_0) r(w).$$

By Lemma 23.18 this result can be applied to compute the probability a particular edge of a general graph is in a matching generated according to some hard-core distribution. In the next section, we use this fact to prove (23.15).

23.4 Comparing the Procedures

To prove (23.15), we consider a fixed arc $e = uv$ of G_γ and condition on a fixed choice of $F_\gamma = F$ and $M_\gamma = M$. If e is incident to an edge of F_γ there is nothing to prove. Otherwise, $e \in G_\gamma^e$ and $e \in H_\gamma^e$. We use e' to denote $\pi^{-1}(e)$ in what follows.

We let M_1 be a random matching in $T(G_\gamma^e, v)$ chosen according to the λ_γ and M_2 be a random matching in $T(H_\gamma^e, v)$ chosen according to the λ_γ. By Lemma 23.18, we can consider these trees rather than the corresponding graphs and hence we need to prove:

23.21 *The probability that:*

$$\begin{aligned} &|\mathbf{Pr}(e' \in M_1 | Q(F,M)) - \mathbf{Pr}(e' \in M_2 | Q^*(F,M))| \\ &> \frac{\mathbf{Pr}(e' \in M_2 | Q^*(F,M))}{2(\log \Delta)^4} \end{aligned}$$

is at most $e^{-\Delta^{\frac{1}{3}}}$.

Remark We highlight for the reader the fact that this statement, which holds for every pair (F, M), concerns the random choices of $Q(F, M)$ and $Q^*(F, M)$ for such a pair.

To simplify matters, we want to consider bounded size trees. Recall that $t' = 8(K+1)^2(\log \Delta)^{20} + 2 = \sqrt{t}$. We let T' be the subtree of $T(G - V(F_\gamma), v)$ within distance t' of $\{v\}$. We let U_1 be the (random) forest formed by the arcs of T' which correspond to edges of G_y which are in G_γ^e (i.e. those which were not activated under REAL). We let U_2 be the (random) forest formed by the arcs of T' which correspond to edges which are in H_γ^e. (i.e. those which were not activated under IDEAL). We note that the component T_1

of U_1 containing v is precisely the subtree of $T(G^e_\gamma, v)$ consisting of those arcs within distance t' of $\{v\}$. Similarly, the component T_2 of U_2 containing $\{v\}$ is precisely the subtree of $T(H^e_\gamma, v)$ consisting of those arcs within distance t' of $\{v\}$.

We let M'_i be a random matching in T_i chosen according to the hard-core distribution given by the λ_γ. We note that by Lemma 22.10 (applied to the appropriate tree), for any choice C of $Q^*(F,M) \cup Q(F,M)$ and $i \in \{1,2\}$ we have:

$$\left(1 - \frac{1}{(\log \Delta)^{20}}\right) \mathbf{Pr}(e' \in M'_i|C) \le \mathbf{Pr}(e' \in M_i|C)$$
$$\le \left(1 + \frac{1}{(\log \Delta)^{20}}\right) \mathbf{Pr}(e' \in M'_i|C),$$

Remark This statement holds for every choice of $Q(F,M)$ and $Q^*(F,M)$, and is simply comparing the choice of a random matching in a tree and in its truncation to the subtree within distance t' of e.

Thus, to prove (23.21) and hence (23.15) we need only show:

23.22 *The probability that:*

$$|\mathbf{Pr}(e \in M'_1|Q(F,M)) - \mathbf{Pr}(e \in M'_2|Q^*(F,M))| > \frac{\mathbf{Pr}(e \in M'_2|Q^*(F,M))}{2(\log \Delta)^5}$$

is at most $e^{-\Delta^{\frac{1}{3}}}$.

Remark Again, we note that this statement holds for every pair (F,M) and concerns the random choices of $Q(F,M)$ and $Q^*(F,M)$ for such a pair.

Now, we know how to compute $\mathbf{Pr}(e' \in M'_1|Q(F,M))$. We apply (A) and (B) of the last section to T_1 with $\lambda = \lambda_\gamma$ thereby defining r_1. Then, $\mathbf{Pr}(e' \in M'_1 = \lambda_\gamma(e)r_1(v)r_1(\{v,e,u\})$. In the same fashion, we can compute $\mathbf{Pr}(e' \in M'_2|Q^*(F,M))$ by applying (A) and (B) to T_2 with $\lambda = \lambda_\gamma$ thereby defining r_2 and set $\mathbf{Pr}(e' \in M'_2|Q^*(F,M)) = \lambda_\gamma(e)r_2(v)r_2(\{v,e,u\})$,

It will be convenient to extend our definition of r_i to all of $V' = V(T')$. So, for every vertex y of V', we let $C(y)$ be the children of y in T', and $C_i(y) = \{x|x \in C(y), xy \in U_i\}$. We define $r_i(y)$ on V' recursively starting at the leaves of T' by setting $r_i(y) = 1$ if $C_i(y) = \emptyset$ and setting $r_i(y) = (1 + \sum_{w \in C_i(y)} \lambda_\gamma(yw)r_i(w))^{-1}$ otherwise. We also define $r_3(y)$ on V' recursively starting at the leaves of T' by setting $r_3(y) = 1$ if $C(y) = \emptyset$ and setting $r_3(y) = (1 + (1-\alpha)\sum_{w \in C(y)} \lambda_\gamma(yw)r_3(w))^{-1}$ otherwise.

For each y in V' and $i \in \{1,2\}$, we let $D_i(y)$ be the event that

$$\left|\sum_{w \in C_i(y)} \lambda_\gamma(yw)r_3(w) - (1-\alpha)\sum_{w \in C(y)} \lambda_\gamma(yw)r_3(w)\right| > \Delta^{-\frac{1}{3}}.$$

Remark $D_i(y)$ is a random event as the $C_i(y)$ are random sets determined by the outcome of our two processes REAL and IDEAL.

We will show:

23.23 $\forall y \in V'$, $i \in \{1,2\}$,*we have:* $\mathbf{Pr}(D_i(y)) \leq e^{-\Delta^{\frac{1}{2}}}$.

Since V' has at most $\Delta^{t'} \leq e^{\Delta^{\frac{1}{10}}}$ nodes this yields:

23.24 $\mathbf{Pr}(D_1(y)$ *and* $D_2(y)$ *fail for all* $y) \leq \frac{1}{2}e^{-\Delta^{\frac{2}{5}}}$.

We will also show:

23.25 *If* $D_i(y)$ *fails for all* $y \in V'$ *and* $i \in \{1,2\}$, *then*

$$|\mathbf{Pr}(e' \in M_1' | Q(F,M)) - \mathbf{Pr}(e' \in M_2' | Q^*(F,M))| < \frac{\mathbf{Pr}(e' \in M_2' | Q^*(F,M))}{2(\log \Delta)^5}.$$

Combining (23.25) and (23.24) yields (23.22), which is the desired result.

Proof of (23.23). We note first that since each edge of $G_\gamma - M_\gamma$ is in U_i with probability $1-\alpha$, if y is incident to no edge of M_γ then the expected value of the first sum in the definition of $D_i(y)$ is equal to the second. Further, r_3 is by definition always below 1 so each term in these sums is at most $\frac{K}{C}$. Since at most one edge incident to y is in M_γ, the expected value of the first sum differs from the second by at most $\frac{K}{C}$.

It remains to prove that the first sum is concentrated. If $i=2$ then the edges incident to y are activated and deleted independently. Since each of the Δ deletions affects the value of the sum by at most $\frac{K}{C}$, we can apply the Simple Concentration Bound to obtain the desired result. If $i=1$ then we need to be a bit more careful. Rather than proving concentration on the sum of $\lambda \times r_3$ on the edges which remain we consider the sum of this product over the edges which are deleted. We again use the fact that deleting an edge causes this new sum to increase by at most $\frac{K}{C}$. We claim that deleting an edge will increase the new sum by at least $\frac{1}{(K+1)C}$. To see this we note that by definition $\frac{1}{r_3(w)}$ is at most $\sum_{f=wx} \lambda(f)+1$ which is at most $K+1$. Furthermore, the fact that the marginal on each edge in the first iteration was $\frac{1}{C}$ implies that each $\lambda(f)$ is at least $\frac{1}{C}$. The claimed result follows. Now, those edges incident to y which are activated and deleted are determined by the choice of the F'_γ for $\gamma' \neq \gamma$. To certify that edges of weight W have been deleted we need only specify one matching containing each deleted edge. So, by our claim, we require at most $WC(K+1)$ matchings to certify that the new sum is W. Also each matching chosen causes us to delete at most one edge and hence affects the new sum by at most $\frac{K}{C}$. Thus, we can apply Talagrand's Inequality to the random variable which is C times the new sum to obtain the desired result. □

Proof of (23.25). The proof is straightforward. We prove by induction on j that if $D_i(y)$ fails for $i \in \{1,2\}$ and every y in U then for each vertex y at distance $t'-j$ from v in T':

$$\left(1+\frac{1}{\Delta^{\frac{1}{4}}}\right)^{-1} r_3(y) \le r_i(y) \le \left(1+\frac{1}{\Delta^{\frac{1}{4}}}\right) r_3(y).$$

Since $\mathbf{Pr}(e \in M_1'|Q(F,M)) = \lambda_\gamma(e) r_1(v) r_1(\{uev\})$ and $\mathbf{Pr}(e \in M_2'|Q^*(F,M)) = \lambda_\gamma(e) r_2(v) r_2(\{uev\})$, (23.25) follows.

To prove our inequality, we note that it holds for leaves of T' as in this case $r_1 = r_2 = r_3 = 1$. So, we can assume that $C(y) \neq \emptyset$. So, we have:

$$\begin{aligned}
r_i(y) &= \left(1+\sum_{w\in C_i(y)} \lambda_\gamma(yw) r_i(w)\right)^{-1} \\
&\ge \left(1+\left(1+\frac{1}{\Delta^{\frac{1}{4}}}\right)\sum_{w\in C_i(y)} \lambda_\gamma(yw) r_3(w)\right)^{-1} \\
&\ge \left(1+\left(1+\frac{1}{\Delta^{\frac{1}{4}}}\right)(1-\alpha)\left(\sum_{w\in C(y)} \lambda_\gamma(yw) r_3(w) + \Delta^{-\frac{1}{3}}\right)\right)^{-1} \\
&\ge \left(\left(1+\frac{1}{\Delta^{\frac{1}{4}}}\right)\left(1+(1-\alpha)\sum_{w\in C(y)} \lambda_\gamma(yw) r_3(w)\right)\right)^{-1} \\
&= \left(1+\frac{1}{\Delta^{\frac{1}{4}}}\right)^{-1} r_3(y).
\end{aligned}$$

The proof that $r_i(y) \le \left(1+\frac{1}{\Delta^{\frac{1}{4}}}\right) r_3(y)$ is symmetric. □

Thus, we have shown that (23.15) and hence (23.11) holds.

23.4.1 Proving Lemma 23.9

To complete the proof of Lemma 23.9, we need to discuss (a) how we obtain the extension (23.10) of (23.11) in which we condition on the choice R of the matchings at distance $t=(t')^2$ from e, and (b) how to combine (23.10) with the Lopsided Local Lemma to show that we can ensure that (P1) and (P2) hold everywhere with positive probability, thereby proving Lemma 23.11.

Strengthening (23.11) to obtain (23.10) is straightforward. The reader can verify that in the proof of (23.11) we bound $\mathbf{Pr}(e \in M_\gamma^e|Q)$ by considering only the intersection of the matchings F_γ and M_γ with the set E' of edges of G at distance at most t' from e. On the other hand, R is determined by the choice of the matching edges at distance at least $t = (t')^2$ from v. Hence, the edges of E' are at distance at least $t - t' >> t'$ from the matching edges determined by R. It follows by Lemma 22.10 that the probability that an edge in E' is in M_γ or F_γ given R is not significantly different from the unconditional probability that it is in M_γ or F_γ. This allows us to use essentially the same proof as above, just carrying through an extra error term as we recursively compute r_1 and r_2. We omit the details.

At first sight, it appears that we cannot use (23.10) to prove Lemma 23.9. The natural first step would be to consider for each edge e, the event A_e that (P1) fails for e and for each vertex v, the event A_v that (P2) fails for v. We would like to deduce from (23.10) that for all $x \in V \cup E$, the probability that A_x holds is small regardless of the outcome of all the events indexed by edges and vertices t-distant from x. Unfortunately, although $Q(e)$ determines whether A_v holds for all v which are t-distant from e, it does not determine whether A_f holds for f which are t-distant from e. This is because the exact probability that $f \in M'_\gamma$ can only be determined by examining all the edges of G'_γ. Indeed it may be that the presence or absence of e in G'_γ will decide if (P1) holds for f or not. Thus, the Lopsided Local Lemma cannot be applied to this set of events. Has our analysis then all been for nought?

Fortunately not! The key point is that, as we used repeatedly throughout the last proof, for any edge e with endpoints u and v, $\mathbf{Pr}(e \in M'_\gamma)$ is within a factor of $1 + \frac{1}{(\log\ \Delta)^{20}}$ of $\mathbf{Pr}((e \in Z'_\gamma)$, where Z'_γ is a random matching in $G'_\gamma \cap S_{<t'}(v)$ chosen using the activity vector λ_γ. So, we define, for each edge e, the new event A'_e that

$$\left| \sum_{G'_\gamma \ni e} \mathbf{Pr}(e \in Z'_\gamma | Q) - \sum_{G_\gamma \ni e} \mathbf{Pr}(e \in M_\gamma) \right| \leq \frac{1}{2(\log\ \Delta)^4},$$

and note that by the above remarks, if the A'_e hold, so do the A_e. But now, since A'_e is defined by the choice of edges of the matchings within t' of e, for any edge f at distance at least $t + t'$ from e, A'_f is determined by $Q(e)$. So, we are in a position to apply the Lopsided Local Lemma to prove that with positive probability all the A_v hold, and all the A'_e and hence all the A_e hold. We omit the routine details.

Part IX

Algorithmic Aspects

In the next two chapters we discuss efficient algorithms for finding many of the colourings whose existence we have demonstrated via the probabilistic method. Once again, Linearity of Expectation plays a crucial role.

In the next chapter we present a simple algorithm, due to Erdős and Selfridge, which finds a proper 2-colouring of a hypergraph H provided the expected number of monochromatic edges in a uniformily random 2-colouring of H is less than one. The core of the algorithm is a procedure which efficiently calculates the expected number of monochromatic edges in a uniformly random completion of a partial colouring using Linearity of Expectation.

We then turn to the much more difficult task of making our applications of the Local Lemma algorithmic. We first present an algorithm due to Beck for finding proper 2-colourings of k-uniform hypergraphs of sufficiently low degree. We then generalize the approach so it can be used to develop algorithms to find the structures guaranteed to exist for a much wider class of applications of the Local Lemma.

The analysis of Beck's algorithm and its generalizations are much more difficult than the analysis of the Erdős-Selfridge algorithm. However, as we shall see, Erdős and Selfridge's simple but powerful idea plays a crucial role in these more complicated algorithms.

24. The Method of Conditional Expectations

Throughout this chapter, we consider hypergraphs for which the expected number of monochromatic edges in a uniformly random 2-colouring is less than one. To begin, we present an efficient algorithm, due to Erdős and Selfridge [46] for finding proper 2-colourings of such hypergraphs.

We then consider a related game in which two players Red and Blue alternately colour vertices of a hypergraph until all the vertices are coloured. If any edge is monochromatically coloured then the first player to have monochromatically coloured an edge wins. Otherwise, the game is a draw.

We shall see that for the hypergraphs under consideration optimal play by both players ensures that the game ends in a draw, and provide an efficiently computable strategy that the players can use to ensure this outcome.

Note that the final position in a drawn game is a proper 2-colouring, so the second result implies the first.

24.1 The Basic Ideas

For the rest of this chapter, we use H to denote a hypergraph with vertex set $\{v_1, \ldots, v_n\}$ and edge set E such that the expected number of monochromatic edges in a uniformly random 2-colouring of H is less than one. This implies that H has a proper 2-colouring, we want to find such a creature. To do so, we will colour the vertices one by one, ensuring that every partial colouring we construct can be completed to a proper 2-colouring.

Choosing the colour of the first vertex v_1 is easy because it is irrelevant, by the symmetry between the two colours.

As the reader may suspect, the best colour to assign the second vertex v_2 is that which is not assigned to v_1. Indeed, we will show that for any assignment of different colours to v_1 and v_2, the expected number of monochromatic edges in a uniformly random 2-colouring completing this assignment is less than one. This implies that we can indeed complete such a partial assignment to a proper 2-colouring.

By symmetry, we can restrict our attention to the case in which v_1 is red and v_2 is blue. To compute the conditional expectation of the number of monochromatic edges we apply Linearity of Expectation. I.e., we compute the conditional probability that each edge is monochromatic and sum these probabilities to obtain the conditional expectation.

If e is disjoint from $\{v_1, v_2\}$ then the conditional probability that e is monochromatic is $2 \times 2^{-|e|}$ as the conditioning is irrelevant. If e contains v_1 but not v_2 then for e to be monochromatic, all the vertices in $V(e) - v_1$ must be coloured red, so the conditional probability that e is monochromatic is $2^{1-|e|}$. If e contains v_2 but not v_1 then for e to be monochromatic, all the vertices in $V(e) - v_2$ must be coloured blue, so the conditional probability that e is monochromatic is $2^{1-|e|}$. If e contains both v_2 and v_1 then e cannot be monochromatic, so the conditional probability under consideration is zero.

Thus, setting $E' = E - \{e | e \in E, \{v_1, v_2\} \subseteq V(e)\}$, we see that the expected number of monochromatic edges given that v_1 is red and v_2 is blue is

$$\sum_{e \in E'} 2^{1-|e|} \leq \sum_{e \in E} 2^{1-|e|} < 1.$$

So there is indeed a proper 2-colouring in which v_1 is red and v_2 is blue.

We will proceed in a similar manner to choose our partial colourings so that for every partial colouring P, the expected number of monochromatic edges in a uniformly chosen completion of P is less than one. In particular, this implies that the final colouring in the sequence has no monochromatic edges.

Crucial to this approach is the fact that we can use Linearity of Expectation to compute these expected values, as we did above. Forthwith the details.

24.2 An Algorithm

As usual, we let $X = X(C)$ be the number of monochromatic edges under a colouring C. If C is a random colouring then X is a random variable. For a partial colouring P, we let $CE_P(X)$ be the expected value of X for a uniformly chosen random completion of P. I.e. $CE_P(X) = \mathbf{E}(X(C))$ where C is the colouring obtained by colouring each vertex uncoloured by P, independently, red with probability $\frac{1}{2}$ and blue with probability $\frac{1}{2}$.

We will iteratively construct partial colourings $P_0, P_1, \ldots, P_n$ where for each i:

(a) the set of vertices coloured under P_i is $\{v_1, \ldots, v_i\}$ (thus P_0 colours no vertices),
(b) P_i and P_{i-1} agree on $\{v_1, \ldots, v_{i-1}\}$, and
(c) $CE_{P_i}(X) < 1$.

Clearly, $CE_{P_n}(X) = X(P_n)$. Since, by (c), $X(P_n) < 1$, it must be zero. Thus P_n is the desired proper 2-colouring.

Having constructed P_{i-1}, there are two possible choices for P_i, we can extend P_{i-1} either by colouring v_i red or by colouring it blue. We denote the first possibility by P_i^r and the second by P_i^b. We shall show:

Observation 24.1 *We can compute $CE_P(X)$ in polynomial time for any partial colouring P.*

and:

Observation 24.2 $\min(CE_{P_i^r}(X), CE_{P_i^b}(X)) \leq CE_{P_{i-1}}(X)$.

With these results in hand, it is easy to iteratively construct $P_1, \ldots, P_n$. Given P_{i-1} we simply compute $CE_{P_i^r}(X)$ and $CE_{P_i^b}(X)$ and choose for P_i one of these possibilities which minimizes $CE_{P_i}(X)$.

The proof of Observation 24.1 is straightforward. To compute $CE_P(X)$ we simply compute the conditional probability for each edge e that e will be monochromatic and sum these values. If e contains vertices of both colours this probability is zero. If e contains no coloured vertices, this probability is $2^{1-|e|}$. Finally, if e contains coloured vertices of only one colour and u uncoloured vertices then this probability is 2^{-u}.

The proof of Observation 24.2 is also straightforward. Actually a stronger fact is true: $CE_{P_{i-1}}(X)$ lies between $CE_{P_i^r}(X)$ and $CE_{P_i^b}(X)$. To gain an intuition as to why this is true, readers should consider the similar statement:

> The average height of the people in a room lies between the average height of the men and the average height of the women.

With this hint, fastidious readers should be able to fill in the details of the proof, which we omit.

24.3 Generalized Tic-Tac-Toe

The English children's game Tic-Tac-Toe will be familiar to many readers (actually, a variant of this game was played in Egypt in 1440 BC, and a related game, renju, in China in 2500 BC). Two players Nought (O) and Cross (X) alternately place their symbol in the squares of a 3x3 grid (see Fig. 24.1). A player can only place his symbol in an unoccupied grid square so the game lasts for at most nine moves. The first player to place his symbol on all the squares in a line (row, column, or diagonal) wins. If all the squares are occupied and no player has covered a line then the game is a draw.

We can reformulate this game in terms of bicolouring hypergraphs. The Tic-Tac-Toe hypergraph has 9 vertices and 8 edges and is depicted in Fig. 24.2. Two players Blue and Red alternately colour an uncoloured vertex. The first player to monochromatically colour an edge wins. If the players complete a proper 2-colouring of the hypergraph then the game is a draw.

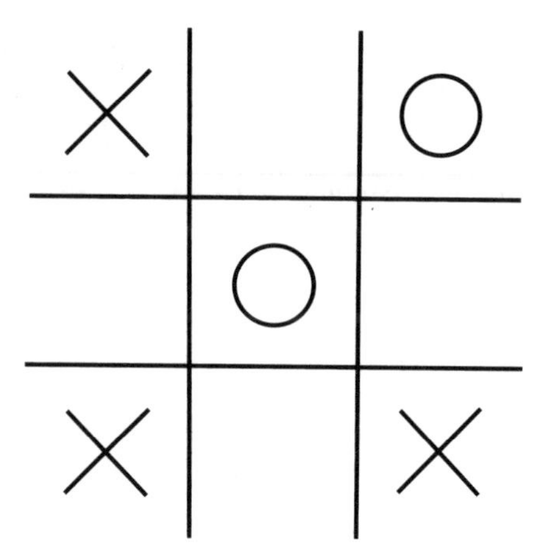

Fig. 24.1.

We can, of course, play this game on any hypergraph H. For some hypergraphs, the first player wins while for others the second player can force a draw. The second player can never win because the game is so symmetric that the first player can steal any winning strategy for the second player (see. Exercise 24.1). In general it is PSpace-complete to determine if the second player can force a draw for an input hypergraph H (cf. [136]). However, as we shall see, the techniques of the previous section can be used to show that for certain hypergraphs, the second player can force a draw.

Lemma 24.3 *If the expected number of monochromatic edges in a uniformly random bicolouring of H is less than 1, then the second player can force a draw in Generalized Tic-Tac-Toe on H.*

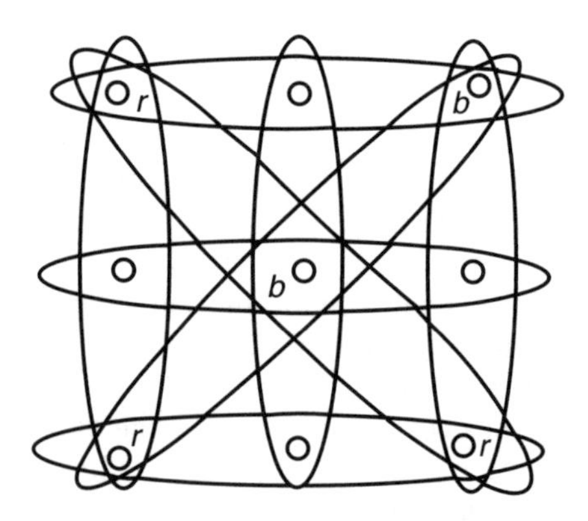

Fig. 24.2. The Tic-Tac-Toe Hypergraph

Remark The natural converse to this statement is false. I.e., there is no lower bound on the expected number of monochromatic edges in a random colouring which guarantees that the first player wins. To see this, consider the hypergraph H_n which consists of n disjoint edges each with two vertices. The expected number of monochromatic edges in a random colouring of H_n is $\frac{n}{2}$. However, the second player can draw by always playing in the edge which the first player just played in.

Remark The bound in Lemma 24.3 is tight, as the following example shows. Let F_n be the hypergraph with $2n + 1$ vertices $a_1, a_2, \ldots a_n, b_1, b_2, \ldots, b_n, c$ whose edge set contains exactly the 2^n subsets of vertices consisting of c and n other vertices all with different indices. The expected number of monochromatic edges in a uniformly random bicolouring of H is exactly one. On the other hand, the first player can win Generalized Tic-Tac-Toe on F_n by first picking c and then always picking a vertex with the same index as that just picked by the second player.

We prove Lemma 24.3 below. First however, we discuss a relationship between Generalized Tic-Tac-Toe and the Local Lemma, and present an intriguing conjecture due to Beck.

We note that two players can cooperate to arrange a draw when playing Generalized Tic-Tac-Toe on F precisely if there is a proper 2-colouring of F such that the number of vertices of each colour differ by at most one. It is an easy matter to modify our earlier proofs of the existence of proper 2-colourings for k-uniform hypergraphs of bounded degree to prove the existence of such special proper colourings for k-uniform hypergraphs of bounded degree (indeed, a bound of $\frac{2^k}{16k}$ will do, see Exercise 24.2). This does not imply that the second player can force a draw on all such graphs however, as the first player may be able to avoid all the drawing positions with optimal play. However, Beck has raised the possibility that if the maximum degree of a k-uniform hypergraph is sufficiently small then the second player can indeed force a draw. Specifically, he proposed

Conjecture 24.4 *There exists a constannt $c > 1$ such that if the maximum degree of a k-uniform hypergraph H is less than c^k then the second player can force a draw when playing Generalized Tic-Tac-Toe on H.*

For more on this open problem and some partial results see [16, 17]

24.4 Proof of Lemma 24.3

To ease the exposition in this section, we assume that the players in a game of Generalized Tic-Tac-Toe continue to alternate turns until the whole graph is coloured, even if one obtains a monochromatic edge.

Lemma 24.3 asserts that a certain condition ensures that the second player can prevent the first player from obtaining a monochromatic edge first. We find it more convenient to prove that this condition allows the second player to prevent the first player from obtaining a monochromatic edge at all. That is, we prove the following strengthening of Lemma 24.3

Lemma 24.5 *Suppose Blue and Red play Generalized Tic-Tac-Toe (to completion) on a hypergraph H for which the expected number of monochromatic edges in a uniformly random 2-colouring is less than one. Then, if Blue plays second, he has an efficiently computable strategy to prevent Red from colouring all of any edge.*

Now, the expected number of all red edges in a uniformly random 2-colouring of a hypergraph is clearly exactly half the expected number of monochromatic edges. Furthermore, any initial move can at most double the expected number of all red edges. Combining these two facts, with the lemma below yields Lemma 24.5.

Lemma 24.6 *Suppose Blue and Red play Generalized Tic-Tac-Toe (to completion) on a hypergraph H for which the expected number of all red edges in a uniformly random 2-colouring is less than one. Then, if Blue plays first, he has an efficiently computable strategy to prevent Red from colouring all of any edge.*

Proof We use $R = R(C)$ to denote the number of all red edges in a colouring C. For any partial colouring P, we use $CE_P(R)$ to denote the expected number of all red edges in uniformly chosen random completion of P. Blue simply colours the vertex v which minimizes the value of $CE_P(R)$ for the resultant partial colouring P.

Clearly, Blue can efficiently compute the value of $CE_P(R)$ for any candidate partial colouring P, using an algorithm similar to that for computing $CE_P(X)$. We claim that no matter what vertex w Red chooses to colour on his next turn, letting P_1 be the partial colour before Blue's turn and P_2 the partial colouring after Red's turn, we have: $CE_{P_2}(R) \leq CE_{P_1}(R)$. Iteratively applying this claim proves Lemma 24.6, as the initial condition implies that for the original colouring P_0, $CE_{P_0}(R) < 1$ and hence this will also be true for the final colouring.

It remains only to prove the claim. To do so, we consider the set E_1 of edges which contain v and contain no blue vertex, and the set E_2 of edges which contain w and contain no blue vertex. We let $E_3 = E_1 \cap E_2$. For every edge e in $E_1 \cup E_2$, we let $u(e)$ be the number of uncoloured vertices in e under P_1.

We note that colouring v blue decreased the conditional expected value of R by $\sum_{e \in E_1} 2^{-u(e)}$. On the other hand, if Blue had coloured w blue he would have decreased the conditional expected value of R by $\sum_{e \in E_2} 2^{-u(e)}$. It follows that

$$\sum_{e \in E_1} 2^{-u(e)} \geq \sum_{e \in E_2} 2^{-u(e)}.$$

Now, $CE_{P_2}(R)$ is clearly $CE_{P_1}(R) - \sum_{e \in E_1} 2^{-u(e)} + \sum_{e \in E_2 - E_1} 2^{-u(e)}$. The desired claim follows by the above inequality. □

Exercises

Exercise 24.1 Prove that there is no hypergraph H on which Generalized Tic-Tac-Toe is a win for the second player, if both players play optimally.

Exercise 24.2 Prove that if H is a k-uniform hypergraph with $2n$ vertices and maximum degree at most $\frac{2^k}{16k}$, then we can 2-colour H so that there are n vertices of each colour and no monochromatic edge.

25. Algorithmic Aspects of the Local Lemma

In this chapter, we discuss finding proper 2-colourings of hypergraphs where each edge intersects a bounded number of other edges. We present the following theorem of Beck [19]:

Theorem 25.1 *There is a deterministic polytime algorithm which will find a proper 2-colouring of any hypergraph H in which each edge has size at least k and intersects at most $d \leq 2^{k/16-2}$ other edges.*

Recall that Theorem 4.2 implies that such a 2-colouring exists, even if we replace the exponent "$k/16-2$" by "$k-3$", but the approach taken here to actually construct the 2-colouring efficiently requires the smaller bound on d.

We will first present a randomized algorithm, and then show how to use the techniques of the previous chapter to derandomize it.

Since there is no bound on the number of edges in our hypergraph, the expected number of monochromatic edges in a uniformly random 2-colouring can be arbitrarily large. So picking such a random colouring will not work, and a direct application of the Erdős–Selfridge approach is also doomed to failure. Instead, we will take a different approach.

As in the previous chapter, we will colour the vertices one-at-a-time, but here our choice of colours is different. We will not specify how to make the colour choices until later in our discussion as how we do so depends on whether we are using a randomized algorithm or a deterministic one. However, the reader should be aware that these choices are not made to avoid monochromatic edges. Rather, we avoid monochromatic edges by permitting ourselves to leave some vertices uncoloured. Specifically, when we consider a vertex v, if it lies in an edge half of whose vertices have been coloured, all with the same colour, then we will not colour v. This ensures that after this first phase, no completely coloured edge is monochromatic. Indeed, if an edge does not contain vertices of both colours then at least half of its vertices are uncoloured.

In our second phase, we will colour the vertices which were passed over in the first phase so that every edge contains a vertex of both colours. Since the only edges with which we need to concern ourselves have at least $k/2$ uncoloured vertices, and since $d \times 2^{1-k/2} < \frac{1}{4}$, a simple application of the Local Lemma (namely Theorem 4.2), ensures that there exists a completion of the partial 2-colouring to a proper 2-colouring.

We have thus reduced our problem to a similar smaller problem which, at first sight, does not seem any easier to solve. It turns out however, that if we make a judicious choice of the colour assignments in our first pass, then the smaller problem will have a very simple structure and so we will be able to quickly find a completion of our colouring in a straightforward manner.

One way to make judicious choices is to simply assign a uniformly random colour to each vertex that we colour. This ensures that the probability of the first $k/2$ vertices of an edge all being given the same colour is at most $2^{-k/2}$. Using this fact, we can show that with sufficiently high probability, the subhypergraph we need to colour in the second phase has very small components. So we can carry out the second phase by using, on each component, an exhaustive search through all possible colourings.

This yields a fairly simple randomized algorithm. To derandomize the algorithm, we must find a deterministic way of carrying out the first phase so that the components of the resulting subhypergraph are all small. We can do so by applying the Erdős–Selfridge technique; when colouring a vertex, we choose the colour which minimizes the conditional expected number of large components.

In this introductory discussion, we have oversimplified the procedure somewhat. For example, we usually repeat the first phase twice to reduce even further the size of the components which we colour using brute force. Also, when derandomizing the algorithm, we do not compute the conditional expected number of large components in the subhypergraph considered in the second phase. Instead, we bound this number by focusing on a larger variable which is simpler to deal with. These and other details will be given more fully in the next section.

25.1 The Algorithm

25.1.1 The Basics

We start by presenting the randomized form of our algorithm in the case where k is fixed. We are given a hypergraph H on n vertices and m hyperedges satisfying the conditions of Theorem 25.1. Since we are doing an asymptotic analysis of the running time, we can assume that m is large.

In the first phase, we arbitrarily order the vertices $v_1, \ldots, v_n$. We go through the vertices in this order, assigning a uniformly random colour to each vertex. After colouring v_i, if an edge e containing v_i has half its vertices coloured, all with the same colour, then we say that e is *bad*, and for each $v_j \in e$ with $j > i$, at step j we will pass over v_j without assigning it a colour.

We let U denote the set of vertices which are not coloured during this first phase, and we let M denote the set of edges which don't yet have vertices of both colours. For each $e \in M$ we define e' to be the set of uncoloured vertices in e. An important consequence of our procedure is that every e' has size at least $\frac{1}{2}k$.

In our second phase, we colour U. No edge outside of M can become monochromatic, no matter how we colour U, so we can ignore all such edges in this phase. Thus, we focus our attention on the hypergraph H' with vertex set U, and edge set $\{e' : e \in M\}$. As we observed, each edge of H' has size at least $\frac{1}{2}k$, and no edge intersects more than $2^{k/16-2}$ other edges. Therefore the Local Lemma implies that there exists a proper 2-colouring of H'. Clearly, using such a 2-colouring to complete the partial colouring formed in the first phase will yield a proper 2-colouring of H.

The main part of our analysis is to show that the components of H' will all be small. In particular, letting m be the number of edges in H, we will prove the following.

Lemma 25.2 *With probability at least $\frac{1}{2}$, every component of H' has at most $5(d+1)\log m$ vertices.*

So we can run Phase I, and if H' has any component with more than $5(d+1)\log m$ vertices, then we start over. The expected number of times we must do this is at most 2. Having obtained such an H', we can find a proper 2-colouring of H' in polynomial time using exhaustive search. For, to find such a 2-colouring of H', we need only find a proper 2-colouring of each of its components. Since $k = O(1)$ we have $d = O(1)$ and so there are only $2^{O(\log m)} = \text{poly}(m)$ candidate 2-colourings for each of these components.

The main step is to prove Lemma 25.2, which we do now. We will bound the expected number of large components of H' by showing that every such component must have many disjoint bad edges. Because disjoint edges become bad independently, the probability that a specific large collection of disjoint edges all turn bad is very small. This will help show that the probability of H' having a big component is small.

As usual, we use $L(H)$ to denote the line graph of H. $L^{(a,b)}(H)$ is the graph with vertex set $V(L(H))$ $(= E(H))$ in which two vertices are adjacent if they are at distance exactly a or b in $L(H)$. We call $T \subseteq E(H)$ a $(1,2)$-*tree* if the subgraph induced by T in $L^{(1,2)}(H)$ is connected. We call $T \subseteq E(H)$ a $(2,3)$-*tree* if the subgraph induced by T in $L^{(2,3)}(H)$ is connected *and* T is a stable set in $L(H)$ (i.e. no two edges of T intersect in H). We call an (a,b)-tree *bad* if it contains only bad edges from H.

Lemma 25.3 *Every component C of H' with ℓ vertices contains a bad $(2,3)$-tree with at least $\ell/(k(d+1))$ edges.*

This lemma follows immediately from two simple facts:

Fact 25.4 *Every component C of H' with ℓ vertices contains a bad $(1,2)$-tree with at least ℓ/k edges.*

Proof Note that every vertex of C lies in a bad edge, by definition of H'. Since each edge contains at most k vertices, the number of bad edges in C is at least ℓ/k.

Let T be a maximal subset of the bad hyperedges of C which forms a $(1,2)$-tree. We show that T contains all the bad hyperedges of C. Suppose the contrary and consider some bad hyperedge $e \notin T$. If e intersects T, then e can be added to T to form a larger $(1,2)$-tree, thereby contradicting the maximality of T. Otherwise, since C is connected, there must be a path from T to e; let P be the shortest such path. Let e_0, e_1, e_2 be the first three hyperedges of P, where e_0 is in T. Consider any v in $e_1 \cap e_2$ and any bad edge f containing v. Now, f does not belong to T, by the minimality of P, but f is of distance at most 2 from T. Thus f can be added to T to form a larger bad $(1,2)$-tree, thereby contradicting the maximality of T. □

Fact 25.5 *For any $(1,2)$-tree with t hyperedges, there is a subset of at least $t/(d+1)$ of these hyperedges which forms a $(2,3)$-tree.*

Proof Consider any $(1,2)$-tree T and a maximal subset T' of the hyperedges of T which forms a $(2,3)$-tree. Consider any hyperedge $e \in T - T'$; we will show that e must intersect some edge in T'. This implies our fact since every hyperedge in T' intersects at most d hyperedges in $T - T'$.

So suppose that e does not intersect any hyperedges in T'. Since T is a $(1,2)$-tree, there must be a path in $L^{(1,2)}(H)$ from T' to e; let P be a shortest such path, and let $e_0 \in T'$ be the first hyperedge in P. If this path has no internal hyperedge then e is at distance 2 in $L(H)$ from e_0 and so e can be added to T' to form a bigger $(2,3)$-tree, thereby contradicting the maximality of T'.

Otherwise P has at least 1 internal edge, so let e_0, e_1, e_2 be the first 3 edges on P, where again $e_0 \in T'$. If e_1 does not intersect any edge in T' then we can add e_1 to T' to form a larger $(2,3)$-tree. Otherwise, since e_2 is at distance at most 2 from e_1, it is at distance at most 3 from an edge in T'. By the minimality of the path, e_2 does not intersect any edges in T', so we can add e_2 to T' to form a larger $(2,3)$-tree, thereby contradicting the maximality of T'. □

These two facts imply Lemma 25.3. Along with Claim 25.6, below, and Markov's Inequality, this yields Lemma 25.2.

Claim 25.6 *The expected number of bad $(2,3)$-trees with at least $\frac{5}{k}\log m$ edges is less than $\frac{1}{2}$.*

Proof We first show that for each $r \geq 1$, H contains fewer than $m \times (4d^3)^r$ different $(2,3)$-trees with r hyperedges. To choose such a $(2,3)$-tree, we will first choose an unlabeled tree T on r vertices. It is well-known that there are at most 4^r choices for T (cf. [76]). We then choose an edge of H to map onto each vertex of T, starting with an arbitrary vertex v_1 of T, and then proceeding through the rest of the vertices of T in a breadth-first order. There are m choices of an edge in H to map onto v_1. For every subsequent vertex v_i of T, we have already specified an edge e' which maps onto a neighbour of v_i. Thus, the edge mapping onto v_i must be to one of the at most d^3 neighbours in $L^{(2,3)}(H)$ of e'. Therefore, there are a total of $4^r \times m \times (d^3)^{r-1} < m \times (4d^3)^r$ such $(2,3)$-trees, as required.

Now consider any such tree. It is easily seen that the probability of a particular edge becoming bad is at most $2^{1-\frac{1}{2}k}$. Furthermore, since no two edges of the tree intersect, the probability that all of them become bad is at most $(2^{1-\frac{1}{2}k})^r$. Therefore, the expected number of bad $(2,3)$-trees of size r is at most $m(4d^3 \times 2^{1-\frac{1}{2}k})^r$. Since $d \le 2^{k/16-2}$, this expected number is at most $m(2^{3k/16-4} \times 2^{1-\frac{1}{2}k})^r < m \times 2^{-5kr/16}$, which is less than $\frac{1}{2}$ for $r \ge \frac{5}{k}\log m$. Of course, if there is a bad $(2,3)$-tree of size at least $\frac{5}{k}\log m$ then there is one of size exactly $\lceil \frac{5}{k}\log m \rceil$, and so this completes our proof. □

25.1.2 Further Details

If k grows with m, or if we want a running time which is near linear, rather than merely polynomial, then we simply perform another iteration of Phase I on each of the components of H', taking advantage of the fact that these components are all small. This time, an edge becomes *bad* if we colour at least half of its at least $k/2$ vertices, and every edge in the resulting hypergraph, H', will have size at least $k/4$. Again, the Local Lemma will show that H'', has a proper 2-coloring, since $d \times 2^{1-k/4} < \frac{1}{4}$.

Recall that every component of H' has at most $5(d+1)\log m$ vertices, and thus has at most $m' = 5(d+1)^2 \log m$ hyperedges, since every vertex lies in at most $d+1$ edges. Mimicking the analysis from Lemma 25.2, we can show that with probability at least $\frac{1}{2}$, every component of H'' has at most $16(d+1)\log(m') < 40d(\log\log m + \log d)$ vertices and at most $40d^2(\log\log m + \log d)$ hyperedges (for m sufficiently large). This time, the analysis uses the fact that $m'(4d^3 \times 2^{1-k/4})^{(16/k)\log m'} < \frac{1}{2}$ which holds since $d < 2^{k/16-2}$. We can ensure that these bounds on the size of the components of H'' hold, by restarting Phase II if they fail. The expected number of times we run Phase II is less than 2.

If d is small in terms of m, then we can again apply exhaustive search to find a successful completion of our partial colouring. For example, if $d \le \log\log m$ then the number of possible 2-colourings for any component of H'' is less than $2^{(\log\log m)^3}$ which is less than $O(m^\epsilon)$ for any $\epsilon > 0$. Thus we can perform an exhaustive search for each component in a total of at most $O(m^{1+\epsilon})$ time.

However, if k, and hence d, grows too quickly with m, then we won't be able to carry out the exhaustive search in polytime. But if $d > \log\log m$, then there at most $40d^2(\log\log m + \log d) < 80d^3$ hyperedges in each component of H'' and so the expected number of monochromatic edges in a uniformly random 2-colouring of a component of H'' is less than $80d^3 \times 2^{1-k/4} < 1$. Therefore, we can simply apply the Erdős–Selfridge procedure from Chapter 24 to find a proper 2-colouring of each component.

This completes the proof that our randomized algorithm runs in near linear expected time. Now we will see that we can derandomize this algorithm using the techniques of the previous chapter.

When we come to a vertex in Phase I, instead of assigning it a uniformly random colour, we give it the colour that minimizes the expected number of bad $(2,3)$-trees with exactly $\frac{5}{k}\log m$ edges that would arise if we were to complete Phase I by using uniformly random choices to colour the vertices.

To show that this is a polytime algorithm, we only need to show that we can compute this expected value quickly. As we saw in the proof of Claim 25.6, the total number of $(2,3)$-trees in H of size $r = \frac{5}{k}\log m$ is at most $m\times(4d^3)^r \leq m \times (4 \times 2^{3k/16-6})^r < m^2$. Furthermore, for any such $(2,3)$-tree, it is very easy to compute the probability that if we continue Phase I using random choices, the tree would become bad. Therefore, we can compute the required expected values very quickly.

The same procedure will work again on H', allowing us to find an appropriate H'' in polytime, thus yielding a deterministic polytime algorithm to find a proper 2-colouring of H.

25.2 A Different Approach

Several people (including, possibly, you reader) have suggested the following very simple approach for finding a 2-colouring of H. Give every vertex a random colour, then uncolour all vertices which lie in monochromatic edges. Repeat until there are no monochromatic edges.

One's intuition might suggest that this approach ought to work. The expected number of monochromatic edges after the first iteration is $2^{1-k}m$ which is fairly small. Furthermore, it seems reasonable to believe that this expected number would decrease quickly with each iteration, and so we would rapidly converge to a proper 2-colouring. However, thus far no one has succeeded in proving this latter statement.

It would be very interesting to determine whether this procedure works to find a 2-colouring guaranteed by Theorem 4.2 with any non-trivial lowering of d. In fact, possibly it works without lowering d at all, at least when k is large.

It is important to note that it is not enough to show that the probability of an edge remaining uncoloured for several iterations is small; this is quite easy to prove. The main issue is dealing with edges which are 2-coloured for several iterations, but then become monochromatic. For example, consider an edge e which has $k-1$ red vertices and 1 blue vertex which lies in an all blue edge. With probability $\frac{1}{2}$, e will become monochromatic during the next iteration.

In [2], Alon provides a somewhat simpler proof of Theorem 25.1 (with 16 replaced by 500), using an algorithm which is somewhere between the one just described and Beck's algorithm. In Alon's algorithm, we assign a random colour to each vertex, and then we uncolour every hyperedge which is close to being monochromatic in the sense that it has fewer than $\frac{k}{8}$ vertices of one colour. Virtually the same analysis as in the last section applies here. Note how this overcomes the issue raised in the previous paragraph. An edge such as e which is 2-coloured, but perilously close to becoming monochromatic will be completely recoloured along with the monochromatic edges.

Alon's approach, while simpler than Beck's, is not nearly as widely applicable. We discuss the robustness of Beck's approach in the next section.

25.3 Applicability of the Technique

Beck's technique does not work just for the particular problem of 2-colouring hypergraphs. Rather, it can be used on a wide range of applications of the Local Lemma. In this section, we present a theorem of Molloy and Reed [120] which roughly describes the type of application for which Beck's technique works well.

In what follows, $\mathcal{T} = \{t_1, \ldots, t_m\}$ is a set of independent random trials. $\mathcal{A} = \{A_1, \ldots, A_n\}$ is a set of events such that each A_i is determined by the outcome of the trials in $T_i \subseteq \mathcal{T}$. Using a 2-phase process similar to Beck's, we can prove:

Theorem 25.7 *If the following holds:*

1. *for each* $1 \leq i \leq n, \mathbf{Pr}(A_i) \leq p$;
2. *each* T_i *intersects at most* d *other* T_j*'s;*
3. $pd^9 < 1/512$;
4. *for each* $1 \leq i \leq n, |T_i| \leq \omega$;
5. *for each* $1 \leq j \leq m$, *the size of the domain of* t_i *is at most* γ, *and we can carry out the random trial in time* τ_1;
6. *for each* $1 \leq i \leq n$, $t_{j_1}, \ldots, t_{j_k} \in T_i$ *and* $w_{j_1}, \ldots, w_{j_k}$ *in the domains of* $t_{j_1}, \ldots, t_{j_k}$ *respectively, we can compute, in time* τ_2, *the probability of* A_i *conditional of the outcomes of* $t_{j_1}, \ldots, t_{j_k}$ *being* $w_{j_1}, \ldots, w_{j_k}$,

then there is a randomized $\mathrm{O}(m \times d \times (\tau_1 + \tau_2) + m \times \gamma^{\omega d \log\log m})$*-time algorithm which will find outcomes of* $t_1, \ldots, t_m$ *such that none of the events in* $\mathcal{A}$ *hold.*

So, for example, if $d, \omega, \gamma = \mathrm{O}(\log^{1/3} m)$, $\tau_1 = \mathrm{poly}(\gamma)$, and $\tau_2 = \mathrm{O}(\gamma^\omega \times \mathrm{poly}(\omega))$, then the running time of our algorithm is nearly linear in m.

Note that the first three conditions are essentially the conditions of the Local Lemma, with a constant bound on pd replaced by a constant bound on pd^9. The remaining ones are just technical conditions which bound the running time of our algorithm. Condition 6 of Theorem 25.7 says that we can compute the conditional probabilities quickly, which we need to do in order to determine which trials to pass over in the first phase. Condition 5 says that we can carry out the random trials efficiently. The bounds in Conditions 4, 5 are used to bound the time required to use exhaustive search to find satisfactory outcomes of the trials, in the final phase.

The exponent "9" in Condition 3 can probably be decreased somewhat, but to get it to 1 would require a significantly different proof. The requirement that pd^{1+x} be bounded for some constant $x > 0$ is the crucial restriction here, and it captures the nature of the applications that are amenable to Beck's technique.

It is worth noting that for many applications of the Local Lemma, a minor adjustment in a parameter (eg. Δ_0) will cause pd^9 to be small enough for Theorem 25.7 to apply. For example, in the vast majority of applications encountered in this book, d is a polynomial in some parameter, usually Δ, and p is exponentially small in that parameter; so for large Δ, pd^x is bounded for any constant x.

To prove Theorem 25.7, we use essentially the same algorithm that Beck did. We consider the hypergraph H which has vertex set $\mathcal{T}$ and where for each event A_i, we have a hyperedge e_i with vertex set T_i. We carry out the random trials one-at-a-time. We say that a hyperedge e_i becomes bad if the conditional probability of A_i ever exceeds $p^{2/3}$; we do not carry out any more trials in T_i after e_i becomes bad. We define H' in a similar manner as before and show that with probability at least $\frac{1}{2}$, all components of H' are small. We repeat the process; this time a hyperedge is bad if the conditional probability of the corresponding event increases from $p^{2/3}$ to $p^{1/3}$. The Local Lemma guarantees that there is a way to complete the remaining random trials so that none of the A_i hold. All components of H'' will be sufficiently small that we can find this guaranteed completion via exhaustive search.

The reader should note that, in the case of 2-colouring hypergraphs, this is the algorithm described earlier in the chapter, except that we say an edge is bad in the first phase if we colour $k/3$ vertices, all the same colour; and in the second phase if we colour another $k/3$ vertices, all the same colour.

The crucial Condition 3 comes in to play twice. First, at the end of the second phase, no event has conditional probability greater than $p^{1/3}$. Condition 3 yields a bound on $p^{1/3} \times d$ which, along with the Local Lemma, implies that with positive probability, a random completion of the trials will result in no event in $\mathcal{A}$ occurring. Second, the term $m(4d^3 \times 2^{1-\frac{1}{2}k})^r$ from the proof of Claim 25.6 becomes $m(4d^3 \times p^{1/3})^r$ here. Our bound on pd^9 yields a bound on this quantity.

There is one detail that we have glossed over. In our analysis of Phase II, we assume that at the end of Phase I, no conditional probability exceeds $p^{2/3}$. However, we have to be careful about this, since we do not freeze an event until after its conditional probability exceeds $p^{2/3}$. It is possible that the outcome of a single trial can cause the conditional probability of an event to increase significantly. Thus, possibly an event A_i whose conditional probability is below $p^{2/3}$ (and hence e_i is not bad) could have its conditional probability jump far higher than $p^{2/3}$. To handle this difficulty, we undo any trial which causes an event to become bad. Thus, we use the following freezing rule. If the outcome of a trial t causes the conditional probability of any A_i to increase above $p^{2/3}$, then we undo t and freeze it along with all other trials in A_i, and we say that A_i is bad. This way we ensure that at the end of Phase I, no conditional probability exceeds $p^{2/3}$, and at the same time, the probability of an edge becoming bad is at most $p^{1/3}$.

It is important to note that our running time depends crucially on ω, γ, the bounds on the number of trials per event, and the size of the domains of the trials. This is because we are bounding the time to do an exhaustive search using the most naive approach. So, a blind application of this theorem only provides a polynomial running time when both of these quantities are bounded.

For example, in most of the applications in this book, Theorem 25.7 implies a polytime algorithm if the maximum degree Δ is bounded by a constant, or by a very slowly increasing function of the size of the graph. But it does not yield a polytime algorithm if Δ can be arbitrarily large. To obtain a polytime algorithm for general Δ, one must be less naive about the final exhaustive search. For example, one might treat events corresponding to vertices of high degree in a different manner, mimicking our treatment of the case $d > \log\log m$ in Sect. 25.1.2.

Furthermore, Theorem 25.7 gives no guarantees that the algorithm can be derandomized. But this is something that can indeed be done for many individual applications.

Thus, in many cases, it is advisable to use Theorem 25.7 as just a guideline as to when the technique should work. Namely, if pd^9 is bounded, then there is a good chance that the argument can be fine-tuned to provide an algorithm that works well for the problem at hand.

25.3.1 Further Extensions

In his original paper [19], Beck provided an algorithmic analogue to Exercise 5.1. More recently, Czumaj and Scheideler [34] obtained an algorithmic analogue to an extension of Theorem 19.2:

Theorem 25.8 *There is a polytime algorithm which will find a proper 2-colouring of any hypergraph H which satisfies for each hyperedge $e \in H$: $|e| \geq 2$ and*

$$\sum_{f \cap e \neq \emptyset} 2^{-|f|/50} < |e|/100.$$

Remark Czumaj and Scheideler actually stated their main theorem in a different, but essentially equivalent form. Their statement is designed to evoke the General Local Lemma rather than the Weighted Local Lemma.

The Weighted Local Lemma implies the existence of a 2-colouring even if the exponent "$-|f|/50$" is replaced by "$-|f|/2$", using the same proof as that used for Theorem 19.2. A more careful application of the General Local Lemma allows the exponent to be replaced by "$-(1-\epsilon)|e|$" for any $\epsilon > 0$ so long as the constant "100" is increased as a function of $1/\epsilon$.

Czumaj and Scheideler's algorithmic result requires the extra multiplicative term "1/50" for the same reasons that Beck required the similar "1/16" in the conditions for Theorem 25.1. In other words, this stronger condition is analogous to Condition 3 of Theorem 25.7.

Czumaj and Scheideler had to modify Beck's technique a bit, in that their algorithm is somewhat different and the analysis is much more difficult. Note that Theorem 25.7 implies their result for the case where the edge sizes are all small enough to satisfy Condition 4, for some suitably small ω. Dealing with very large edges, and in particular with the way that they interact with small edges, was the main difficulty in proving Theorem 25.8. See [34] for the details; see also [35] for further extensions.

In [139], Salavatipour strengthened the arguments in [35] to prove the following discrepancy version of Theorem 25.8:

Theorem 25.9 *For any $0 < \alpha < 1$, there exists $1 > \beta, \gamma > 0$ and a polytime algorithm which does the following. Suppose that H is a hypergraph and that for each $e \in H$ we have:*

$$\sum_{f \cap e \neq \emptyset} 2^{-\beta|f|} < \gamma|e|.$$

Then the polytime algorithm will find a 2-colouring of H such that for every edge e, the difference between the number of red vertices and the number of blue vertices in e is at most $\alpha|e|$.

A somewhat weaker result was provided in [35].

25.4 Extending the Approach

All the results described thus far in this chapter require something analogous to Condition 3 of Theorem 25.7. In this section we present another approach, introduced by Molloy and Reed in [120]. It often applies in situations where pd is bounded by a constant, but pd^{1+x} is not for any fixed $x > 0$.

To begin, we study the problem of colouring a k-uniform hypergraph of maximum degree Δ. As usual, our goal is that no edge is monochromatic. Typically, k can be small and Δ will be large, so we will require many more than two colours.

25.4.1 3-Uniform Hypergraphs

We begin by considering the case $k = 3$. It is trivial to use our standard greedy algorithm to $(\Delta+1)$-colour any hypergraph with at least 2 vertices per edge, so that no edge is monochromatic. The Local Lemma implies that if every edge has at least 3 vertices, then we can do substantially better – we only need $O(\sqrt{\Delta})$ colours. To see this, suppose that we assign to each vertex a uniformly random colour from a list of C colours, where as usual, each choice is independent of that for other vertices. For each edge e, we let A_e be the event that e is monochromatic. Clearly, $\mathbf{Pr}(A_e) \leq p = \frac{1}{C^2}$. Furthermore, each event is mutually independent of all but at most $d = 3(\Delta - 1)$ other events. Thus, so long as $C \geq 2\sqrt{3} \times \sqrt{\Delta}$, we have $pd < \frac{1}{4}$ and so with positive probability no edge is monochromatic. In this section, we will prove an algorithmic analogue of this simple result:

Theorem 25.10 *There is a polytime algorithm which properly colours any hypergraph with maximum degree Δ and minimum edge-size at least 3, using $O(\sqrt{\Delta})$ colours.*

We will only present the randomized version of the algorithm here, leaving it to the reader to verify that it can be derandomized (see Exercise 25.1). We can assume that the hypergraph is 3-uniform, since if it is not, then by deleting $|e|-3$ vertices from each edge e of size greater than 3 we obtain a 3-uniform hypergraph any proper colouring of which yields a proper colouring of the original hypergraph.

Note that in our proof of the existence of such a colouring, if $C = O(\sqrt{\Delta})$ then pd^9 is of order Δ^7 and so we cannot apply Theorem 25.7. If we wish to bound pd^9 by a constant then we need $C \geq \Delta^{4.5}$ which results in a very uninteresting theorem since we have already noted that a trivial algorithm colours the hypergraph with $\Delta + 1$ colours. So in order to prove Theorem 25.10, we need to deviate somewhat from the procedure of the earlier sections, using a variation developed in [120].

The intuition behind the modification is straightforward. Consider any particular vertex v. When we assign random colours to $V(H) - v$, the probability that a particular edge containing v has two vertices of the same colour is of order $\Omega(1/\sqrt{\Delta})$ and hence the expected number of such edges is $\Omega(\sqrt{\Delta})$. Thus, if we were to freeze each v as soon as it lies in such an edge, then we would expect to freeze the majority of the vertices. So we cannot mimic the arguments used to prove Theorem 25.1. Instead we only freeze v if it lies in many such edges. As a result, we freeze far fewer vertices – few enough for the analysis to work. However, this requires significant modifications in the procedure. Forthwith the details.

As usual, we can assume that $\Delta \geq \Delta_0$ for some large constant Δ_0 since for $\Delta < \Delta_0$ the trivial greedy algorithm will use only $\Delta_0 = O(1)$ colours. For Δ sufficiently large, our algorithm will use $C = 10\sqrt{\Delta}$ colours.

As before, the first phase of our algorithm will entail colouring the vertices one-at-a-time. The difference is that this time, if at least half, i.e. two, of the vertices on a hyperedge receive the same colour, c, then rather than leaving the remaining vertex until the next phase, we instead forbid that vertex from receiving the colour c. More specifically, for each vertex v, we maintain a list L_v of permissible colours for v. When we come to a vertex v, we assign it a uniformly random colour from L_v. Initially, $L_v = \{1, \dots, C\}$ for each v, and for every hyperedge $\{u, v, w\}$, if u, v both get the same colour c then c is removed from L_w. This ensures that no edge will become monochromatic.

We have to take care to ensure that no L_v loses too many colours. For example, since $C = O(\sqrt{\Delta})$ and v lies in up to Δ edges, it is possible that every colour will get deleted from L_v. We prevent this as follows. If $|L_v|$ ever decreases to $9\sqrt{\Delta}$ for some uncoloured v, then we say that v is *bad* and we *freeze* v and all uncoloured vertices in those hyperedges containing v. When we reach a frozen vertex, we pass over it without colouring it. This ensures that at the end of the phase, every L_v contains at least $9\sqrt{\Delta}$ colours. More formally, our procedure runs as follows:

First Phase:
The input is a hypergraph H with vertices $v_1, \dots, v_n$.

0. We initialize $L_v = \{1, \dots, 10\sqrt{\Delta}\}$ for each v.
1. For $i = 1$ to n, if v_i is not frozen then:
   a) Assign to v_i a random colour c chosen uniformly from L_{v_i}.
   b) For each edge $\{v_i, u, w\}$ where u has colour c and w is uncoloured:
      i. Remove c from L_w.
      ii. If $|L_w| = 9\sqrt{\Delta}$ then w is bad and we freeze w and all uncoloured vertices in those hyperedges containing w.

H' is the hypergraph induced by all edges containing at least one uncoloured vertex at the end of the First Phase.

The Local Lemma implies that the partial colouring produced by the First Phase can be completed to a colouring of H, as follows. Consider assigning to each uncoloured v a uniformly random colour from L_v. If an edge e has either 2 or 3 coloured vertices, then the probability that e becomes monochromatic is 0, because of the removal of colours from the lists in the first phase. If e has either 0 or 1 uncoloured vertices, then this probability is at most $1/(9\sqrt{\Delta})^2$. Thus, the proof follows as in the argument preceding the statement of Theorem 25.10.

As in the earlier sections, we will show that the components of H' are so small that we can find this completion quickly.

Lemma 25.11 *With probability at least $\frac{1}{2}$, every component of H' has at most $8\Delta^3 \log n$ vertices.*

This lemma, proven below, allows us to complete the proof of Theorem 25.10 in the same way that we proved Theorem 25.1. We repeat essentially the same procedure on each component of H'. This time, if v is assigned colour c then we remove c from L_u for every edge $\{v, u, w\} \in H$ where w was previously assigned c in either the First or Second Phase. We freeze v and its neighbours if L_v ever drops to size $8\sqrt{\Delta}$. The same analysis as that used to prove Lemma 25.11 will show that this reduces the maximum component size to $8\Delta^3 \log(8\Delta^3 \log n) < 9\Delta^3 \log\log n$. Once again, the Local Lemma proves that the partial colouring can be completed, and we can quickly find the completion on each individual component. Before describing exactly how to find these colourings, we prove our main lemma. The proof is very similar to that of Lemma 25.2.

Proof of Lemma 25.11 Again, we focus on (a, b)-trees. This time, it is vertices that go bad, rather than edges, so it is more convenient to define these trees in terms of H itself, rather than the line graph of H. $H^{(3,5)}$ is the graph with vertex set $V(H)$ in which two vertices are adjacent iff they are at distance exactly $3, 4$ or 5 in H. A $(3, 5)$-*tree* is a connected induced subgraph of $H^{(3,5)}$ containing no two vertices of distance less than 3 in H. A $(3, 5)$-tree is *bad* if it contains only bad vertices.

A proof nearly identical to that of Lemma 25.3 yields that every component of H' with at least ℓ vertices must contain a bad $(3, 5)$-tree with at least $\ell/(8\Delta^3)$ vertices. The maximum degree in $H^{(3,5)}$ is at most $(2\Delta)^5$, and so it follows as in the proof of Claim 25.6 that every vertex lies in at most $(4 \times 32\Delta^5)^r$ different $(3, 5)$-trees. Thus, Lemma 25.11 follows from the following:

Claim 25.12 *The probability that a particular* $(3, 5)$*-tree of size* r *becomes bad is at most* $(e/9)^{r\sqrt{\Delta}}$.

It is not hard to show that the probability of a particular vertex becoming bad is at most $(e/9)^{\sqrt{\Delta}}$. It is tempting to say that Claim 25.12 follows since two vertices at distance at least 3 in H have different neighbourhoods and so the events that they become bad are independent. However, this is not true. The colour choice for a vertex v can affect the list L_u for any neighbour u of v, and so the colours assigned to u and v are not independent. Similarly, the colours assigned to v and to a neighbour w of u are not independent, and so on. This chain of dependency spreads throughout the graph so that for two vertices of arbitrary distance, the choice of colours assigned to them are not independent, let alone the events that they become bad. So instead, we proceed as follows.

Consider r vertices which form a $(3, 5)$-tree, T. For T to become bad, there must be, for each $v \in T$, a set of $\sqrt{\Delta}$ edges containing v such that the two other vertices in each edge get the same colour. This gives a set of $r\sqrt{\Delta}$ disjoint monochromatic pairs of vertices. There are $\binom{\Delta}{\sqrt{\Delta}}^r$ choices for these pairs. When we come to the second vertex of a pair, regardless of what colours were assigned to earlier vertices in the procedure, the probability that it gets the same colour as its partner is at most $1/(9\sqrt{\Delta})$. So the probability that the tree becomes bad is at most:

$$\left(\frac{\Delta}{\sqrt{\Delta}}\right)^r \times \left(\frac{1}{9\sqrt{\Delta}}\right)^{r\sqrt{\Delta}} \leq \left(\frac{e\Delta}{\sqrt{\Delta}} \times \frac{1}{9\sqrt{\Delta}}\right)^{r\sqrt{\Delta}} = \left(\frac{e}{9}\right)^{r\sqrt{\Delta}}.$$

Therefore, the expected number of bad $(3,5)$-trees of size r is at most $n \times (4 \times 32\Delta^5)^r \times (e/9)^{r\sqrt{\Delta}} < n \times (\frac{1}{3})^r$ for Δ sufficiently large, and so the expected number of components in $H^{'}$ with ℓ vertices is at most $n \times (\frac{1}{3})^{(\ell/8)\Delta^3}$. This establishes Lemma 25.11. □

As we said earlier, the same analysis yields the following bound on the component sizes of $H^{''}$:

Lemma 25.13 *With probability at least $\frac{1}{2}$, every component of $H^{''}$ has at most $9\Delta^3 \log\log n$ vertices.*

It only remains to show how to complete the colouring on $H^{''}$. First, we consider the case $\Delta \geq (\log\log n)^3$. Claim 25.12 with $r = 1$ implies that for any component X of $H^{''}$, the expected number of bad vertices in X after the Second Phase is at most $O(\Delta^3 \log n (e/9)^{\sqrt{\Delta}}) = o(1)$. So for any such component, by repeating the Second Phase a constant expected number of times, we obtain $H^{''} = \emptyset$.

Thus, we can assume that $\Delta < (\log\log n)^3$, and so the number of vertices in any component X of $H^{''}$ is at most $O((\log\log n)^{10})$, and $C = O\left((\log\log n)^{3/2}\right)$. Using exhaustive search through all the colourings of X takes $O(C^{|X|})$ time which is $o(n^\epsilon)$ for any $\epsilon > 0$.

25.4.2 k-Uniform Hypergraphs with $k \geq 4$

For any $k \geq 2$, if H is a k-uniform hypergraph with maximum degree Δ, then the Local Lemma implies that H can be coloured with $O\left(\Delta^{\frac{1}{k-1}}\right)$ colours. The general proof is the same as for the 3-uniform case – this time if there are C colours, then the probability that an edge becomes monochromatic is $1/C^{k-1}$. Theorem 25.10 extends to such hypergraphs as follows:

Theorem 25.14 *For each fixed $k \geq 2$, there is a polytime algorithm which properly colours any k-uniform hypergraph with maximum degree Δ using $O\left(\Delta^{\frac{1}{k-1}}\right)$ colours.*

Note that this is trivial for $k = 2$.

The proof is along the same lines as that of Theorem 25.10, except for one complication. As before, we remove a colour c from L_v if c is assigned to every other vertex in an edge containing v. Again, we want to apply the Local Lemma to show that the partial colouring produced by the First Phase can be extended to a colouring of H. However, if an edge e has exactly two coloured vertices and they both get the same colour, then the probability that the edge becomes monochromatic in a random completion is $1/C^{k-2}$ rather than $1/C^{k-1}$. If there are more than $O\left(\Delta^{\frac{k-2}{k-1}}\right)$ such edges intersecting a particular edge e, then the Local Lemma will not apply.

There are two ways to deal with this complication. The simplest is to use a fresh set of colours for the Second Phase. This ensures that any edge for which even one vertex was coloured in the First Phase cannot become monochromatic in the Second Phase, thus overcoming our complication. Similarly, we use a third set of colours in the Third Phase. This has an effect of tripling the total number of colours used, which is absorbed by the $O\left(\Delta^{\frac{1}{k-1}}\right)$ notation.

For the case where H is linear, there is an alternate, more difficult, way to deal with this complication. We describe it here because it generalizes to other situations where the first method does not work. Consider, for example, the case $k = 4$, and set $C = 10\Delta^{1/3}$ to be the number of colours used. At any stage of the procedure, for each vertex v we denote by T_v, the number of edges e containing v in which two vertices other than v have the same colour. If T_v ever reaches $\Delta^{2/3}$ then we say that v is *bad* and we freeze v along with all the uncoloured vertices in the edges containing v. Of course, we do the same if L_v ever drops to size $9\Delta^{1/3}$.

It is easy to see that the Asymmetric Local Lemma now ensures that the partial colouring produced in the First Phase can be completed. Furthermore, the expected size of T_v at the end of the first iteration is at most $\Delta \times 3 \times \frac{1}{9\Delta^{1/3}} = \Delta^{2/3}/3$. So to bound the probability of v becoming bad, it is enough to show that T_v is concentrated.

If H is linear, then it is not difficult to show that T_v is concentrated since the choice of colour assigned to any vertex other than v can affect T_v by at most 1. Furthermore, a proof along the lines of that of Lemma 25.11 proves that with probability at least $\frac{1}{2}$, all the components of H' are small.

As usual, we repeat the phase, this time freezing if some T_v reaches $2\Delta^{2/3}$ or if some $|L_v|$ drops to $8\Delta^{1/3}$. The Asymmetric Local Lemma guarantees that the partial colouring we obtain can be completed, as follows. We consider assigning to each v a uniform colour from L_v, and we use A_e to denote the event that e becomes monochromatic. For each edge e, the sum over all f intersecting e of $\mathbf{Pr}(A_f)$ is at most $4\Delta \times (1/8\Delta^{1/3})^3 + \sum_{v \in e} T_v \times (1/8\Delta^{1/3})^2 < \frac{1}{12}$, since each $T_v \leq 2\Delta^{2/3}$. Furthermore, with sufficiently high probability, all the components of the remaining hypergraph are very small. We omit the details which should, by now, be straightforward.

This technique extends easily for general fixed k. For each v and $2 \leq i \leq k-2$, one keeps track of the number of edges intersecting v which have exactly i coloured vertices, all with the same colour, and by freezing when necessary, ensures that this number never exceeds $2\Delta^{\frac{k-2}{k-1}}$.

25.4.3 The General Technique

The algorithm for colouring k-uniform hypergraphs discussed above is an example intended to illustrate an approach which handles problems for which pd is bounded by a constant but pd^9 is not. We close this chapter, and the main body of the book, by elucidating the main ideas behind this new approach, which was introduced in [120].

Suppose that we have a set of independent random trials $\mathcal{T} = \{t_1, \ldots, t_m\}$ and a set of events $\mathcal{A} = \{A_1, \ldots, A_n\}$ where each A_i is determined by the outcome of the trials in $T_i \subseteq \mathcal{T}$. Suppose further that we have $\mathbf{Pr}(A_i) \leq p$ for all i and that each T_i intersects at most d other T_j. As before, this defines a hypergraph whose vertices are the trials and whose edges are the T_i.

In both the new and the old approach, we carry out a sequence of iterations. In each iteration we have a probability distribution on the remaining trials and we carry out a trial using the corresponding distribution.

In both approaches, when choosing the outcome of a trial, we ensure that we will still be able to apply the Local Lemma to show that we can choose outcomes for the remaining trials in such a way that none of the A_i hold. If there is a *local danger* which might lead to this condition failing in the future then we freeze all the trials near this danger.

We need to ensure that we freeze so few vertices that with high probability the components of the frozen hypergraph are very small. This allows us to use exhaustive search to choose outcomes for the frozen trials so that none of the A_i hold.

Recall that when pd^9 is bounded by, say $\frac{1}{512}$, we define local danger in a straightforward manner. Namely, we freeze the trials in T_i if the probability of A_i conditional on the outcomes of the trials already carried out becomes too high (above $p^{2/3}$ in the first iteration, $p^{1/3}$ in the second), and then we undo the trial which caused this probability to be too high. Now, our bound on pd^9 ensures that $p^{1/3}$ is less than $\frac{1}{4d}$. So, since at the end of the second iteration we have that the conditional probability of any event becoming bad is at most $p^{\frac{1}{3}}$, the Local Lemma does indeed imply that we can choose outcomes for the frozen trials in such a way that none of the A_i hold. Furthermore, the probability that T_i is frozen in, e.g., the first iteration is at most $\frac{p}{p^{2/3}} = p^{1/3}$ which is very small by our bound on pd^9. So, we expect to freeze very few trials and it is true that with high probability the components of the frozen hypergraph are all small.

The weakness of this approach is that it requires freezing the trials in T_i before the conditional probability of A_i rises above $\frac{1}{4d}$ (or more precisely $\frac{1}{ed}$). If $p = \Omega(\frac{1}{d})$, this could force us to freeze many trials. So many, in fact, that the components of the frozen hypergraph would be too large. To avoid this difficulty, we use the following new idea:

Idea 1: *We take advantage of the fact that the degrees in the hypergraph tend to decrease.*

At any point in the process, for each event A_i, we define $H(A_i)$ to be the sum over all j such that $T_i \cap T_j \neq \emptyset$ of the conditional probability of A_j. We note that if at the end of a phase, each $H(A_i)$ is at most $\frac{1}{4}$ then the Asymmetric Local Lemma implies that we can still successfully complete the trials.

When we had $pd^9 < \frac{1}{512}$, by preventing any conditional probabilities from exceeding $p^{1/3}$, we implicitly bounded each $H(A_i)$ by $p^{1/3}d$ which was much less than $\frac{1}{4}$. This is a very loose bound, however, since the conditional probabilities of the A_i tend to vary widely, and most of these probabilities go to zero, thus causing the degrees in H' to also vary. When $pd = \Omega(1)$, we must focus more closely on $H(A_i)$, and by freezing trials if necessary, keep it from becoming too large.

For example, in Sect. 25.4.2 for the case $k = 4$, we initially had for each edge e, $H(A_e) \leq 4\Delta \times (1/C)^3 = \frac{1}{250}$. To obtain a strong bound on $H(A_e)$ as the procedure progressed we considered T_v the number of edges containing v and a pair of vertices disjoint from v which have been assigned the same colour. Throughout the procedure, $H(A_e)$ was bounded by $4\Delta \times (1/9\Delta^{1/3})^3 + \sum_{v \in e} T_v \times (1/9\Delta^{1/3})^2$. Thus, the key to keeping $H(A_e)$ small was to keep T_v small. By freezing before T_v exceeded $\Delta^{2/3}$, in the first phase, we kept $H(A_e) \leq \frac{4}{9^3} + 4 \times \frac{1}{9^2} < \frac{1}{19}$. During the second phase, we kept $H(A_e) < \frac{1}{12}$, which is more than small enough to guarantee the existence of a good completion.

Note that by focusing on an overall bound on $H(A_i)$, we no longer need to enforce an overall bound of $\frac{1}{4d}$ on the conditional probabilities of the events. I.e., we can allow some of these probabilities to become quite high, so long as each $H(A_i)$ remains small.

Of course, we have to avoid freezing too many trials. In the proof of Theorem 25.7 we required the probability of an event becoming bad to be at most $\frac{1}{8d^3}$ (which implies that the probability that we freeze a trial is $\frac{k}{8d^3} = o(\frac{1}{8d^2})$). In order to mimic this proof in the general situation, we need to obtain, for each i, a bound of the same order on the probability that the trials in T_i are frozen because $H(A_i)$ becomes too large.

However, if $pd = \Omega(1)$ and we make no other modifications to the procedure then the probability of $H(A_i)$ becoming too large is much higher than $\frac{1}{8d^3}$. To see this note that if A_i occurs during our process, then $H(A_i)$ will become at least 1, which is too large. and so we will freeze the trials in T_i (and undo the trial which caused A_i to go bad). Now, intuitively, the probability that A_i will occur during our process is roughly $p = \Omega(\frac{1}{d})$, which is much larger than $\frac{1}{8d^3}$. So, we need to introduce another new idea:

Idea 2: *We modify the distributions of the trials.*

For example, in Subsections 25.4.1 and 25.4.2, for each vertex v we had a trial t_v which initially consisted of assigning v a colour from $\{1, \ldots, C\}$ chosen from the uniform distribution. As the procedure progressed, we occasionally modified that distribution, setting $\mathbf{Pr}(c) = 0$ for certain colours c by removing c from L_v, and thereby increasing $\mathbf{Pr}(c')$ for each colour c' remaining in L_v. Note that the colours deleted were chosen so as to decrease the probability that any event "containing" v occurred, and hence lowered the values of the corresponding $H(A_i)$. Modifying the lists in this way allowed us to keep the $H(A_i)$ low without freezing too many trials.

Idea 2 is not just a necessary technical modification which is needed in the implementation of Idea 1. It is a powerful tool in its own right. For example, in Sect. 25.4.1, by modifying the distributions we ensured that all the conditional probabilities remained so low that Idea 1 was not needed at all!

The main difficulty encountered when we try to apply this technique is the following, which, though obvious, is worth stressing.

Complication: The T_i intersect.

Thus, if we modify the distribution for some trial to ensure that some A_i does not occur, this may increase $H(A_j)$ for some $i \neq j$. We need to avoid situations in which we skew our distributions so much that other $H(A_j)$ become too high. One way to do this is to set limits on the amount by which a trial's probability distribution can change, and resort to freezing when some trial's distribution approaches its limits. For example, in Sect. 25.4.1 our limit was that we would only set $\mathbf{Pr}(c) = 0$ for at most $\sqrt{\Delta}$ colours c, i.e. we kept $|L_v| \geq 9\sqrt{\Delta}$. Generally, the amount of freezing required to enforce the limits is much less than the amount of freezing that would take place if we did not modify the distributions. This is what makes the approach so powerful.

Of course, we need to bound the amount of freezing which occurs because of these limits. In Sect. 25.4.2 this amounted to proving that T_v was highly concentrated.

Clearly, the precise manner in which we implement these ideas depends on the specific application. Thus, we have provided here only an illustrative example and a roadmap for this technique.

In the same way, this book presents some elucidating examples of and a roadmap to the probabilistic method. We hope that they will guide readers to new and interesting applications of this powerful technique.

Exercises

Exercise 25.1 Show how to derandomize the algorithm from Sect. 25.4.1. Start by being more precise in the proof of Lemma 25.11 and showing that in fact the expected number of bad $(3,5)$-trees of size r is less than $\frac{1}{2}$ for $r = \log n/\sqrt{\Delta}$.

Exercise 25.2 Find an algorithm corresponding to Theorem 19.3. Specifically, for every constant β, show that there is a polytime algorithm which will find a β-frugal colouring of any graph G with maximum degree Δ sufficiently large, using $O(\Delta^{1+1/\beta})$ colours.

References

1. Ajtai M., Komlós J., Szemeredi E. (1981) A dense infinite Sidon sequence. Europ. J. Comb. 2:1-11
2. Alon N. (1991) A Parallel Algorithmic Version of the Local Lemma. Random Structures & Algorithms 2:367–379
3. Alon N. (1992) The Strong Chromatic Number of a Graph. Random Structures & Algorithms 3:1–7
4. Alon N.(1993) Restricted Colourings of Graphs. In: Surveys in Combinatorics, Proc. 14th British Combinatorial Conference, Cambridge University Press, Cambridge, 1–33
5. Alon N. (1996) Independence Numbers of Locally Sparse Graphs and a Ramsey Type Problem. Random Structures & Algorithms 9:271-278
6. Alon N. Personal communication
7. Alon N., Krivelevich M., Sudakov B. (1997) Subgraphs with Large Cochromatic Number. Journal of Graph Theory 25:295–297
8. Alon N., McDiarmid C., Reed B. (1991) Acyclic Colouring of Graphs. Random Structures & Algorithms 2:277–288
9. Alon N., Sudakov B. and Zaks A. (2001) Acyclic Edge Colourings of Graphs. J. Graph Th. 37:157–167
10. Alon N., Spencer J. (1992) The Probabilistic Method. Wiley, New York
11. Andersen L. (1992) The Strong Chromatic Index of a Cubic Graph Is at Most 10. Discrete Math. 108:231–252
12. Appel K., Hakken W. (1976) Every Planar Map Is Four Colorable. Bull. Amer. Math. Soc. 82:711–712
13. Arora S., Lund C. (1996) Hardness of Approximation, in Approximation Algorithms for NP-hard problems (Ed. D. Hochbaum), PWS Publishing, Boston
14. Azuma K. (1967) Weighted Sums of Certain Dependent Random Variables. Tokuku Math. Journal 19:357–367
15. Baik J., Deift P., Johansson K. (1999) On the Distribution of the Length of the Longest Increasing Subsequence of Random Permutations. J. Amer. Math. Soc. 12:1119–1178
16. Beck J. Multi-dimensional Tic-Tac-Toe. J. Combinatorial Th. (A) (to appear).
17. Beck J. The Erdős-Selfridge Theorem in Positional Game Theory, manuscript
18. Beck J. (1978) On 3-Chromatic Hypergraphs. Discrete Math. 24:127–137
19. Beck J. (1991) An Algorithmic Approach to the Lovász Local Lemma I. Random Structures & Algorithms 2:343–365
20. Behzad M. (1965) Graphs and Their Chromatic Numbers. Ph.D. Thesis, Michigan State University
21. Berge C. (1963) Perfect Graphs. In: Six Papers on Graph Theory, Indian Statistical Institute, Calcutta, 1–21
22. Beutelspacher A., Hering P. (1984) Minimal Graphs for Which the Chromatic Number Equals the Maximal Degree. Ars Combinatorica 18:201–216

23. Bollobás B. (1978) Chromatic Number, Girth and Maximal Degree. Discrete Math. 24:311–314
24. Bollobás B. (1984) Random Graphs. Academic Press, London-New York
25. Bollobás B. (1988) The Chromatic Number of Random Graphs, Combinatorica 8:49–55
26. Bollobás B., Catlin P., Erdős P. (1980) Hadwiger's Conjecture Is True for Almost Every Graph. Europ. J. Comb. 1:195–199
27. Bollobás B., Harris A. (1985) List-Colourings of Graphs. Graphs and Combinatorics 1:115–127
28. Bondy J.A., Murty U.S.R. (1979) Graph Theory and Related Topics. Academic Press, Toronto
29. Borodin O., Kostochka A. (1977) On an Upper Bound on a Graph's Chromatic Number, Depending on the Graphs's Degree and Density. J. Combinatorial Th. (B) 23:247–250
30. Brooks R.L. (1941) On Colouring the Nodes of a Network. Proc. Cambridge Phil. Soc. 37:194–197
31. Catlin P. (1978) A Bound on the Chromatic Number of a Graph, Discrete Math. 22:81–83
32. Catlin P. (1979) Hajös' Graph Colouring Conjecture: Variations and Counter-Examples. J. Combinatorial Th. (B) 26:268–274
33. Chung F.R.K., Gyárfás A., Trotter W.T., Tuza Z. (1990) The Maximum Number of Edges in $2K_2$-free Graphs of Bounded Degree. Discrete Math. 81:129–135
34. Czumaj A., Schiedeler C. (2000) Coloring Nonuniform Hypergraphs: A New Algorithmic Approach to the General Lovász Local Lemma. Random Structures & Algorithms 17:213–237
35. Czumaj A., Schiedeler C. (2000) An Algorithmic Approach to the General Lovász Local Lemma with Applications to Scheduling and Satisfiability Problems. Proceedings of the 32nd Annual ACM Symposium on Theory of Computing (STOC), 38–47
36. Dirac G.A. (1952) A Property of 4-Chromatic Graphs, and Some Remarks on Critical Graphs. J. London Math. Soc. 27:85–92
37. Edmonds J. (1965) Maximum Matchings and a Polytope with 0-1 Vertices. J. Res. Nat. Bur. Standards Sec. B 69b:125–130
38. Emden-Weinert S., Hougardy S., Kreuter B. (1998) Uniquely Colourable Graphs and the Hardness of Colouring Graphs of Large Girth. Combinatorics Probability & Computing 7:375–386
39. Erdős P. (1959) Graph Theory and Probability. Canadian J. of Math. 11:34–38
40. Erdős P. (1964) On a Combinatorial Problem II. Acta Math. Hung. Acad. Sci. 15:445–447
41. Erdős P., Fajtlowicz S. (1981) On the Conjecture of Hajós. Combinatorica 1:141–143
42. Erdős P., Gimbel J. (1993) Some Problems and Results in Cochromatic Theory. Ann. Discrete Math. 55:261–264
43. Erdős P., Gimbel J., Kratsch D. (1992) Some Extremal Results in Cochromatic and Dichromatic Theory. J. Graph Th. 15:579–585
44. Erdős P., Lovász L. (1975) Problems and Results on 3-Chromatic Hypergraphs and Some Related Questions. In: "Infinite and Finite Sets" (A. Hajnal et. al. Eds), Colloq. Math. Soc. J. Bolyai 11, North Holland, Amsterdam, 609–627
45. Erdős P., Rubin A., Taylor H. (1979) Choosability in Graphs. Congr. Numeratum 26:125–157
46. Erdős P., Selfridge J. (1973) On a Combinatorial Game. J. Combinatorial Th. (A) 14:298–301

47. Erdős P., Szekeres G. (1935) A Combinatorial Problem in Geometry. Composito Math. 3:463–470
48. Fajtlowicz S. (1978) On the Size of Independent Sets in Graphs. Proceedings of the 9th South-East Conference on Combinatorics, Graph Theory and Computing, 269–274
49. Fajtlowicz S. (1984) Independence, Clique, and Maximum Degree. Combinatorica 4:35–38
50. Farzad B. (2001) When the Chromatic Number is Close to the Maximum Degree. M.Sc. thesis, Dept. of Computer Science, University of Toronto.
51. Farzad B., Molloy M., Reed B. In preparation.
52. Faudree R.J., Gyárfás A., Schelp R.H., Tuza Z. (1989) Induced Matchings in Bipartite Graphs. Discrete Math. 78:83–87
53. Faudree R.J., Gyárfás A., Schelp R.H., Tuza Z. (1990) The Strong Chromatic Index of Graphs. Ars Combinatorica 29-B:205–211
54. Fellows M. (1990) Transversals of Vertex Partitions in Graphs. SIAM J. Discrete Math. 3:206–215
55. Fernandez de la Véga W. (1983) On the Maximum Density of Graphs Which Have No Subcontractions to K^S. Discrete Math. 46:109–110
56. Fernandez de la Véga W. (1983) On the Maximum Cardinality of a Consistent Set of Arcs in a Random Tournament. J. Combin. Th. (B) 35:328–332
57. Fiege U., Kilian J. (1998) Zero Knowledge and the Chromatic Number. J. Computer and System Science 57:187–199
58. Fleischner H., Steibitz M. (1992) A Solution to a Colouring Problem of P. Erdős. Discrete Math. 101:39–48
59. Fleischner H., Steibitz M. Some Remarks on the Cycle Plus Triangles Problem. manuscript
60. Frankl P., Rödl V. (1985) Near Perfect Coverings in Graphs and Hypergraphs. Europ. J. Comb. 6:317–326
61. Frieze A. (1991) On the Length of the Longest Monotone Increasing Subsequence in a Random Permutation. Ann. Appl. Prob. 1:301–305
62. Frobenius G. (1917) Uber Zerlegbare Determinanten. Sitzungsber. Konig Preuss. Akad. Wiss. XVIII:274–277
63. Furer M. (1995) Improved Hardness Results for Approximating the Chromatic Number. Proceedings of the Thirty-Sixth Annual IEEE Symposium on Foundations of Computer Science, 414–421
64. Galvin F. (1995) The List Chromatic Index of a Bipartite Multigraph. J.Combinatorial Th.(B) 63:153–158
65. Garey M., Johnson D., Stockmeyer L. (1976) Some Simplified NP-complete Graph Problems. Theor. Comp. SCi. 1:237–267
66. Godsil C., (1981) Matchings and Walks in Graphs. Journal of Graph Theory 5:285-297
67. Goldberg A. (1973) On Multigraphs of Almost Maximal Chromatic Class. Diskret. Analiz. 23:3–7
68. Grimmett G. (1999) Percolation. 2nd Edition. Springer Verlag, Berlin
69. Grimmett G., Stirzaker D. (1992) Probability and Random Processes. 2nd Edition. Oxford University Press, Oxford
70. Grimmett G., Stirzaker D. (2001) One Thousand Exercises in Probability. 1st Edition. Oxford University Press, Oxford
71. Grimmett G., Welsh D. (1986) An Introduction to Probability. 1st Edition. Oxford University Press, Oxford
72. Grunbaum B. (1973) Acyclic Colouring of Planar Graphs. Israel J. Math. 14:390–408

73. Hadwiger H. (1943) Uber eine Klassifikation der Streckencomplexe. Vierteljarsch Naturforsch Ges. Zurich 88:133–142
74. Häggkvist R., Chetwynd A. (1992) Some Upper Bounds on the Total and List Chromatic Numbers of Multigraphs. J. Graph Th. 16:503–516
75. Häggkvist R., Janssen J. (1997) New Bounds on the List Chromatic Index of the Complete Graph and Other Simple Graphs. Combinatorics, Probability & Computing 6:273–295
76. Harary F., Palmer E. (1973) Graphical Enumeration, 1st Edition, Academic Press
77. Haxell P. A Note on Vertex List-colouring. Combinatorics, Probability & Computing (to appear)
78. Hind H. (1990) An Improved Bound for the Total Chromatic Number of a Graph. Graphs and Combinatorics 6:153–159
79. Hind H. (1994) Recent Developments in Total Colouring. Discrete Math. 125-211–218
80. Hind H., Molloy M., Reed B. (1997) Colouring a Graph Frugally. Combinatorica 17:469–482
81. Hind H., Molloy M., Reed B. (1998) Total Colouring with Δ+poly($\log \Delta$) Colours. SIAM J. Computing 28:816–821
82. Holyer I. (1981) The NP-completeness of Edge Coloring. SIAM J. Computing 10:718–720
83. Horák P. (1990) The Strong Chromatic Index of Graphs with Maximum Degree Four. Contemporary Methods in Graph Theory, R. Bodendiek, editor, B.I. Wissenschaftsverlag, 399–403
84. Horák P., Qing H., Trotter W.T. (1993) Induced Matchings in Cubic Graphs. J. Graph Th. 17:151–160
85. Jensen T., Toft B. (1995) Graph Coloring Problems. Wiley, New York
86. Johansson A. (1996) Asymptotic Choice Number for Triangle Free Graphs. DIMACS Technical Report 91-5
87. Johansson A. The Choice Number of Sparse Graphs. Unpublished manuscript
88. Kahn J. (1992) Coloring Nearly Disjoint Hypergraphs with $n + o(n)$ Colors. J. Combinatorial Th. (A) 59:31–39
89. Kahn J. (1996) Asymptotically Good List-Colorings. J. Combinatorial Th. (A) 73:1–59
90. Kahn J. (1996) Asymptotics of the Chromatic Index for Multigraphs. J. Combinatorial Th. (B) 68:233–255
91. Kahn J. (2000) Asymptotics of the List-Chromatic Index for Multigraphs. Random Structures & Algorithms 17:117–156
92. Kahn J., Kayll M. (1997) On the Stochastic Independence Properties of Hard-core Distributions. Combinatorica 17:369–391
93. Karp R. (1972) Reducibility among Combinatorial Problems. In Complexity of Computer Computations, Plenum Press, 85–103
94. Kilakos K., Reed B. (1993) Fractionally Colouring Total Graphs. Combinatorica 13:435–440
95. Kim J.H. (1995) On Brooks' Theorem for Sparse Graphs. Combinatorics, Probability and Computing 4:97–132
96. Kim J.H. (1996) On Increasing Subsequences of Random Permutations. J. Combinatorial Th. (A) 76:148–155
97. Kostochka A. (1980) Degree, Girth, and Chromatic Number. Met. Diskret. Analiz. 35:45–70
98. Kostochka A. (1982) The Minimum Hadwiger Number for Graphs with a Given Mean Degree of Vertices (in Russian). Met. Diskret. Analiz. 38:37–58

99. Kostochka A. (1984) Lower Bound on the Hadwiger Number of Graphs by Their Average Degree. Combinatorica 4:307–316
100. Kostochka A. Personal Communication.
101. Kostochka A., Mazurova N. (1977) An Inequality in the Theory of Graph Colouring. Met. Diskret. Analiz. 30:173–208
102. Lawrence J. (1978) Covering the Vertex Set of a Graph with Subgraphs of Smaller Degree. Discrete Math. 21:61–68
103. Lee C. (1990) Some Recent Results on Convex Polytopes. Contemporary Mathematics 114:3-19.
104. Lesniak L., Straight H. (1977) The Cochromatic Number of a Graph. Ars Combinatorica 3:34–46
105. Logan B., Shepp L. (1977) A Variational Problem for Young Tableaux. Adv. Math. 26:206–222
106. Lovász L. (1968) On Chromatic Number of Finite Set-Systems. Acta Math. Acad. Sci. Hung. 19:59–67
107. Lovász L., Plummer M. (1986) Matching Theory. North Holland, Amsterdam
108. Lund C., Yannakakis M. (1993) On the Hardness of Approximating Minimization Problems. In: Proc. 25th ACM Symposium on the Theory of Computing, ACM, 286–293
109. Mader W. (1968) Homomorphiesätze für Graphen. Math. Ann. 178:154–168
110. Maffray F., Priessmann M. (1996) On the NP-completeness of the k-Colourability Problem for Triangle-Free Graphs. Discrete Math. 162:313–317
111. Matula D., Kŭcera L. (1990) An Expose-and-Merge Algorithm and the Chromatic Number of a Random Graph. In Random Graphs, 1987, eds. J. Jaworski, M. Karoński and A. Ruciński, Wiley, Chichester 175–188
112. McDiarmid C. (1989) On the Method of Bounded Differences. In: Surveys in Combinatorics, Proc. 14th British Combinatorial Conference, Cambridge University Press, Cambridge 148–188
113. McDiarmid C. (1997) Hypergraph Coloring and the Lovasz Local Lemma. Disc. Math. 167/168: 481-486
114. McDiarmid C. (1998) Concentration. In: Probabilistic Methods for Algorithmic Discrete Mathematics, (Habib M., McDiarmid C., Ramirez-Alfonsin J., Reed B., Eds.), Springer, Berlin 195–248
115. McDiarmid C. Concentration for Independent Permutations. Manuscript.
116. McDiarmid C., Reed B. (1989) Building Heaps Fast. J. Algorithms 10:352–365
117. McDiarmid C., Reed B. (1993) On Total Colourings of Graphs. J. Combinatorial Th. (B) 57:122–130
118. Molloy M., Reed B. (1997) A Bound on the Strong Chromatic Index of a Graph. J. Combinatorial Th. (B) 69:103–109
119. Molloy M., Reed B. (1998) A Bound on the Total Chromatic Number. Combinatorica 18:241–280
120. Molloy M., Reed B. (1998) Further Algorithmic Aspects of the Local Lemma. Proceedings of the 30th ACM Symposium on Theory of Computing, 524–529
121. Molloy M., Reed B. (1998) Colouring Graphs Whose Chromatic Number Is Near Their Maximum Degree. Proceedings of Latin American Theoretical Informatics, 216–225.
122. Molloy M., Reed B. (1999) Graph Colouring via the Probabilistic Method. In: Graph Theory and Computational Biology, Gyarfas A. and Lovasz L. editors, J. Bolyai Math. Soc., 125–155
123. Molloy M., Reed B. (2000) Near-optimal List Colourings. Random Structures & Algorithms 17:376–402

124. Molloy M., Reed B. (2001) Colouring Graphs When the Number of Colours is Almost the Maximum Degree. Proceedings the 33rd ACM Symposium on Theory of Computing.
125. Motwani R., Raghavan P. (1995) Randomized Algorithms. Cambridge University Press, Cambridge
126. Nesetřil J., Rödl V. (1979) A Short Proof of the Existence of Highly Chromatic Hypergraphs without Short Cycles. J. Combinatorial Th. (B) 27:225–227
127. Padberg M., Rao R. (1982) Odd Minimum Cutsets and b-matchings. Mathematics of Operations Research 7:67–80
128. Perkovic L., Reed B. Edge Colouring in Polynomial Expected Time. Unpublished manuscript
129. Pippenger N., Spencer J. (1989) Asymptotic Behavior of the Chromatic Index for Hypergraphs. J. Combinatorial Th. (A) 51:24–42
130. Rabinovich Y., Sinclair A., Widgerson A. (1992) Quadratic Dynamical Systems. Proceedings of the Thirty-Third Annual IEEE Symposium on Foundations of Computer Science 304–313
131. Radhakrishnanz J., Srivinasan A. (2000) Improved Bounds and Algorithms for Hypergraph 2-colouring. Random Structures & Algorithms 16:4–32
132. Reed B. (1998) ω, Δ, and χ. J. Graph Th. 27:177–212
133. Reed B. (1999) The List Colouring Constants. J. Graph Th. 31:149–153
134. Reed B. and Sudakov B. Asymptotically, the List Colouring Constants are 1. Manuscript
135. Reed B. and Sudakov B. Acyclic Edge Colouring. Manuscript
136. Reich S. (1980) Go-Bang ist PSPACE-vollständig, Acta Informatica 13:59-66
137. Robertson N., Seymour P., Thomas R. (1993) Hadwiger's Conjecture for K_6-free Graphs, Combinatorica 13:279–361
138. Rödl V. (1985) On a Packing and Covering Problem. Europ. J. Comb. 5:69–78
139. Salavatipour M. In preparation.
140. Scheinerman E., Ullman D. (1997) Fractional Graph Theory, Wiley, New York
141. Sedgewick R., Flajolet P. (1996) An Introduction to the Analysis of Algorithms. Addison-Wesley, Boston
142. Seymour P. (1979) On multi-colourings of Cubic Graphs and Conjectures of Fulkerson and Tutte. Proc. Lond. Math. Soc. 38:423–460
143. Shearer J. (1985) On a Problem of Spencer. Combinatorica 3:241–245
144. Spencer J. (1984) Six Standard Deviations Suffice. Trans. Amer. Math. Soc. 289:679–706
145. Spencer J. Applications of Talagrand's Inequality. Unpublished manuscript.
146. Steele M. (1997) Probability Theory and Combinatorial Optimization. SIAM NSF-CBMS Series Vol. 69
147. Tait P. (1878–1880) Remarks on the Colouring of Maps. Proceedings of the Royal Edinborough Society, 10:729
148. Talagrand M. (1995) Concentration of Measure and Isoperimetric Inequalities in Product Spaces. Institut Des Hautes Études Scientifiques, Publications Mathématiques 81:73–205
149. Thomason A. (1984) An Extremal Function for Contractions of Graphs. Math. Proc. Camb. Phil. Soc., 95:261–265
150. Thomason A. The Extremal Function for Complete Minors. J. Combinatorial Th. (B) 81:318–338
151. Thomassen C. (1992) The Even Cycle Problem for Directed Graphs. J. Amer. Math. Soc. 5:217–229
152. Toft B. (1975) On Colour-critical Hypergraphs. Colloq. Math. Soc. Janos Bolyai 10:1445-1457

153. Veršik A., Kerov C. (1977) Asymptotics for the Plancheral Measure of the Symmetric Group and a Limiting Form for Young Tableaux. Dokl. Akad. Nauk USSR 233:1024–1027
154. Vizing V. (1964) On the Estimate of the Chromatic Class of a p-graph. Met. Diskret. Analiz. 3:25–30
155. Vizing V. (1968) Some Unsolved Problems in Graph Theory. Russian Math Surveys 23:125–141
156. Vizing V. (1976) Colouring the Vertices of a Graph with Prescribed Colours. Met. Diskret. Analiz. 29:3–10
157. Vu V. (1999) On Some Degree Conditions which Guarantee the Upper Bound of Chromatic (Choice) Number of Random Graphs. J. Graph Th. 31:201-226
158. Vu V. A General Upper Bound on the List Chromatic Number of Locally Sparse Graphs Combinatorics, Probability & Computing (to appear).
159. Wagner K. (1937) Uber eine Eigenschaft der ebenen Komplexe. Math. Ann. 114:570–590
160. Zykov A. (1949) On Some Problems of Linear Complexes. Mat. Sbornik N.S. 24:163–188. English translation in Amer. Math. Soc. Transl. 79 (1952)

Index

Printing: Mercedes-Druck, Berlin
Binding: Stürtz AG, Würzburg